AN INTRODUCTION TO MODERN GENETICS

DONALD I. PATT / **GAIL R. PATT** / Boston University

AN INTRODUCTION TO MODERN GENETICS

ADDISON-WESLEY PUBLISHING COMPANY

Reading, Massachusetts · Menlo Park, California
London · Amsterdam · Don Mills, Ontario · Sydney

This book is in the
ADDISON-WESLEY SERIES IN THE LIFE SCIENCES

Consulting Editor
Johns Hopkins III

Second printing, May 1976

ISBN 0-201-05743-3
ABCDEFGHIJ-HA-79876

To Stephen, Andrew, Jonathan, and Anthony

Preface

Forty years ago, the term "genetics" called to mind patterns of inheritance in corn and fruit flies. Today it presents a spectrum that runs from molecules and highly technical biochemical and serological techniques all the way to mathematically sophisticated studies of populations. Every teacher and every author prefers his or her own mix of the various aspects of modern genetics—of the molecular, the cytogenetic, the classical, and the populational. Predictably, we have our own preferences. Ours have been derived from our personal experiences teaching a one-semester course in introductory genetics to a heterogeneous group of college students including biology majors, premedical students, and others motivated solely by intellectual interest. This text is meant for such a course and for such a mixture of students. It presupposes a college course in introductory biology, but not ones in chemistry or mathematics.

This book emphasizes some of the experimental techniques that have been used so successfully to explain genetic principles, especially with bacteria and viruses. This is in response to our experience that students frequently have difficulty understanding the manipulation of the invisible. Over the years, we have learned how distorted a student's conceptualization of an experimental procedure—and consequently its inferences—can be, simply because too many little things were taken for granted. We hope this book will promote an appreciation for the marvels of insight and inductive reasoning that lie between some deceptively simple experimental procedures and the highly sophisticated hypotheses that are derived from them. For similar, although unrelated, reasons, no steps are omitted in mathematical derivations. Some students are facile with algebra; others are not. It seemed to be worth some tedium to ensure that no one would flounder with elementary algebra simply because intermediate equations were lacking.

Because of the emphasis on methodology and analysis, illustrations play a vital role in the explanation of many techniques and concepts. They are not secondary in any sense, for they are designed both to supplement and to augment the text.

Some topics are taken up in more depth than others, and the emphasis shifts from topic to topic. The reasons for this obvious unevenness stem from what we perceive to be the interests and capabilities of beginners in genetics, as well as from our own feelings about the relative impact that each of the topics has had and likely will have on genetics as a science and on individuals and society. In each case, we have tried to stress principles and concepts, from which the text fans out in different directions. Even a cursory reading will reveal that molecular aspects and mechanisms and developmental genetics have been treated in far more detail than either the classical or populational aspects. We have done this because the current mainstream of genetics is at the molecular level, and introductory courses in biology are preparing students better and better each year to handle molecular concepts. We believe that the major advances in the future will derive from existing knowledge of molecular biology, and that some of these advances will have a profound impact upon society. For example, we wish to feel that students who have used this textbook will be better able to evaluate progress and projections regarding techniques of genetic engineering and will have a solid theoretical base upon which to build moral and political judgments.

By way of contrast, population genetics is at a level that we feel typical college students are ill-equipped to handle, and we are loath to interrupt the flow of the text with a primer on complex statistical methods that would permit an adequately detailed treatment of the subject. Advanced computer technology is required by many population geneticists to solve for all the variables in their model systems, many of which are still at a very tentative stage of development. For these reasons, this book emphasizes the principles of population genetics rather than the methodology,

as well as the more immediate societal implications that arise from the study of the human gene pool and its changes.

Classical transmission genetics has still a different emphasis. Beginning with a very brief introduction in Chapter 1 and continuing in later chapters, the historical aspects of the concepts are stressed. Although many of these concepts are not required for an understanding of molecular genetics, they were required as a stimulus for molecular discoveries and have intellectual merit in their own right. They are also important in agricultural and clinical genetics. The historical treatment extends into the future with a rather detailed discussion on non-Mendelian inheritance.

The order of the chapters is to some degree arbitrary. Other arrangements can be used according to the pedagogical bent of the instructor. For example, in a molecularly oriented course, Chapters 1 and 2 might be followed directly by Chapters 7 through 10, at which point the remainder of the book could be used in normal order.

There are few footnotes in the body of the text; references are keyed to a bibliography at the end of the book. This has been done in order to make the text more enjoyable to read, but more importantly to enhance the usefulness of the book as a reference source for students who wish to delve more deeply into particular aspects of genetics.

There are a number of people and institutions whose help was invaluable in the preparation of this book. The President and Fellows of Clare Hall, Cambridge University, most graciously provided us with living space, working space, and friendship during the year when much of the book was written. Professional encouragement was extended by Professor Thoday, Head of the Department of Genetics at that same institution. Mrs. Moira Hutchinson, Librarian of the Department of Genetics, deserves special thanks, as does Patricia Allen, Secretary of the Science Department, Boston University.

Harvard, Massachusetts   D. I. P.
January 1975   G. R. P.

Contents

1

History of Genetic Theory

Variability is a fundamental characteristic of life; it is the raw material of adaptation and of evolution by natural selection. The science of genetics began with attempts to discover the physical basis of biological variation, the mechanisms by which variations arise, and the principles or laws that govern their transmission from one generation to another. As with any science, once some of the basic principles pertaining to it have been discovered and explained, ways are found to apply these principles to human advantage. Genetics is no exception. The virtually explosive growth of knowledge of hereditary mechanisms to which the scientific world has been witness in the twentieth century has provided mankind with powerful tools of applied genetic theory that have had a profound effect upon humankind both economically and politically. In the so-called area of "genetic engineering," the probability of even more fundamental effects in the future seems inevitable.

This book attempts to trace the development of the major concepts in genetics and to show their relevance to human affairs.

With the work of Gregor Mendel (1822–1884), the science of genetics began. However, in order to understand the revolutionary nature of Mendelian ideas, it is important to place Mendel and his work in the proper historical context. Prior to his time theories of inheritance proposed a direct link between parent and offspring in the sense that each part of the parental body was believed to produce some kind of substance or "essence" that became incorporated into the gametes and was transmitted intact to the offspring. Such theories were propounded by Hippocrates in the fourth century B.C. and, with minor modification, held sway until shortly before the close of the nineteenth century. Darwin's ideas of heredity were not materially different from those of Hippocrates.

The major challenge to the notion of direct inheritance came with the work of August Weismann (1834–1914). He removed the tails from newborn mice for twenty-two successive generations. If inheritance were direct, the adult tailless mice should have tailless progeny. Because such surgery was not able to alter the trait in any way, Weismann reasoned that inheritance must be indirect, namely, that there must be a distinction between the unspecialized reproductive cells (germ plasm) and the other cells of the body (somatoplasm). Alteration of the somatoplasm could not alter the germ plasm because the gametes were isolated and preserved intact in each generation.

Another quite separate line of work was being undertaken by botanists. Then, as now, an increase in agricultural output was considered desirable. Basic discoveries about the function of the pollen grain and methods of plant reproduction were being used to create hardier and more productive hybrid plants. The garden pea plant was a favorite object for study because of the existence of distinct and recognizable strains and also because the plant is self-pollinating and therefore more amenable to experimental procedures. However, because the major aim of research in plant hybridization was an increase in agricultural productivity, these studies had little effect on scientific thoughts concerning heredity.

It was in this context that Mendel worked. He repeated and extended much of the previous work on the hybridization of the garden pea but, unlike his predecessors, he used the techniques of hybridization specifically as a tool to probe the mysteries of heredity.

Mendel's contributions to the science of genetics are known today as his first and second laws of inheritance. The first law, the Law of Segregation, holds that hereditary characteristics are determined by particulate factors, now called *genes*, that these factors occur in pairs in the cells of the individual, and that members of pairs segregate during the production of gametes so that any given gamete receives and transmits only one member of each pair. Mendel's second law, the Law of Independent Assortment, states that during the formation of gametes, pairs of factors segregate inde-

pendently of one another. A pair of factors, say, for character A are transmitted to the gametes uninfluenced by the transmission of any other pair of factors, such as those for character B. The principles of segregation and of independent assortment

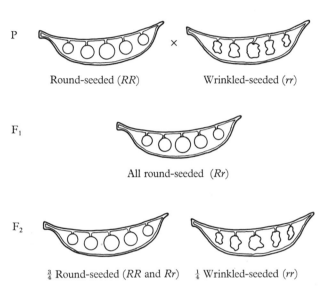

P   Round-seeded (RR)   ×   Wrinkled-seeded (rr)

F_1 All round-seeded (Rr)

F_2 ¾ Round-seeded (RR and Rr)   ¼ Wrinkled-seeded (rr)

Fig. 1–1.   Segregation of genes. Assuming that each plant cell (with the exception of the gametes) contains two genes for the trait of seed shape, the round-seeded and wrinkled-seeded parent plants (P) could be designated genetically as RR and rr, respectively. The round-seeded parent would produce gametes, all of which would carry one R gene; the gametes of the wrinkled-seeded parent would all contain one r gene. Hence, the F_1 generation would be the recipient of both variations of the seed-shape gene. They would appear round-seeded because roundness is dominant to wrinkledness. Half of the F_1 gametes would carry one R gene; the other half would carry one r gene. If each pollen nucleus has an equal probability of fertilizing any egg, then ¼ of the F_2 plants would be RR, ¼ would be rr, and ½ would be Rr. In terms of appearance, ¾ of the F_2 would be round-seeded and ¼ would be wrinkled-seeded. When the F_2 plants are allowed to self-pollinate, the wrinkled-seeded plants would breed true as did the original rr parents, but the round-seeded plants, not all carrying the R gene uniformly, would produce a mixture of the two types of plants.

are represented schematically in Figs. 1–1 and 1–2.

Contrary to the picture that is sometimes portrayed of him as an introspective plant hobbyist, Mendel was in fact a widely read and well-trained student of science who apparently accepted the religious calling, at least in part, because it was an academic calling. As a priest, he was also a teacher and scholar. He was sent to the University of

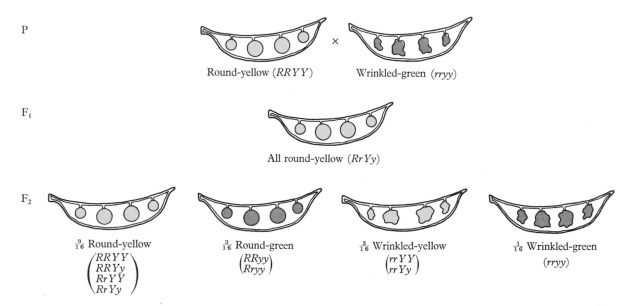

P

Round-yellow (*RRYY*)   ×   Wrinkled-green (*rryy*)

F₁

All round-yellow (*RrYy*)

F₂

$\frac{9}{16}$ Round-yellow

$\begin{pmatrix} RRYY \\ RRYy \\ RrYY \\ RrYy \end{pmatrix}$

$\frac{3}{16}$ Round-green

$\begin{pmatrix} RRyy \\ Rryy \end{pmatrix}$

$\frac{3}{16}$ Wrinkled-yellow

$\begin{pmatrix} rrYY \\ rrYy \end{pmatrix}$

$\frac{1}{16}$ Wrinkled-green

(*rryy*)

Fig. 1–2. Independent assortment. The traits of round seeds and yellow seeds are dominant to wrinkledness and greenness. Hence, the pure-breeding parents can be designated as *RRYY* and *rryy*. Each gamete must contain one gene for seed shape and one for seed color. The gametes of the *RRYY* parent will all carry one *R* and one *Y* gene; those of the *rryy* parent will all carry one *r* and one *y* gene. After fertilization, the F₁ plants so produced will have both variations of both genes and can be represented as *RrYy*. During F₁ gamete formation, ¹/₂ of the gametes will carry an *R* gene and ¹/₂ will contain an *r* gene. Similarly, ¹/₂ will carry a *Y* gene and ¹/₂ a *y* gene. Since the color genes are assorted into gametes independently of the shape genes, the halves referred to above need not coincide. By the laws of chance, the F₁ gametes will be as follows:

$$^1/_4 \; RY$$
$$^1/_4 \; Ry$$
$$^1/_4 \; rY$$
$$^1/_4 \; ry$$

Random combination of any gamete with any other will yield the 9:3:3:1 ratio for the appearance of the F₂ generation, in that ⁹/₁₆ will have at least one dominant gene for each trait; the ⁶/₁₆ classes will have at least one dominant gene for one trait and two recessive genes for the other; and ¹/₁₆ will have both recessive genes for each trait.

Vienna by his order for further training in mathematics and science, including botany and microscopy. Upon his return to the monastery and until his election to Abbot in 1868, he spent a great deal of time deriving 34 pure strains of pea plants, devising and carrying out experiments in hybridization. In keeping with his training, he applied mathematical analysis to his findings. The paper that presented the results of his researches was read and discussed at two evening meetings of the Brünn Natural Sciences Society in early 1865. It was published the following year in the Society's proceedings and widely disseminated. It has been specifically established that copies of the proceedings and reprints of his paper were requested by scientists and were sent to Berlin, the United States, and England. Finally, the fact that Mendel's paper was read by others is apparent from the fact that it was included in bibliographies on plant hybridization. And yet, Mendel's work was not considered important and it seemingly ''died on the vine'' for close to 40 years.

Why was the importance of Mendel's work not immediately recognized? Of critical importance here is the fact that, had Darwin been aware of Mendel's work, the weak points pertaining to heredity as propounded in Darwin's *Origin of Species* could have been bolstered to a very considerable extent. It is known that Mendel had read Darwin, but the reverse probably did not occur. The reasons that are most widely accepted by historians have the ring of truth that often goes with triviality. Most biologists were poorly trained in mathematics, and similarly, most mathematicians had little knowledge or demonstrated interest in biology. In brief, Mendel constituted a bridge across a gulf that no one was interested in crossing. Perhaps more important, however, was Mendel's failure to follow up his brilliantly conceived principles with generalizations about other plants or animals. His important paper, therefore, remained for years an isolated item of interest only to experimenters who worked with peas. In all fairness,

Mendel was unable to extend his conclusions because, unluckily, he was persuaded to try to apply them to hawkweed, a plant with unusual reproductive patterns.

THE CHROMOSOME THEORY

Between the publication of Mendel's paper in 1866 and the opening years of the twentieth century, evolutionists were fighting their battles and Mendel died quietly of Bright's disease in 1884. Nevertheless, behind the backdrop of Darwinian controversy, some fundamental advances were made in cytology that led many scholars to take the same experimental aproach as Mendel had. With advances in microscopy, the significance of the fusion of the nuclei of egg and sperm during fertilization was grasped and the idea developed that chromosomes, which had been observed for many years but whose significance was not understood, might be hereditary determinants. It was recognized by 1900 that the number of chromosomes is a species-specific character, that chromosomes divide during cell division, and that a halving of chromosome number occurs during the special kind of cell division that precedes the formation of gametes.

By 1900 Mendel's experiments with garden peas had been independently duplicated by Carl Correns and Erich von Tschermak. Hugo De Vries had also carried out similar experiments with maize, primroses, poppies, and many other flowering plants. Each of these men independently recognized the 3:1 ratio that characterized second-generation offspring, as Mendel had. Then in 1900, when each of them was set to publish his work, they discovered Mendel's old paper. They provided the follow-up work that Mendel had neither the time, the opportunity, nor the good luck to do. The universality of Mendel's principles became even more apparent in the following two years when Bateson, Saunders, and Cuenot extended genetic observation to animals.

Exceptions to Mendelian rules posed problems, especially since they were so numerous. The exceptions were of three kinds, namely, they pertained (1) to characters that showed continuous variation such as height, weight, or intensity of coloration, (2) to pairs of characters that did not seem to assort independently, and (3) to certain other characters that even later could not be explained by the chromosome theory of inheritance.

Within the first decade of the twentieth century, continuous variation of quantitative characters was shown to be attributable to the cumulative expression of many pairs of genes, each acting according to Mendelian rules. In addition, the notion of dominance was modified, because certain gene pairs act as though neither alternative is truly dominant to the other. The matter of certain traits appearing not to be governed by chromosomally-borne genes or factors, although now understood in essence, is still a subject of active investigation, and it is well recognized that such traits are indeed determined extrachromosomally.

In spite of the exceptions, recognition of the role played by the chromosomes in inheritance clarified the mechanism of Mendel's laws and provided an explanation for nonindependent assortment that strengthened, rather than weakened, Mendel's argument. By the late 1800's, the essential features of mitosis had been established. Meiosis was recognized as a different process that consists of two rather modified divisions that result in reduction of the chromosome number in the gametes. In the opening years of the twentieth century, Montgomery and Sutton independently concluded that chromosome pairing is an important feature of meiosis and that, furthermore, the elements in each pair have a different origin. That is, one member of a pair of chromosomes is of maternal and the other of paternal origin.

In 1903 Sutton clearly associated Mendelian factors with chromosomes (as their vehicle) by pointing out the resemblances in behavior between the two. The segregation of pairs of factors that Mendel postulated during gamete formation is paralleled by the separation of homologous chromosomes during meiosis. Mendel's ratios and statistical predictions implicate independent segregation of chromosome pairs during meiosis, which is the basis of independent assortment.

The chromosome theory of inheritance was confirmed by the discovery of the sex chromosomes and by demonstration that each chromosome contains not one, but many genes. That genes were "linked" together in one chromosome was a critical discovery, because the backlog of experimental data that contradicted the principle of independent assortment could now be explained. It became possible to think of traits that are always transmitted together as a manifestation of the fact that the genes that determine them reside on the same chromosome.

A major source of confusion stemmed from the observation that some traits show such linkage most of the time but not all of the time. In 1911, Thomas Hunt Morgan suggested that during meiosis chromosomes can exchange parts. Linkage, therefore, need not be absolute and, depending upon the location of the sets of genes in question, genes governing different traits will show independent assortment if they are located on different pairs of chromosomes, or linkage in varying degrees if they are located on the same pair of chromosomes. The degree of variation was thought to be a function of how close together (or far apart) they may be on that pair of chromosomes.

GENETICS AND EVOLUTION

During the early years of the twentieth century, another significant line of genetic theory began to develop that not only added more support to Mendelism, but related it to evolutionary theory and, consequently, to philosophy. The background was laid by the scientific writings of Charles Darwin and Alfred Russel Wallace.

Darwin had sailed from England in 1831 aboard the *H.M.S. Beagle* as a theology student of little promise and even less reputation. In all honesty, but with not much knowledge, he had undertaken the voyage in order to prove the validity of the immutability of species as set forth in Genesis. Wallace left England several years later (1847) for a collecting trip to Brazil. For Wallace, a man of little wealth but with a declared interest in the origin of species, the voyage was a bonanza because, by selling duplicate collections of Brazilian flora and fauna, he was able to meet his expenses while at the same time pursuing his intellectual goals. Darwin by the 1840's and Wallace by the 1850's had arrived at similar conclusions from their experiences: Similarities between organisms are a result of descent from a common ancestral species, whereas differences represent divergences owing to natural selection. Their theory supposed that individuals of a species show random variation, much of which is heritable. Members of a particular species differ in their reproductive potential, partly because of such randomly occurring heritable variations. Some characteristics increase reproductive "fitness" in a particular environment while others decrease it. Species undergo change because competition results in natural selection for those individuals that are more fit, resulting in a progressive decline in the less reproductively fit.

This theory differed from previous views in a number of important ways. First of all, it provided another counterview to the ideas of the immutability of species. Secondly, it held that evolution is not a process linked with goals or progress in any predictable direction. Hereditary variation is random rather than environmentally induced. Other scientists, notably Lamarck (1744–1829), believed in evolutionary change, but Lamarck's views on change centered around increasing complexity toward "desirable" goals, such complexity being caused by the direct action of heat and gravity on the bodily fluids. That organisms differ in different parts of the world was not surprising to Lamarck, because climatic variation would cause heritable changes in bodily organization. Lamarck was concerned with progressive heirarchy in the living world; Darwin and Wallace were trying to answer questions concerning the conversion of varieties into full-fledged, reproductively separable species.

Although hereditary variation was one of the assumptions in Darwin's view of evolution, he and Wallace maintained that the argument of natural selection could stand alone, irrespective of the causes and mode of transmission of hereditary traits. They agreed that variation was not a product of environmental change and that selection within an environment elevated previously existing variation to new heights. But Darwin did attempt a theory of inheritance to underlie his work (direct inheritance by granules or *gemmules*) which indeed did cloud the issue of natural selection, especially after the acceptance of Weismann's work.

In 1908 the synthesis of the inheritance of traits in individuals and the evolution of groups began with the independent analyses of G. H. Hardy and W. Weinberg. Hardy, a mathematician, and Weinberg, a geneticist, demonstrated that in the absence of mutation and selection, Mendelian principles could explain genetic stability in populations over time. On the basis of this equilibrium, mutation and selection could be measured and given a physical reality. By the 1920's developments in statistical methods had advanced enough to permit the Russian geneticist Chetverikov to show that variation can be hidden by genetic means in a population, and that this variation is so great as to allow evolution even when selection pressures are of low order and environmental change is slight. In the late 1920's and early 1930's, Sewall Wright, J. B. S. Haldane, and R. A. Fisher introduced the statistical methodology necessary for a rigorous and temporal treatment of Mendelian principles at the population level. By 1937 the synthesis of genetics and evolution was sufficiently complete to allow Theodosius Dobzhansky to publish *Genetics and the Origin of Species*. The linkage of the transmission genetics of Mendel and his successors with Darwin's and Wallace's theory of natural selection changed the subsequent history of both fields.

THEORIES OF HOW GENES ACT:
ONE GENE—ONE ENZYME

With solutions to questions of gene transmission in sight, interest naturally turned to the next order of business: How do genes act? Today we speak of the gene or genes for blue eyes, not fully realizing the intricacy involved in producing not only eyes, but those of a very definite type. If genes are, in fact, particulate nuclear substances, they must work in a physically understandable way. As early as 1909, the link between the concept of the gene and the realization of a physical characteristic was seen to be enzymatic.

As a result of a number of years of study, Archibald Garrod (1902), an English physician at Oxford, demonstrated that the rare human disease alkaptonuria was caused by a biochemical failure. In the normal metabolism of the amino acids phenylalanine and tyrosine, the benzene rings of these molecules are degraded. In individuals suffering from alkaptonuria, this degradation does not occur. Bateson and Saunders (1902) pointed out that Garrod's data seemed to indicate that the disease was heritable; it appeared to be transmitted by a single recessive gene. Garrod suggested that alkaptonurics probably lack a special enzyme that is required to split the benzene ring, and Bateson (1909) generalized the concept by concluding that "the consequences of their [mutant genes] presence is in so many instances comparable with the effects produced by ferments [enzymes], that with some confidence we suspect that the operations of some units [genes] are in an essential way carried out by the formation of definite substances acting as ferments."

This view of genetic activity received support from those who studied characteristics that appeared to be due to two or more genes. As an example, Bateson, Saunders, and Punnett (1905) crossed a pure-breeding pink-flowered strain of *Salvia horminum* with a pure-breeding white variety. They obtained an F_1 generation of all purple flowers and an F_2 consisting of 225 purple, 92 pink, and 114 white. This ratio, approximately 9:3:4, appeared to be derived from the standard 9:3:3:1 ratio expected for the independent inheritance of two genes in the F_2 generation, except that the last two classes were combined. Flower color, in this case, was regulated by two genes probably acting sequentially. The parallel between stepwise, sequential activity of genes and the same characteristic of enzymes was too apparent to miss (refer to Fig. 1–3).

By the 1940's it was possible to create biochemical mutants by X or ultraviolet irradiation and study the inheritance of the new traits. Beadle and Tatum (1941) devised a technique whereby new varieties (mutants) of the breadmold *Neurospora crassa* could be produced by irradiation and recovered for study. Their work, and that of several others as well, involved careful breeding experiments that revealed the mutation of at least seven different genes, all resulting in the production of varieties lacking the ability to synthesize the amino acid arginine. However, each of the different mutants could be treated in different ways to permit growth, that is, mutant no. 1 could produce arginine if substance A was added to its food. Mutant no. 2 required either substance A or B. Mutant no. 3 needed either A, B, or C, and so forth. Correlation of the mutants with the biochemical pathway for arginine synthesis led to the conclusion that each mutant had a defective enzyme in the series of enzymes necessary for arginine production. If the "blockage" were bypassed by providing the mutant *Neurospora* with the normal product of the blocked enzyme, or any succeeding product, then the rest of the biochemical process could proceed. This led Beadle to conclude in 1945 that every gene directs a specific enzymatic reaction, and that specific genes act by directing the synthesis of specific enzymatic proteins (Fig. 1–4).

Fig. 1–3. Hypothetical metabolic pathway in (a) purple *Salvia*, (b) pink *Salvia*, and (c) white *Salvia*. According to the cross done by Bateson, Saunders, and Punnett, a pure-breeding pink (*aaBB*) was crossed with a pure-breeding white (*AAbb*). The F₁ was purple (*AaBb*), having derived the *A* and *b* genes from the white parent and the *a* and *B* genes from the pink parent, as follows.

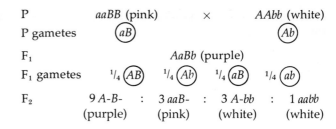

P	*aaBB* (pink)	×	*AAbb* (white)
P gametes	(*aB*)		(*Ab*)
F₁		*AaBb* (purple)	
F₁ gametes	¹/₄ (*AB*) ¹/₄ (*Ab*)	¹/₄ (*aB*)	¹/₄ (*ab*)
F₂	9 *A-B-* : 3 *aaB-* :	3 *A-bb* :	1 *aabb*
	(purple) (pink)	(white)	(white)

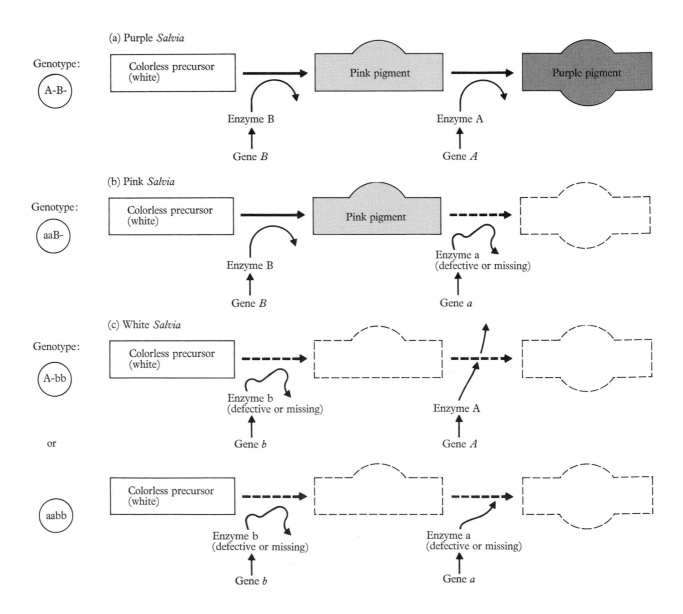

How would you deduce that the P generation white plant was neither *aabb* nor *Aabb*?

A dash instead of an indicated gene (as in *aaB-*) means that the appearance would not be altered by one or the other variant. Plants that are *A-B-* are purple. Some of these are *AABB*, others are *AaBB*, others are *AaBb*, and still others are *AABb*. However, because of dominance they all have a similar appearance.

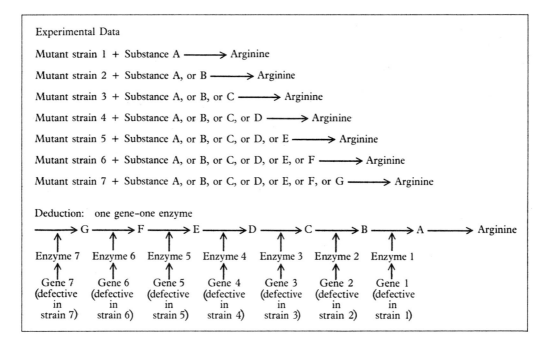

Experimental Data

Mutant strain 1 + Substance A ⟶ Arginine

Mutant strain 2 + Substance A, or B ⟶ Arginine

Mutant strain 3 + Substance A, or B, or C ⟶ Arginine

Mutant strain 4 + Substance A, or B, or C, or D ⟶ Arginine

Mutant strain 5 + Substance A, or B, or C, or D, or E ⟶ Arginine

Mutant strain 6 + Substance A, or B, or C, or D, or E, or F ⟶ Arginine

Mutant strain 7 + Substance A, or B, or C, or D, or E, or F, or G ⟶ Arginine

Deduction: one gene—one enzyme

⟶ G ⟶ F ⟶ E ⟶ D ⟶ C ⟶ B ⟶ A ⟶ Arginine

Enzyme 7	Enzyme 6	Enzyme 5	Enzyme 4	Enzyme 3	Enzyme 2	Enzyme 1
Gene 7 (defective in strain 7)	Gene 6 (defective in strain 6)	Gene 5 (defective in strain 5)	Gene 4 (defective in strain 4)	Gene 3 (defective in strain 3)	Gene 2 (defective in strain 2)	Gene 1 (defective in strain 1)

Fig. 1–4.   Different nutritionally deficient strains of *Neurospora crassa* require one of several precursor substances in order to manufacture the amino acid arginine that is necessary for growth. Because the various precursors form a metabolic chain or pathway, Beadle and Tatum deduced that each of the mutants lacked one of the enzymes required in the pathway, and that each mutant had one deficient gene for the appropriate enzyme. Hence mutant strain 1, defective for the last enzyme in this chain, must be supplied with substance A to circumvent the block. Strain 2, unable to catalyze the reaction of C to B, must be supplied with either B or A. Strain 7 is unable to manufacture substance G. Since the rest of its enzymes are functional, strain 7 requires substance G or any of the other substances that follow the faulty enzymatic reaction.

THEORY OF THE
CHEMICAL NATURE OF THE GENE

When the interest in genes centered on their passage to the next generation, it was sufficient to have a simple conceptual definition of the gene. Transmission genetics could progress if genes were visualized as beads on a chromosomal string. However, with the new biochemical direction in which genetics was heading, such a concept was inadequate. It was now necessary to account for the transmission of characteristics in light of these characteristics being end products of biochemical pathways. More specificity was needed.

Results of chemical studies during the last third of the nineteenth century had revealed that the basic constituents of cell nuclei are nucleic acids in combination with proteins. Analyses of chromosomes performed at a later date confirmed these earlier findings. The question then was which of these substances represents the gene—protein or nucleic acid? For a number of logical reasons, proteins were in favor and retained this favor until the decade beginning in 1943. Then the evidence for the genetic role of nucleic acids, specifically deoxyribonucleic acid (DNA) in most organisms, began to accumulate.

The shift in attention from protein to DNA brought with it another kind of shift in genetics. Mendel had worked with higher plants as a result of his background in agricultural hybridization. The major findings in chromosome research occurred when geneticists turned their attention to the fruit fly, *Drosophila melanogaster*. The link to biochemistry was closely tied to studying the fungi. Then geneticists and biochemists turned their attention to bacteria and viruses and their successes were due in great measure to the organisms with which they worked. In addition to the advantages inherent in the size and speed of multiplication of these organisms, their environment can be controlled to an extent not possible with higher animals or plants. Furthermore, the concentration upon prokaryotic cells (bacteria and blue-green algae) and viruses has led to one of today's major thrusts in genetics, which is to answer the question, Can theories of prokaryotic genetics be extended to more complex organisms? Are bears and bacteria really alike?

By the early 1950's, it was apparent that *the biological problem to be solved was the structure of DNA*. If this molecule has a structure that could allow certain processes to occur, the protein versus nucleic acid argument could be settled in a way that would also explain how genes work. Specifically, the genetic material must be relatively stable chemically, but must allow for a low rate of change as well. In addition, the fundamental ability to duplicate must be explainable within the confines of the molecular structure of the gene. Finally, the structure must be one that allows the gene to carry out its task of specifically regulating the synthesis of innumerable enzymes.

Two major groups of people set to work to decipher the structure of DNA. According to James Watson, an American postdoctoral student who played a key role with the English group, the rivalry between them and Linus Pauling and his coworkers in California was keen. Pauling had come up with a specific three-dimensional shape, the alpha-helix, for protein molecules, and both he and those of the English group intuitively believed that this configuration would be important in deciphering the structure of DNA.

Based on the crystallographic evidence of Maurice Wilkins and Rosalind Franklin, the structure that Watson and Crick (1953) finally proposed clearly did meet the necessary molecular criteria and, in relatively short order, the phrase "genetic code" became at least a schoolroom if not a household phrase (Fig. 1–5). Masses of data and thousands of man-hours provided proof that DNA, arranged as a double helix, could and would be stable but mutable, could replicate itself, and could account for the regulation of enzyme synthesis, a feature that is central to the regulation of cells and organisms.

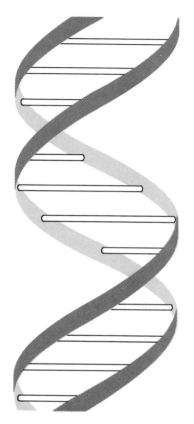

Fig. 1–5. The Watson-Crick model of the double helix of DNA. (A detailed analysis of this model is represented in Fig. 7–9.)

THEORY OF GENETIC CONTROL

The last conceptual synthesis that has come of age in genetics concerns the matter of gene regulation. However, in this relatively new area of investigation, the accepted concepts are still tentative and we do not know if they are universally applicable.

It has been known for decades that organisms are capable of going through stages that differ with respect to metabolic needs and enzymatic requirements. Such differentiation is as true for bacteria as it is for more complex forms. However, a paradox arises when one tries to reconcile the fact of genetic identity in all cells of a multicellular organism with the fact of differentiation.

Solutions to the paradox were forthcoming, at least with respect to bacteria, when it was realized that great variations in amounts of various enzymes occur not only from cell to cell, but also from time to time in an individual cell. Fluctuating environmental conditions seemed to go hand in hand with similar fluctuations in internal chemistry. Not all of these variations could be explained as examples of enzymatic feedback or other types of control at the enzymatic level. It was necessary to postulate a control mechanism that worked at a different level in the cell. Somehow, certain features of environmental fluctuation must have a direct effect on the genes themselves so that certain genes could be active at some times and dormant at other times.

In the early 1960's, Francois Jacob and Jacques Monod proposed that not all genes function in the production of metabolic enzymes. Genes that work in the classical sense were called structural genes because they dictate the molecular structure of enzymatic proteins. They found that in some cases, a group of structural genes corresponding to a group of sequentially acting enzymes was under the guidance of an *operator* in much the same way that all the electrical outlets in one room can be under the control of a master switch for that room. A set consisting of an operator and its structural genes was termed an *operon*. Another variety of control gene was postulated. This gene, the *regulator*, would be sensitive to the changing metabolites in the bacterial environment. If particular enzymes were not needed because the substrate for those enzymes was lacking, the regulator would so inform the operator and this would lead to cessation of function of the structural genes under that operator's control. The analogy of the electrical outlets can be extended here to include the man who turns the switch on or off according to what work he is going to do in the room (Fig. 1–6).

In the past few years, evidence to substantiate the operon theory has been found in bacteria. Geneticists are beginning to understand just how

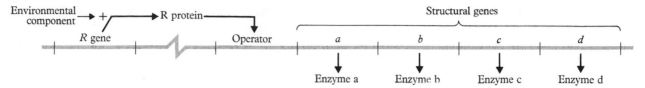

Fig. 1–6.   Schema of the relationship between a group of structural genes and the genes that regulate their activity. Gene *R* directs the synthesis of a protein substance which, under the influence of specific environmental influences, either stimulates or inhibits the operator. When the operator is stimulated, all the structural genes under its control are activated, and their corresponding enzymes are synthesized.

operator and regulator genes work and how they interact with structural genes. Because of the difficulties inherent in their complexity, multicellular organisms provide answers that are less clear and work with them has proceeded at a slower pace. However, work with bacterial operons has provided a model of genetic control that may apply to all living things (Fig. 1–7). How circular this path may be, or to what degree a change in metabolism and structure can affect the environment, remains one of the tantalizing questions of interest today. The genetic mechanism of differentiation, the interplay of cells and environment, remains one of the most absorbing genetic problems.

Fig. 1–7.   The metabolic roles of genes include self-replication (DNA) and directing the synthesis of proteins (DNA → proteins). Proteins, in turn, comprise part of the basic structure of cells as well as act enzymatically to regulate cellular function (proteins → structure and function). On an organismal scale, the workings of collections of cells can alter the environment which then acts as a regulator of genes.

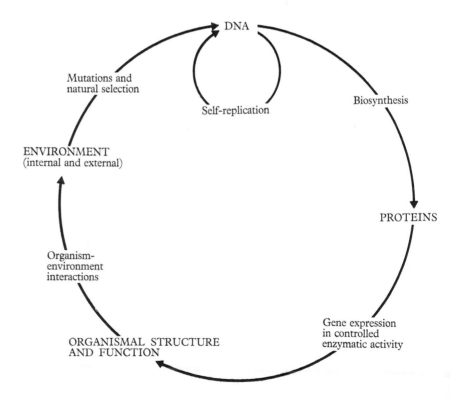

UNANSWERED QUESTIONS

Although the following list is not exhaustive, a few subareas of genetics constitute the forefront of modern research. In each of these fields, facts are being collected at an ever-increasing pace, but not enough is currently known to provide the basis for encompassing theories.

1. Evolution of Genes. In his theory of evolution, Darwin assumed that organisms have always existed; he did not enter into the realm of prebiotic theorizing. But of course, if genes are formed by the replication of other genes, the most natural question is, How did genes begin? Ingenious experiments have been performed in synthetic environments that are thought to resemble those of the early earth. By energizing mixtures of various gases, scientists have been able to isolate amino acids, and nucleic acid precursors formed spontaneously. But the question still remains, How did the informational and replicative mechanisms of nucleic acids and proteins arise?

2. Genes and Behavior. No one disputes the role of genes in determination of physical and metabolic traits such as eye color and enzyme production. Many disputes arise over the role of genes in normal and abnormal behavior, because it is not clear how particular behavior patterns relate to such concrete elements as neuronal pathways and synaptic junctions. To what extent can behavior be ascribed to physical elements in the nervous system? To what extent do these physical elements under genetic control interact? What is the role of nucleic acids in learning and memory? All of these questions lack firm answers and all are presently being pursued with vigor.

3. Genes and Cancer. Cancer cells multiply at a rate faster than that of their normal counterparts. In addition, cancer cells can migrate and exist in areas of the body where the normal cells from which they are derived would not normally reside. Of the three major theories to explain the origin of human cancers, all are basically genetic. One theory holds that accumulation of somatic mutations leads directly to cancer. A second hypothesis is that cancers are caused by the activity of certain viruses, themselves little more than gene packages. The third theory is that cancerous changes are really examples of irreversible differentiation. Some supporters of the last idea believe that human cancer was less widespread in the past because life was too short to permit such changes to occur. In this regard, however, one wonders about leukemia and other malignant growths in children.

GENETIC RELEVANCE

A student of genetics asked, "Why should Arthur Kornberg want to get genes to reproduce in a test tube?" The answer was given by another student, an answer that could apply equally as well to Picasso painting or Beethoven composing, "For that very reason—he wanted to." Obviously, geneticists study genetics because they like it for one reason or another. But the question has relevance in light of the technology that might be built upon genetic discoveries.

The world impact of genetics can be divided into two areas based on what genetics has accomplished and what is theoretically, but not yet practically, possible. On the one hand, discoveries in genetics have accelerated our mechanistic views of man and the universe. After Copernicus, Gallileo, and Newton, the heavens were never so close and the angels lost some of their magic. With continuing emphasis on smaller and smaller genetic entities and with deciphering the structure and role of genes, western man has increasingly wondered about the power of genes. It might be comforting to think that "bad" mental states, such as Mongolian idiocy or perhaps even schizophrenia, were the result of gene action. But now we must wonder about the role of genes in "good" things, such as high intellectual ability or psychological stability. If you add up the bad, the good, and include the neutral that might be under genetic control, not much else is left. Legal cases of individual responsibility are beginning to be questioned because of chromosomal abnormalities. With more accurate techniques available for detection of genetic error and with abortion more widely accepted (or at least tolerated), the social costs of genetically defective individuals have become an area of concern. Philosophical and theological doctrines of free will and behavioral responsibility are being rethought as well. Perhaps the mechanistic push of genetics can best be measured by the counter-reaction of many who do not find solace in the notion that their looks and possibly even actions are a result of particular arrangements of carbon, hydrogen, oxygen, nitrogen, and phosphorus. However, wishes will not make DNA disappear and feelings of cosmic impotence will not keep the molecule from carrying out its mechanistic tasks.

For many individuals, even those who can accept the burden of being no more than genetic packages of chemicals, genetic research is a mixed blessing. One can be excited, even awed, by the incredible complexity of the living cell and by the full realization of the way that this complexity is organized, channeled, and made to respond to change. But if genes are so definite, so tangible, and so capable of being isolated, these same people look with fear at the possibility of genetic engineering. It might be considered moral to tamper with genes in order to cure genetic diseases instead of merely treating the symptoms. However, it might not be acceptable to contemplate the decisions that would haunt us. Who would take the responsibility of deciding what should be changed or left alone? And if we could cure hemophilia, then perhaps we could "cure" the kinds of intelligence that lead to revolutionary ideas as well.

One of the most cogent arguments for increasing the numbers who study modern science, including genetics, rests not with the need for more scientists. Rather it has to do with the impact of living in a society that has come to a new plateau of technological superabundance. Only with widespread understanding of modern theory can society as a whole distinguish between technological progress and human progress. Even more importantly, people must be able to decide if the choice is necessary, and if it is, when or how it shall be made.

QUESTIONS

1. What mechanisms of heredity were postulated by Hippocrates, Aristotle, and Charles Darwin? Compare their relative merits and inadequacies.

2. Discuss the validity of mutilation experiments, such as those performed by August Weismann, as genetic tests. Would experiments that might enhance function (e.g., artificially elevated levels of certain hormones such as the sex hormones, adrenalin, or growth hormone) constitute more valid tests? Why?

3. What were the three major generalizations that Mendel derived from his experiments with garden peas? What observations led him to these principles, and how did he test them?

4. What are the deductive links between the concepts of dominance, gene segregation, and independent assortment?

5. When Bateson crossed red-flowered snapdragons with white-flowered snapdragons, he obtained progeny that were uniformly pink-flowered. When pink snapdragons were crossed with each other, their progeny assorted into three classes that approximated the ratio $\frac{1}{4}$ red: $\frac{1}{2}$ pink: $\frac{1}{4}$ white. What modification of Mendelian principles do these observations suggest?

6. Why is it necessary to provide a theoretical link between the theory of evolution and genes?

7. Hardy and Weinberg showed how Mendelian principles could explain genetic stability under constant environmental conditions. How does this relate to evolution?

8. What is the relationship between the one-gene, one-enzyme concept and the production of a physical trait such as brown eye color?

9. What are the advantages that accrue from many genes being organized in one operon?

2

Chromosomes and Cell Division

Chromosomes had been observed and counted for some years before their genetic significance was fully appreciated. In 1903, Walter Sutton first clearly identified Mendel's genetic units with the chromosomes, pointing out the direct correspondence between them with respect to pairing, segregation, and independent assortment. Eight years later, Thomas Morgan showed that a particular gene (white eye) in *Drosophila melanogaster* was carried by a particular chromosome (the X chromosome), corroborating the Chromosome Theory of Inheritance. Subsequent work by the remarkable group of geneticists at Columbia University (Sturtevant, Morgan, and Bridges) and others demonstrated that not only do particular genes reside on particular chromosomes, they occupy very specific places or *loci* on them so that they can be mapped. Geneticists now viewed genes as finite entities that occupy specific places on specific chromosomes, rather like beads on a string. This concept of genes was easily grasped and facilitated much imaginative research. The Chromosome Theory of Inheritance, therefore, emerged as a well-articulated concept which held, in effect, that the chromosomes are the repositories of hereditary potentialities, namely, the genes. Further, this theory held that by the kind of cell division known as meiosis, a single set of chromosomes, and therefore of genes, is segregated into each germ cell. At fertilization, two sets of chromosomes—one of paternal and one of maternal origin—are included within a single cell, the zygote. Then by mitosis, each pair of chromosomes is transmitted with neither loss nor gain to each of the cells that ultimately comprise a new, individual organism.

An understanding of chromosome structure and the details of meiosis and mitosis is essential to any understanding of how genes are transmitted from one generation to the next, how they are disseminated throughout the cells of each generation, how both genetic diversity and genetic uniformity can be provided for, and how certain

15

kinds of abnormalities or "genetic accidents" can occur. Underlying the principles of mitosis and meiosis are the molecular mechanisms of gene replication and recombination, which are discussed below.

CHROMOSOMES

Chromosomes are finite bodies that appear in *eukaryotic* cells during cell division and are readily visible in appropriately fixed and stained cells with the aid of a compound microscope, or in living cells by phase-contrast microscopy. Although the precise arrangement of a chromosome's chemical components is still far from being clearly understood, it is known that they consist largely of

Fig. 2–1.   Scanning electron micrograph of a metaphase chromosome from a cell of a Chinese hamster. The middle constricted zone marks the location of the centromere. The white bar represents 1μ. [Courtesy of Dr. C. K. Yu (1972), *Canadian J. Genet. Cytol.* **13**.]

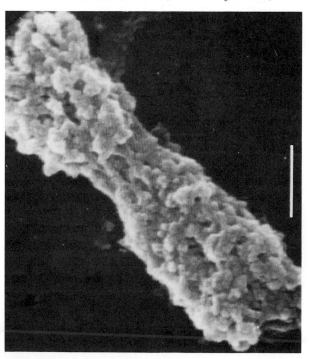

DNA, RNA, and certain kinds of proteins. The DNA occupies the interior or core of the chromosomes in a tightly coiled state, and so the chromosomes are rather compact bodies (Fig. 2–1). All chromosomes are characterized as having a *centromere* (also known as a *primary constriction* or *kinetochore*), inside of which the DNA is not so tightly coiled, RNA is scanty or lacking, and one or two minute granules called *kinetochore granules* are often present (Fig. 2–2). The centromere is the region of attachment of the spindle fiber during cell division. Chromosomes are classified according to their length and the location of their centromere, both of which are constant features for any given chromosome at certain times in the cell cycle. If the centromere is at or very near one end, the chromosome is classified as *acrocentric;* if it is in or near the middle, it is called *metacentric.*

Certain regions of chromosomes are concerned with the production of nucleolar material and are known as *nucleolar organizers* accordingly. In most higher forms of animals and plants, these regions have become concentrated and restricted to a specific site in each member of a pair of homologous chromosomes (Fig. 2–2). At this site, short segments of the DNA of the chromosome are proliferated (specific DNA amplification) at a certain time in the cell cycle to facilitate rapid synthesis of nucleolar RNA, as described in Chapter 7.

In *haploid* eukaryotic cells, such as those that occur among the green algae and molds, chromosomes occur singly—that is, there is only one of a given kind of chromosome in terms of the complex of genes that it carries. In *diploid* cells, such as the somatic cells of most higher animals and plants, the chromosomes occur in identical pairs called *homologous chromosomes* or *homologues.* Consequently, in diploid cells there are two of every gene distributed among the pairs of homologous chromosomes, whereas in haploid cells there is only one. An exception to paired gene representation in diploid cells is found in those species that have sex chromosomes, as described in Chapter 5. In these cases, the cells that have the XY or XO

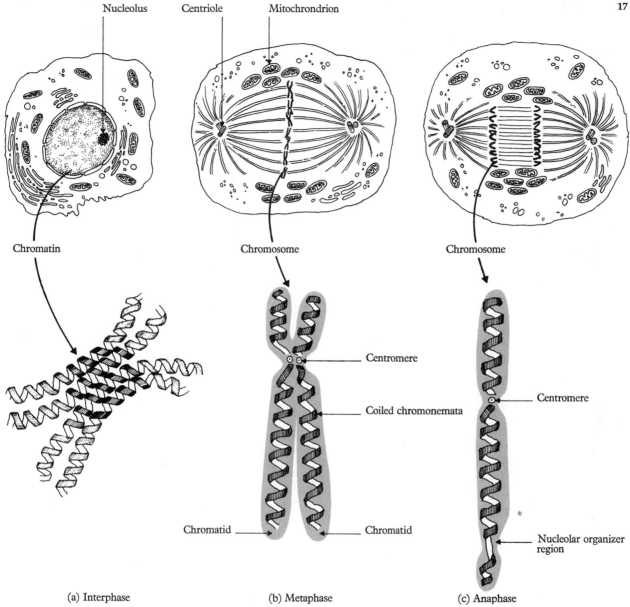

Nucleolus   Centriole   Mitochrondrion

Chromatin

Chromosome

Chromosome

Centromere

Coiled chromonemata

Chromatid

Chromatid

Centromere

Nucleolar organizer region

(a) Interphase   (b) Metaphase   (c) Anaphase

Fig. 2–2. Chromosome structure. (a) During *interphase,* the chromosomal material appears as dispersed clumps of stainable material called "chromatin" located within the nucleus. By electron microscopy the clumps appear as aggregations of coiled strands and granules that are comprised chemically of DNA, RNA, and protein. (b) In the *metaphase* stage of mitosis, the chromosomes are fully formed into pairs of distinct rod-shaped bodies (chromatids) that are joined at the centromere and include a double set of coiled strands (chromonemata). An acrocentric chromosome is shown, enlarged. Compare with Fig. 2– 1. (c) During *anaphase* of mitosis, the chromatids separate and become independent chromosomes that move to opposite poles of the cell. A metacentric chromosome that includes a nucleolar organizer region is shown, enlarged.

constitution have only one of the genes that are carried on the X or the Y, but two of the genes that are carried by the other chromosomes.

The formation of chromosomes during cell division appears to represent an adaptive device for facilitating the uniform distribution of genetic material between the daughter cells being formed. The practicality of this can be appreciated if one considers the physical state of the genetic material—the DNA—during interphase when the cell is not dividing. During interphase the DNA is largely uncoiled, and it appears as a number of extremely long, delicate strands throughout much of the nucleus. The reason for their being uncoiled during interphase will become evident when DNA function is considered. But surely in this state the numerous DNA strands could easily become tangled and/or broken by the very mechanics of cell division, and cell division would likely be a hit-or-miss process, which clearly it is not.

When a cell divides to become two cells, it is essential that every gene be represented in each of the two new cells, called the *daughter cells*. Inasmuch as the genes are distributed among the several chromosomes of a cell, it is therefore essential that each of the chromosomes be represented in each of the daughter cells. If genes *A* and *B*, say, are located in duplicate on a pair of acrocentric chromosomes, and genes *P* and *Q* in duplicate on a pair of metacentric chromosomes, then each daughter cell, if it is to be diploid, must have both the genetic constitution *AABB PPQQ* and the chromosomal complement of two acrocentrics and two metacentrics (Fig. 2–3b). Similarly, if a cell is haploid and has only one of each of these genes and, correspondingly, only one acrocentric and one metacentric chromosome, its daughter cells must have identical composition in these respects (Fig. 2–3a). The mechanism by which the requisite genetic and chromosomal constancy is regulated and achieved in called *mitosis*.

In organisms that reproduce sexually, that is, by combining the genetic information of two different individuals, the sex cells (*gametes*) must

Fig. 2–3. Mitosis assures transmission of the same number and kinds of chromosomes from one cell generation to the next, irrespective of whether the parent cell is (a) haploid (1n) or (b) diploid (2n). In metaphase of mitosis, each chromosome appears as a pair of chromatids that have been produced by replication of DNA. Pairs of chromatids are joined by the centromere, which does not split until the end of metaphase, and are aligned on the metaphase plate (a plane perpendicular to the spindle axis and midway between its poles) randomly and independently of other pairs of chromatids. After the centromeres split, the chromatids—now daughter chromosomes—move to opposite sides of the cell as the cell divides in two between them. Thus diploid cells produce diploid cells, and haploid cells produce haploid cells. The symbols *A*, *B*, *P*, and *Q* designate hypothetical "gene" locations. In the diploid cells, the members of pairs of genes are distinguished by contrasting shades. Genes *A* and *B* are represented as occupying acrocentric chromosomes (centromere in a terminal or subterminal position) and *P* and *Q* as occupying metacentric chromosomes (centromere in or near the middle).

satisfy two primary requirements. The first is that these cells must be haploid, and the second is that every gene—and therefore every chromosome—must be represented. In order to be haploid, it follows that each gene and, correspondingly, each chromosome must be represented singly rather than in pairs. Therefore, if the cell from which gametes are to arise is diploid, the cell division by which they arise must be organized somewhat differently than in mitosis. For example, the germ cells produced by a cell with genetic composition (*genotype*) *AABB PPQQ* and chromosomal composition (*karyotype*) of two acrocentric and two metacentric chromosomes, as cited above, must have the genotype *AB PQ* and the karyotype of one acrocentric and one metacentric. The mechanism by which the reduction in numbers of genes and chromosomes is regulated and achieved is called *meiosis*.

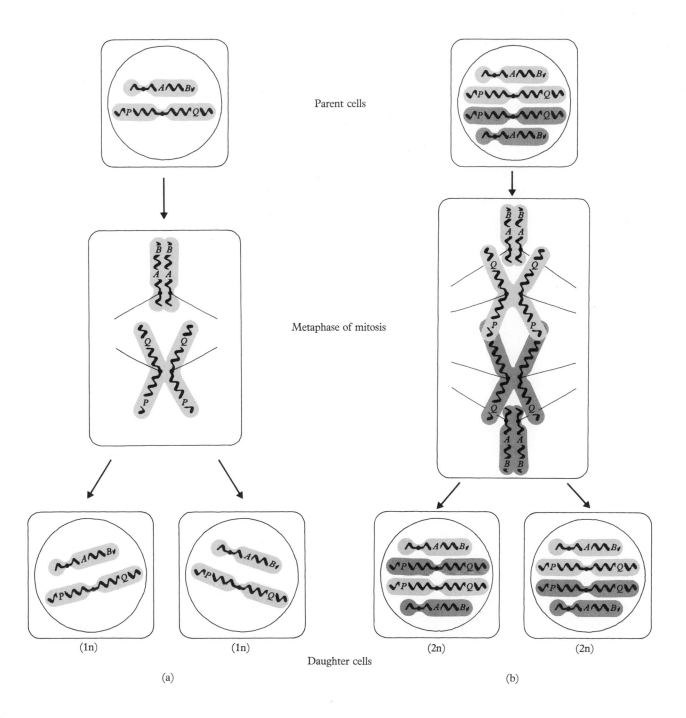

Parent cells

Metaphase of mitosis

Daughter cells

(1n)   (1n)   (2n)   (2n)

(a)   (b)

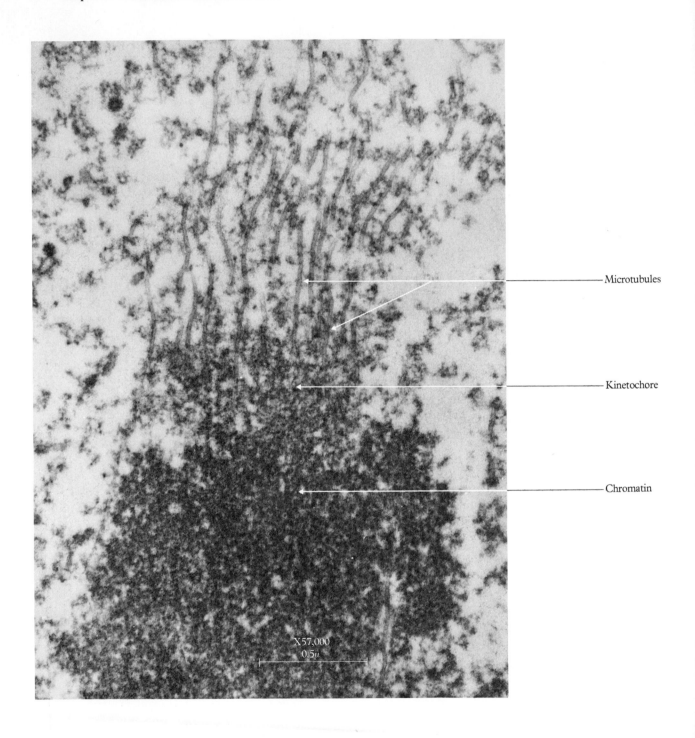

Microtubules

Kinetochore

Chromatin

MITOSIS

Mitosis may be defined as the mechanism of cell division in which a mitotic spindle is formed and by which the genetic and chromosomal composition of a cell is faithfully reproduced in each of the two daughter cells resulting from that division. Mitosis is accomplished by a remarkable process of chemical engineering that involves not only packaging the genes of DNA into chromosomes, but also the formation of the *mitotic spindle* referred to above.

The Mitotic Spindle

The mitotic spindle is a unique system of microtubules (usually referred to as spindle fibers) that both guide and apparently "pull" the products of chromosome replication to opposite poles of the cell. The spindle forms by the assembly of specific protein subunits into microtubules, which are hollow strands that measure about 250 A in diameter (varying with the fixation procedures followed) and anywhere from a few millimicrons to several microns in length. The microtubules become oriented parallel to each other to appear collectively as a spindle-shaped body, from which it derives its name.

The site of origin of spindle microtubules is not clear. Some of them appear to originate from the kinetochore (Fig. 2–4). These form relatively thick bundles that comprise the "traction fibers" referred to by light microscopists. It is believed that the kinetochore granules might represent specific gene products that are directly concerned with the polymerization of these microtubules (Luykx, 1970). Other spindle microtubules appear to be assembled by or in close association with the centrioles in those cells that have centrioles (de Harven, 1968). Still others appear without any apparent reference to particular organelles.

The actual division of the cell during mitosis involves both the movement of chromosomes toward opposite poles of the spindle and segregation of the cell's contents into two daughter cells. The cell is pinched in two by progressive constriction along a girdling cleavage furrow (animal cells) or by the development of a new cell membrane or "cell plate" that cuts across the cell (plant cells). After cell division is accomplished, the spindle microtubules are disassembled and disappear. These phenomena are discussed further in the next section.

◀ **Fig. 2–4.**   A segment of a chromosome of a plant, *Lilium longiflorum*, during meiotic metaphase I showing the kinetochore and microtubules in longitudinal section. [J. P. Braselton and C. C. Bowen (1971), *Caryologia* (Firenze) **24**:49–58. By permission.]

Chromosomal Events and the Cell Cycle

The physical and metabolic activities of growing cells are cyclic—that is, they constitute a sequence of events that is regular and repetitive as long as the cell is growing and dividing. This *cell cycle*, as it is known, is characterized by four more or less distinct stages or phases that have been designated, in order, the G_1-, S-, G_2-, and M-phases (Fig. 2–5).

The G_1-Phase.   With the possible exception of cells undergoing early cleavage (cf. Chapter 11), cells newly formed by mitosis enter a metabolic phase of RNA and protein synthesis that is concerned with that cell's biological maintenance and, if it has any, specialized function. From the standpoint of cell growth, that is, cell division, this phase represents a kind of "gap" in the cell cycle, and so it is called the first gap or G_1-phase. Surely a more generally descriptive designation could have been devised, but G_1 has the advantage of brevity and it conveys as much to cell biologists as any other term would. This is the most prolonged phase in most somatic cells of adult organisms—in fact, most highly differentiated cells such as skeletal muscle fibers and neurons remain permanently in this phase, which might be eighty years or more in some individuals. In embryonic tissues, this is the phase during which the cells increase in volume by imbibing water and nutrients and building new protoplasm and cytoplasmic organelles. The latter may include principally endoplasmic reticulum, Golgi membranes, ribosomes, mitochondria,

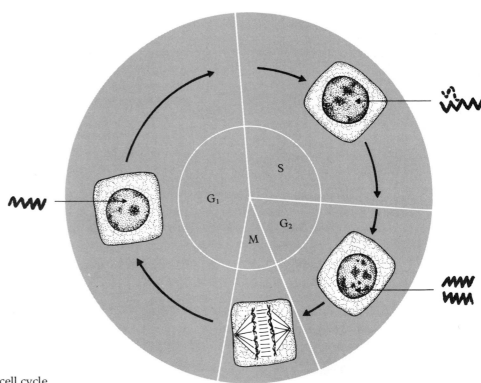

Fig. 2–5. Diagram of the cell cycle. Phases G_1, S, and G_2 constitute interphase, and M is the stage of cell division. The appearance of the cell and the condition of the DNA strands characteristic of each phase are indicated. The dashed line symbolizes DNA strands in the process of replication. (See text for further explanation.)

chloroplasts, and/or cilia. Secretion granules, storage granules or droplets, cell-wall materials, and the like may also be elaborated during this phase according to the particular cell's requirements and functions. So, although this phase may represent merely a "gap" in the reproductive life of a cell, this gap constitutes the essence of a cell in many instances.

The S-Phase. The S-phase is perhaps the most critical phase of the cell cycle, for it is during this interval that the major parts of DNA synthesis and replication take place. (It is called the S-phase because this *synthesis* is its most significant activity.) This can be demonstrated clearly by the uptake and incorporation of specific DNA precursor substances at this time. If a cell is administered molecules of the precursor *thymidine,* which is utilized specifically for DNA synthesis, this substance will be incorporated in significant amounts only during the S-phase. This can be shown by using thymidine in which *tritium*, a radioactive isotope of hydrogen (^{3}H) has been substituted for some of the hydrogen atoms in the molecule's structure. Cells that have been so treated are prepared in thin films by histologic section or "squashing," washed to remove unincorporated thymidine, covered with a thin photographic emulsion, and stored in darkness at low temperature. After an extended period of time dictated by the rate of decay of the tritium atoms, the preparation is developed photographically. The physical location of the molecules of tritiated thymidine can be determined by the appearance of developed silver grains in the overlying emulsion when the film is observed with a compound microscope. If the treated cell was in the S-phase of the cell cycle at the time of thymidine administration, the silver grains are seen concentrated over the nucleus and their pattern of distribution is indicative of strand structure (Fig. 2–6). Moreover, if cells treated with tritiated thymidine during the S-phase are permitted to enter mitosis before they are fixed and examined, the distribution of silver grains corresponds exactly with the chromosomes. If, on the other hand, tritiated thymidine is administered at any other time in the cell cycle, it is not incorporated in any significant amount within the cell and it is washed out when the preparation is washed. Consequently, any silver grains that are seen are few in number and no pattern is evident (Fig. 2–6).

The G$_2$-Phase. After the S-phase and as a consequence of it, the cell is usually committed to division, although cases of G$_2$-arrest have been described in the literature. The second "gap" phase (G$_2$) represents the interval during which the cell's primary activities are concerned with synthesizing the RNA and proteins needed for chromosome synthesis and for the mitotic spindle, although some chromosomal protein is synthesized in the S-phase. The G$_2$-phase is generally of comparatively short duration, as indicated in Fig. 2–5.

These phases of the cell cycle are distinguishable biochemically but they are not distinguishable morphologically. To the microscopist they constitute the nonmitotic stage called *interphase.*

The M-Phase. This is the phase during which the visible events of mitosis or meiosis occur. It begins when the chromosomes become structurally distinct bodies, and it terminates with the formation of nuclear membranes in the new daughter cells that are produced as the consequence of division. The M-phase is the most complex phase in terms of visible morphological phenomena, and so it is subdivided into a series of stages to facilitate its description. These stages are known as *prophase, metaphase, anaphase,* and *telophase.* Where one stage ends and another begins is rather arbitrary because the stages themselves are arbitrarily designated segments of a continuous process, almost as arbitrary as the terms "infancy," "childhood," and "adolescence" in the continuous process of human development.

a) Prophase. The beginning of prophase is marked by the appearance of the chromosomes as slender threads in the nucleus. The most conspicuous

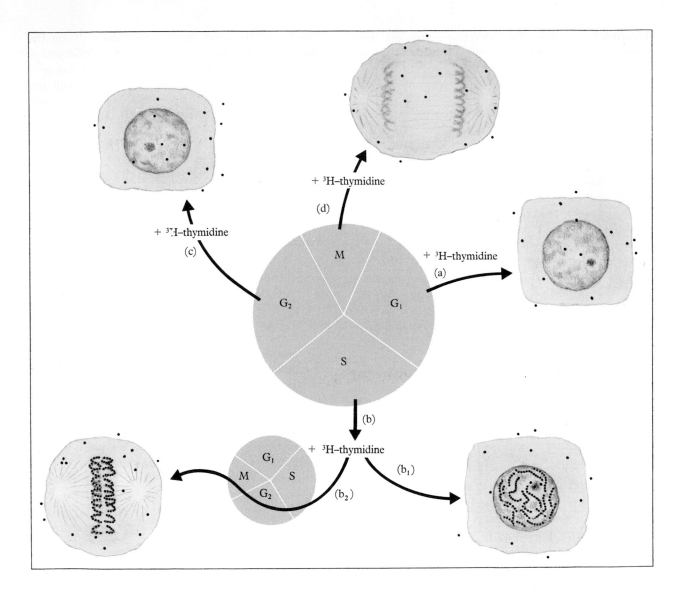

Fig. 2–6.   The occurrence of major DNA synthesis and replication can be demonstrated by the incorporation of radioactive thymidine (^{3}H-thymidine). This diagram shows the distribution of radioactive particles in cells that have been fed labeled thymidine in each of the four phases of the cell cycle. Only when the label is administered to cells in the S-phase (b) is it incorporated in significant amounts. If cells treated during S-phase are fixed while still in this phase, the label is found concentrated in the nucleus (b_1). If such cells are permitted to proceed into the M-phase before fixation, the label is seen to be concentrated in the chromosomes (b_2).

parts of the threads in early prophase are the pairs of strands or *chromonemata* that form their core, one pair per chromosome. The chromonemata are loosely coiled in early prophase and, consequently, they appear long and thin (Fig. 2–7). Cytochemical staining and tritium labeling show that the chromonemata contain RNA and protein as well as DNA. Certainly a single chromonema could not represent a single helical molecule of DNA, for the cross-sectional diameter of a molecule of DNA (approximately 20 A in electron microscopic preparations) is far beyond the resolving power of a compound microscope, whereas chromonemata are readily observable. It has been proposed that the chromonema may include multiple strands of DNA (Moses and Coleman, 1964), but this view is generally regarded as presenting more problems than it solves from the standpoint of genetic analysis. Moreover, Laird (1971) found that mouse chromatids contain only single rep-

Fig. 2–7. Mitosis in (a) animal and (b) plant cells compared. (See text for explanation.) The stages of mitosis are prophase, metaphase, anaphase, and telophase. Note the presence of centrioles in the animal cells, as well as division of the cytoplasm (cytokinesis) by constriction of the cell membrane (arrows). The cells of higher plants lack centrioles, and cytokinesis is effected by the formation of a cell plate which expands centrifugally (arrows).

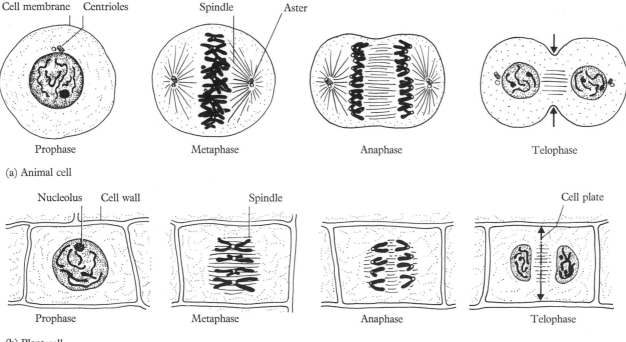

Cell membrane   Centrioles   Spindle   Aster

Prophase   Metaphase   Anaphase   Telophase

(a) Animal cell

Nucleolus   Cell wall   Spindle   Cell plate

Prophase   Metaphase   Anaphase   Telophase

(b) Plant cell

resentatives of most gene components, which argues against the multiple-strand hypotheses. In any event, there is little question that the chromonemal strands are DNA and RNA complexed with certain nuclear proteins (Busch et al., 1964). The role of the proteins in chromosome structure is still not understood, but they are believed to be concerned both with suppressing RNA synthesis during mitosis and with giving structural support to the chromosomes.

As prophase progresses, the chromonemata become more and more tightly coiled and the chromosomes become correspondingly thicker in cross-sectional diameter and shorter in length. Distinguishing separate chromosomes becomes easier and their number can often be counted at this time. Concurrent with chromosomal condensation are progressive disintegration of the nucleolus and the nuclear membrane and the development of the mitotic spindle. In the cells of animals and lower plants, the spindle originates between pairs of centrioles that have come to lie on opposite sides of the nucleus (Fig. 2–7). Emanations of other microtubules around the centrioles constitute the *asters* of animal cells. Centrioles are lacking in cells of higher plants but the spindle develops nevertheless, except that asters are not present.

The end of prophase is marked by completion of nuclear membrane disintegration and spindle formation. The chromosomes are condensed nearly to their maximum and their centromeres are becoming aligned along the metaphase plate. The nucleolus has either disappeared or is represented by a scattering of nucleolar fragments.

b) **Metaphase.**   When alignment of the centromeres on the metaphase plate is completed, the metaphase stage is reached. This stage, although not as prolonged in time as prophase, persists longer than the subsequent stages. The chromosomes become maximally condensed and the chromonemata in each chromosome separate from

each other except at the centromeres, thereby becoming quasi-independent bodies called *chromatids* (Fig. 2–2). Spindle microtubules attach firmly to kinetochores on opposite sides of the centromeres, and then the centromere replicates. Replication of the centromere constitutes completion of chromosome duplication and the chromatids become sister chromosomes. The genetic composition of sister chromosomes is identical because the DNA in each has arisen during the S-phase by DNA replication, which, as will be shown in Chapter 7, produces identical replicas.

c) **Anaphase.**   Replication of the centromeres enables the sister chromosomes to move apart as the spindle elongates. This is the stage of mitosis called anaphase (Fig. 2–7). As the chromosomes separate, division of the cytoplasm and parceling out of cytoplasmic constituents takes place. In animal cells, this is initiated during anaphase by constriction of the cell membrane in the plane of the metaphase plate, thereby pinching the daughter cells apart. In plant cells, which are characterized by a relatively rigid cell wall, cytoplasmic division is effected by the synthesis of a *cell plate* in the center of and in the same plane as the metaphase plate. The cell plate is a flat disc of cytoplasmic membranes and the products of their synthetic activities, the primordia of a new cell wall (see Fig. 2–7). Beginning at the center of the metaphase plate, the cell plate expands peripherally until it meets and fuses with the lateral cell walls, whereupon division of the cytoplasm is completed. This process is also initiated in anaphase.

The role of the spindle microtubules in the dynamics of anaphase can be demonstrated by treating mitosing cells with substances that are known to interfere with microtubule formation or function. The alkaloid drug colchicine is such a substance. It prevents the assembly of protein subunits that comprise the microtubules and anaphase does not occur. The chromosomes remain in the middle of the cell, the nuclear membrane is reestablished, and the cell enters the G_1-phase with double the normal chromosome complement.

d) Telophase.   Once the chromosomes have completed their journey to opposite poles of the cell, the telophase stage is reached. During this stage the nuclear membrane reassembles around the tight cluster of chromosomes at each pole, the chromonemata unwind, chromosomal protein is evidently dispersed, and division of the cytoplasm, which was initiated in anaphase, is completed. As the chromosomal protein becomes uncoupled from the DNA of the chromonemata, the DNA of the nucleolar organizers becomes activated and the new nucleolar material is synthesized. Meanwhile, without any clear-cut event of transition, the cell enters the G_1-phase of the cell cycle and mitosis is completed.

In summary, mitosis results in the production of two daughter cells that are identical with respect to chromosomes and chromosomal genes. The original chromosomes are positioned one-by-one on the metaphase plate independently of one another, and their chromatids, which are exact duplicates, are separated so that one example of each chromosome goes into each daughter cell. The mechanism is so precise and so well regulated that misadventure within the limits of a normal environment occurs with extreme rarity—how often is not known, but its frequency is probably less than one in a hundred thousand times.

BINARY FISSION

Lest the student infer that mitosis is the only mechanism by which genetically identical daughter cells are produced by cell division, it should be pointed out that *prokaryotic* cells also do this, but neither chromosomes nor a mitotic spindle is involved. As implied earlier, the DNA of prokaryotic cells exists as a single, double-helical molecule (Fig. 10–13). Preparatory to cell division, this molecule replicates just as the individual DNA molecules in eukaryotic cells do, but then the two copies separate from each other directly without benefit of a mitotic spindle and are included in separate daughter cells by a relatively more straightforward kind of cell division known as *binary fission*. This is possible because only a single molecule of DNA is involved. In spite of the fact that prokaryotic DNA does not become packaged into chromosomes, the term "chromosome" is often used as a synonym for the DNA molecule in these organisms. Although technically incorrect, the usage is convenient. It should also be mentioned that the reproduction of plasmids (extrachromosomal replicating bodies) is nonmitotic and like that of prokaryotes. The details of cell division in prokaryotes will be considered later in the context of DNA replication.

MEIOSIS

Meiosis resembles mitosis in many ways. It is preceded by the same sequence of phases in the cell cycle (G_1, S, G_2) and all or nearly all DNA replication occurs in the S-phase. Chromosomes are formed and cell division is effected by a mitotic spindle. The major differences are that (1) the M-phase involves two cell divisions even though preceded by only one DNA replication, (2) the first division is characterized by a more complex and prolonged prophase which includes the pairing of homologous chromosomes (synapsis), (3) with rare exceptions, segmental exchange (genetic recombination or crossing-over) takes place between homologues, and (4) although every gene is represented in each of the daughter cells, these cells are generally not genetically identical. The reasons for these differences become clear when one considers the difference in function between the cells produced by the two kinds of cell division, especially in diploid multicellular organisms—mitosis assures genetic uniformity and diploidy throughout the body cells of the individual, whereas meiosis provides genetic diversity and haploidy among the reproductive cells or gametes.

Because meiosis involves two cell divisions which differ in significant ways, Roman numerals I and II are appended to the stages of the first and second divisions, respectively.

a) Prophase I.   The first meiotic prophase is prolonged in time and is divisible into several well-defined substages. As in other kinds of cell division, the process is a continuous one and there is

granules, *chromomeres*, that appear at irregular intervals along their length. Studies with radioactive tracers indicate that there are two chromonemata in each chromosome, just as in mitotic chromosomes, but they cannot be distinguished clearly by light microscopy. There is yet no evidence of spindle formation and the nucleolus is intact, as it usually is during the major part of prophase I.

Zygotene.   This is a critical stage in meiosis. Homologous chromosomes seek each other out by mechanisms that are not yet understood and pair

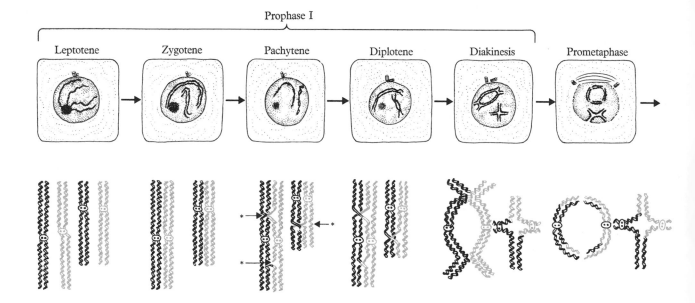

an element of arbitrariness in substage designation. These substages are *leptotene, zygotene, pachytene, diplotene,* and *diakinesis* (Fig. 2–8).

Leptotene.   This is the stage of assembly of the chromosomes. As in the early prophase of mitosis, the chromosomes appear as slender threads, but these threads are characterized by clusters of

up side-by-side, chromomere-by-chromomere. This pairing, called *synapsis,* has been shown to be very precise and, with the exception of polytenic chromosomes which will be described later, it occurs usually during meiosis only. Although the mechanism by which homologues "find" each other and commence pairing is not known, it is known that they become attached next to each other at well-defined points on the nuclear membrane (Comings and Okada, 1970) and that a system of protein microfibrils, the *synaptonemal complex*, forms between them. This complex consists of two parallel lateral elements and a thin central element lying between them (Fig. 2–9). The space between lateral elements and the central element is bridged by microfibrils, and other microfibrils bush out from the lateral elements in all directions. The chromosomes appear to become attached to these bushy fibers at intervals, and these intervals are believed to correspond to the chromomeres

(Comings and Okada, 1970). Synapsis is believed to be a two-step process consisting of (1) chromosome pairing, during which the synaptonemal complex draws the homologues into approximate association with each other, and (2) molecular pairing, during which gene-by-gene matching takes place. The first step is probably initiated during leptotene, because of the appearance of chromomeres at this stage, and completed in zygotene when all of the chromomeres of homologues are visibly brought together. The second step, which is doubtless initiated during zygotene, is probably

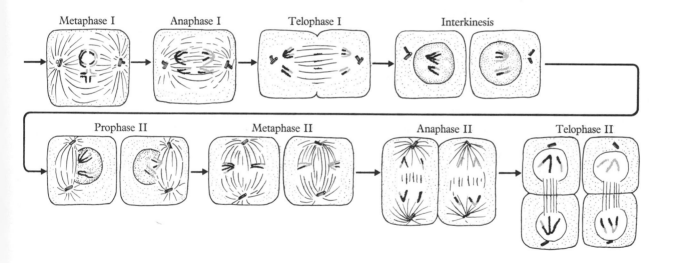

Fig. 2–8. The stages of meiosis in a schematized animal cell. The physical states of the chromosomes and of their chromonemata in each stage of prophase I are diagrammed separately to show more clearly the phenomena of synapsis, recombination, and chiasma formation. Points of breakage and exchange are indicated by *. (See text for description.)

not completed until the next morphological stage, pachytene. Although the synaptonemal complex is not always essential for chromosome pairing, it is always present in those situations where crossing-over occurs, and so it appears to be essential for this process (Moses and Coleman, 1964).

stage. Each of the chromosomes, however, consists of two chromatids which can be distinguished in carefully prepared specimens (Fig. 2–8). Both genetic and cytological evidence indicate that during pachytene, physical breaks in the chromatids occur and are repaired, but in a statistically pre-

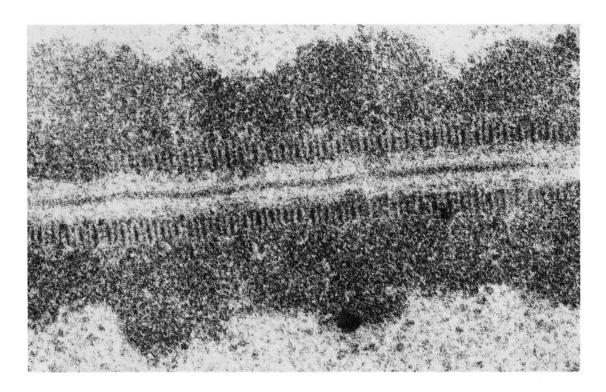

Fig. 2–9. Electron micrograph of a synaptonemal complex at pachytene in the mold *Neottiella*. (See text for explanation.) Magnification X154,000. [Courtesy of Dr. D. von Wettstein (1971), *Proc. Nat. Acad. Sci.* **68**:851.]

Pachytene. Shortening of strands by internal coiling during leptotene and zygotene results in the chromosomes appearing considerably thickened in pachytene (Greek, "thick strand"). The two homologues are in such intimate synapsis that it is difficult to resolve them microscopically at this

dictable number of cases the repair entails crossing-over between homologous chromatids (Fig. 2–8) and consequent formation of a cross or *chiasma* (pl., *chiasmata*) at the point of exchange. The precision with which the breaks and crossing-over occur in terms of location between corresponding parts of the DNA molecules involved is presumably facilitated by the synaptonemal complex. The DNA repair that is implicit in this process could account for the small amount of DNA that is synthesized during pachytene, as measured by tritiated thymidine incorporation (cf. Hotta and Stern, 1971).

Crossing-over between homologous chromosomes constitutes one of the most important events in meiosis, for it is the mechanism by which the genes that constitute a homologous pair of chromosomes can be "shuffled" and exchanged, thereby effecting a variety of genetic combinations that is potentially almost limitless. The amount of genetic variability that can be provided by shuffling whole chromosomes, which is the basis of Mendel's Law of Independent Assortment, is limited by the number of chromosome pairs, which in some species is rather small. Moreover, independent assortment is of limited effectiveness in shuffling genes of related functions, for these tend to be clustered on the same pair of homologous chromosomes. Crossing-over, in conjunction with mutation, represents the most effective means of providing the kinds and amounts of genetic variability that were essential to the relatively explosive evolution that plants and animals have undergone.

The chiasmata that are formed as a consequence of crossing-over are also important. They are believed to help hold the complex of chromatids together until they are aligned on the metaphase plate, thereby reducing the probability of unequal distribution of the chromosomes during anaphase. For reasons not yet understood, aged oocytes do not form many chiasmata and the probability of unequal distribution of chromosomes during meiosis increases accordingly, especially for the short chromosomes. This has been proposed to be one of the reasons for the correlation that exists between the incidence of Down's Syndrome and other similar congenital abnormalities and the age of the mother (cf. Chapter 9).

Diplotene. This stage is characterized by cytologists as the stage in which the pairs of chromatids separate to the extent that they are visible by light microscopy as a pair of strands (Fig. 2–8). The chiasmata are clearly evident in diplotene and perhaps constitute its major diagnostic feature. Chromatid separation implies that the synaptonemal complex is no longer functional and starts to disperse. Whether or not this is so has yet to be ascertained.

Diakinesis. This stage is marked by the appearance of rather unique configurations of the chromatids, and also by disintegration of the nucleolus and the beginning of spindle formation. The chromatid configurations are a consequence of the repulsion of chromatid pairs, which appears to start from the centromeric region and spread out along the length of the chromatid strands (Fig. 2–8). The chiasmata are believed to impede repulsion but not to nullify it, for the chiasmata are forced toward the ends of the chromatids as repulsion proceeds. If you hold your arms out in front of you so that they cross at about the elbows, and then slowly move them apart until the fingers of one hand touch those of the other, you will imitate the way in which a chiasma slides to the ends of the chromatids.

b) Metaphase I. During the early part of metaphase I, the nuclear membrane disintegrates and the spindle microtubules move the chromosomes into position on the metaphase plate. Metaphase I in diploid cells differs from mitotic metaphase in three striking ways (compare Figs. 2–7 and 2–8). First, the diameter of the spindle at the metaphase plate is considerably less in metaphase I. This is because the chromatids are clustered in groups of four rather than groups of two, thereby halving the number of positions on the metaphase plate. Second, the centromeres are positioned on both sides of the metaphase plate rather than directly on it, whereas the opposite is true of the chromatid ends (Fig. 2–8). And third, the centromeres do not "split" in metaphase I as they do in mitotic metaphase, and so the chromatids remain chromatids.

c) Anaphase I.   If it were possible to label maternal and paternal centromeres differentially, one would see them migrating toward opposite poles during anaphase I. This does not represent a breakup of the marriage that brought them together at fertilization, however, because the maternal centromere of one set of chromatids may go toward one pole and the paternal centromere of another set may go to the same pole. Moreover, crossing-over has resulted in a shuffling of maternal and paternal segments of chromatids, so that each daughter cell produced after anaphase I receives a good mix of both maternal and paternal genes.

d) Telophase I.   The chromatids are bunched at the poles in telophase I and division of the cytoplasm takes place. In some species a nuclear envelope is reestablished; in others it is not. If it is reestablished, the daughter cells pass into a brief transitional stage called *interkinesis,* which resembles morphologically an abbreviated interphase because the chromatids tend to "fade out." With or without interkinesis, the daughter cells then pass into prophase II.

e) Prophase II.   This stage resembles mitotic prophase with respect to duration and formation of a new spindle. It differs from mitotic prophase in diploid organisms in that only half the number of chromatids is represented (Fig. 2–8).

f) Metaphase II.   In metaphase II, the centromeres are aligned on the metaphase plate and spindle microtubules attach to them from opposite poles. Then the centromeres replicate and the two cells immediately enter anaphase II.

g) Anaphase II and Telophase II.   The chromosomes migrate toward opposite poles and division of the cytoplasm follows. The four daughter cells produced are haploid and, if the parent cell possessed genetic heterogeneity and crossing-over has occurred, they are all genetically different from one

another (Fig. 2–8). A nuclear membrane forms in each cell and meiosis is completed.

In the foregoing description of meiosis, emphasis was placed upon the nuclear phenomena which are essentially alike in the production of both male and female gametes. However, what becomes of the nuclei and what becomes of the cytoplasm differ not only between the sexes, but between animals and plants. These differences are largely irrelevant to the basic principles of genetics except in two important respects. One of these concerns the apportionment of extrachromosomal (cytoplasmic) "genes" such as chloroplasts and mitochondria, and the other concerns the formation of a specialized tissue in seed-plant reproduction, the endosperm. These differences are best understood if they are described in the context of reproductive cycles, discussed where appropriate in the chapters that follow.

QUESTIONS

1. How does the centromere differ in structure from the rest of the chromosome? What is its functional significance?

2. Suppose you had a photograph of the karyotype of a cell in a particular organism. How would you determine whether the cell was diploid or haploid?

3. What requirements of continued reproduction are served by mitosis? meiosis?

4. What differences can you discern between the spindle microtubules and their attachments in plants and animals (see Fig. 2–4)?

5. Why is each stage of the cell cycle prerequisite to the next?

6. Distinguish between the events of the G_1- and G_2-phases of the cell cycle.

7. Outline the events that occur duing prophase, metaphase, anaphase, and telophase of mitosis.

8. If a cell has four chromosomes at the end of mitotic prophase, how many are present at the end of metaphase?

9. Distinguish between a chromosome and a chromatid.

10. Is the role of the spindle considered to be an active one or a passive one during mitosis? Why?

11. What are the major differences between binary fission and mitosis?

12. Outline the similarities and differences between mitosis and meiosis.

13. Starting with a cell containing two chromosomes, draw the subsequent division stages of mitosis and meiosis.

14. What are the major chromosomal events that occur during prophase I of meiosis?

15. What is the structure and supposed function of the synaptonemal complex?

16. How does crossing-over increase genetic variability? What is the basic cause of variability? In addition to crossing-over, in what other way is variability increased during meiosis?

17. What relationship is thought to exist between the structural events of chiasmata formation and the functional events of crossing-over?

18. Correlate the principles of Mendel with the specific events of meiosis.

19. How many chromosomally different daughter cells could arise from the meiotic division of a cell containing four chromosomes? (Consider the maternal and paternal members of one pair as being "different.") How many different daughter cells could be produced by meiosis of a cell in which $2n = 6$?

20. In which events (phases) do the laws of chance operate during meiosis? Explain your answer(s).

21. A diploid cell with four chromosomes is observed to be undergoing division. How could you determine if the division was mitotic or meiotic?

3

Segregation, Independent Assortment, and Alleles

Most of what is known at the present time about how genetic information is transmitted from one generation to the next was learned before the genetic material itself was either identified or described. As stated in Chapter 1, Gregor Mendel placed the postulate of particulate genetic factors, now called genes, on a base that was supported by repeatable, scientific evidence. It is appropriate to consider some of the evidence now, as well as the other concepts to which it gave rise.

One of the pairs of characters studied by Mendel pertained to the shape of seeds in garden peas. He chose parental plants from pure-breeding lines or varieties that differed in this trait, one variety consistently producing round seeds and the other consistently producing wrinkled seeds. An experimental cross of these parents (the P generation) always yielded first-generation progeny (the F_1, or first filial generation) that produced round seeds. These results were independent of which way the cross was made, that is, whether the round-seeded parental plant donated or received the pollen. Consequently, it was deduced that each parent must have contributed equally to the offspring, but the contribution of the round-seeded parent was dominant to that of the wrinkled-seeded parent (Fig. 3–1a).

SEGREGATION

The concepts of equal parental contribution and dominance were supported by observation of the progeny produced when this F_1 generation was allowed to reproduce by self-fertilization, which is the normal method of reproduction in garden peas. The F_2 generation produced in this way consisted of 5474 round-seeded plants and 1850 wrinkled-seeded plants (Fig. 3–1a). If this ratio (2.96 to 1.0) is interpreted as a chance variation of 3:1, then these results could be accounted for by assuming that each plant contained two genes for seed shape, but that reduction and separation (segregation) of these genes occurred in the formation of the sex cells or gametes. In order to test these hypotheses, Mendel backcrossed the F$_1$ "hybrids" with pure-breeding wrinkled-seeded plants. The progeny resulting from this cross approximated a 1:1 ratio, round-seeded to wrinkled-seeded, which was consistent with the hypotheses (Fig. 3–1b). Further confirmation came from permitting the F$_2$ to self-fertilize. The wrinkled-seeded plants produced only their own kind, and thus indicated they contained only the "recessive" genes for this character. Two-thirds of the round-

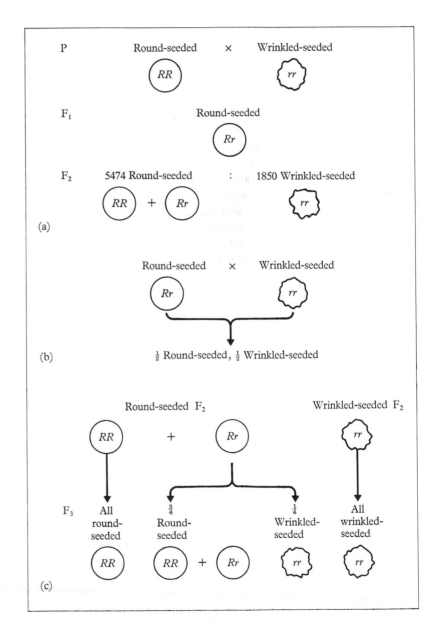

Fig. 3–1. Schematic representation of one of Mendel's experiments. (See text for explanation.)

seeded F_2 plants, when self-fertilized, again produced both types of progeny in a 3:1 ratio. The remaining one-third of the round-seeded F_2 plants produced only round-seeded progeny when self-fertilized (Fig. 3–1c).

That these principles were not limited to factors that determine seed shape was demonstrated by testing parental varieties that differed in other true-breeding characteristics, specifically seed color (yellow vs. green), position of the inflorescence on the plant (terminal vs. axial), and the length of the stem at maturity (tall vs. dwarf). In each case, Mendel found that the principles of segregation and dominance applied and the same characteristic ratios were obtained.

INDEPENDENT ASSORTMENT AND ALLELES

Mendel's second principle, the Law of Independent Assortment, was arrived at by observing the consequences of experimental crossing of two parental varieties that differed in two characters. He crossed one variety that combined the characters of round seed shape and yellow seed color by another variety having the contrasting characters of wrinkled green seeds. The F_1 derived from this cross was uniformly round and yellow-seeded, but, to quote Bateson's translation of Mendel (1909), "The plants raised therefrom [the F_2] yielded seeds of four sorts, which frequently presented themselves in one pod. In all, 556 seeds were yielded by 15 plants, and of these there were: 315 round and yellow, 101 wrinkled and yellow, 108 round and green, and 32 wrinkled and green." From the results of this and other experiments that gave comparable results approximating a 9:3:3:1 ratio, Mendel concluded that ". . . the relation of each pair of different characters in hybrid union is independent of the other differences in the two original parental stocks" and that during production of the gametes, the factors for them assort independently. The results of this cross are illustrated in Fig. 3–2.

As will be shown in subsequent chapters in this book, genes are concerned indirectly with the expression of all the physical and functional manifestations of any cell or organism, and yet it is impossible to distinguish the individual components of the gene-expression complex unless a gene changes. Changes in genes are known as *mutations*. By the effect that mutations have, the normal function of the gene is deduced and so, in a sense, the gene identifies itself. It is because of this that genetics is said to be concerned more with differences than with similarities. Genetics is by no means unique in this respect. The field of endocrinology, to name one other, was also built on knowledge gained by observing the effects pursuant to dysfunctional or malfunctional components of a system.

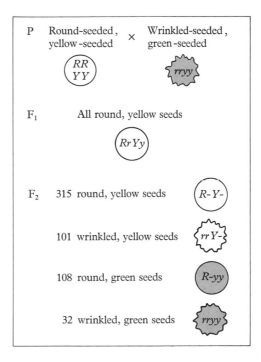

Fig. 3–2. Independent assortment of two separate characteristics. Individually, each produces a 3:1 F_2 monohybrid ratio (423 round : 133 wrinkled : 416 yellow : 140 green). Considered together, a 9:3:3:1 F_2 dihybrid ratio results.

It is mutation that brings into being the alternative form or forms of a gene by which, in turn, that gene can be recognized. Except when the nature of a mutation is the physical loss or absence of a gene, a mutant gene is basically the same gene it was before mutating—just modified to a greater or lesser extent. Genes that have changed and thus occur in two or more different forms are called *allelomorphs* (Greek: *allo* = different, *morpho* = form), a term that has been shortened to *alleles*. Thus the mutant form of a gene is an allele of its normal counterpart, and similarly the normal form is an allele of its mutant counterpart.

The term "normal" has a certain implication that may be misleading, for it immediately suggests that any alternative to it is abnormal. This further implies inferiority, which may not be the case at all. Mutant alleles are not, *ipso facto*, inferior to their "normal" alleles—in fact, in some situations identifying which allele is the normal and which is the mutant may be quite arbitrary. The difficulty is partly avoided by use of the term "wild type" to designate any allele that would ordinarily be considered as the predominating form of the gene in natural or "wild" populations. For example, the kind of robin that you are most likely to see in the park has a pattern of black, brown, russet (red), and white feathers that make it look as a robin should look. This coloration is called wild type, and any and all genes that contribute to the development of this coloration are also called wild type. Occasionally, a colorless or albino robin is encountered (Fig. 3–3). Most cases of albinism in animals or plants can be traced to a difference in a single gene. This gene represents a change from the wild type—a mutant—and so we speak of the mutant gene for albinism and of its wild-type allele.

Not all cases are as readily resolved as is albinism in robins. Anyone who was born and reared in Denmark before the periods of extensive world exploration would have considered blue eyes and blond hair to be wild-type characters for people. At the same time a native of Japan would

have argued for dark eyes and dark hair. Who would have been right? Who would be right now? Nobody knows for certain, but since it is generally believed that the human species is tropical in origin, and since dark pigmentation of both eyes and hair characterizes the overwhelming majority of people indigenous to the tropics, geneticists regard the dark pigmentation as wild type. No geneticist, or anyone else for that matter, would consider either blue eyes or blond hair, *per se*, abnormal. Of course they would be quite unusual in certain contexts, such as in Indonesian, Polynesian, or African peoples.

Fig. 3–3.   Albinism in robins. (Photograph courtesy of Dr. Alphonse Avitabile, Assistant Professor of Biology, University of Connecticut.)

When one considers human blood groupings, any attempt to draw a distinction between wild type and mutant type, or normal and abnormal, would be completely futile. In such cases as this, the alleles are designated according to the blood group they determine. The genetics of human blood groups will be discussed later in this chapter.

Let us now pursue the trait of albinism a little further. For practical purposes, albinism may be regarded as the consequence of an inability to synthesize the pigment *melanin*. Melanin is one end product of a complex of metabolic pathways involving the amino acid *tyrosine* (Fig. 8–5). The amount and distribution of melanin varies among animals and are precisely regulated by many genes as well as certain environmental factors, such as temperature and sunlight, to varying extents. However, when the specific enzyme *tyrosinase* is faulty or is lacking in an individual, no melanin can be produced at all, and albinism results. The synthesis of tyrosinase is regulated by a specific gene. A mutant allele of this gene, which lacks the ability to do this, is responsible for albinism.

In diploid organisms such as Mendel's peas, genes occur in pairs. If both members of a pair are alike, the individual is said to be *homozygous*. If they are different, the individual is called *heterozygous*. Two classes of homozygous individuals are possible—an individual may be either homozygous for the wild-type allele or homozygous for the mutant allele. An individual who has the wild-type allele for tyrosinase synthesis is able to produce melanin. Consequently, individuals who are either homozygous wild type or heterozygous for this allele will be pigmented. Only those individuals who are homozygous mutant will be incapable of developing pigmentation.

This is the kind of situation that Mendel encountered in garden peas, and he called it *dominance*. This is the situation in which an allele is expressed fully, regardless of whether it is in homozygous or heterozygous condition, and such alleles are called *dominant*. Correspondingly, a gene, the expression of which is suppressed or masked by its allele, is termed *recessive* to that allele. Accordingly it can be said that the wild-type gene for tyrosinase synthesis is dominant and its mutant allele, which we may call the allele for albinism, is recessive.

There are three shorthand systems of notation that are used to designate the genetic composition or *genotype* of diploid organisms. In one system, the first letter of either the dominant or recessive trait is used. The choice of which is used is often a matter of convenience. For example, with respect to the gene being discussed, one could use the symbol W for wild type or A to refer to the trait of albinism. Since every malfunction has a wild-type alternative, use of the letter that suggests the mutant effect is usually more appropriate, in this case A. Whether an allele is dominant in its expression or recessive can be indicated by using the upper case letter for one and the lower case of the same letter for the other. For example, the three possible genotypes for pigmentation could be written:

Homozygous wild type	$= A/A$
Homozygous mutant (albino)	$= a/a$
Heterozygous	$= A/a$

The second system of shorthand, one that is used interchangeably with the first, uses a plus sign (+) to represent the wild-type allele. As before, the alternative allele is represented by a letter—lower case if it is recessive, upper case if it is dominant. The third system utilizes an exponent + over the symbol for the mutant allele to designate wild type. Accordingly, the three genotypes under discussion could be written any of these ways:

Homozygous wild type	$= A/A$, or $+/+$, or a^+/a^+
Homozygous mutant (albino)	$= a/a$
Heterozygous	$= A/a$, or $+/a$, or a^+/a

Because diploid organisms contain two of each gene, as mentioned earlier, the genotype as designated symbolically by any of the foregoing systems represents a deduction that is based upon the individual's visible or biochemical appearance, which is known as the *phenotype* (Greek: *pheno* = apparent). In some cases the line of deduction is clear; in other cases it is not clear or must be arrived at after a more or less extensive breeding program. For instance, the genotype of an albino individual must be *a/a* because either of the other two genotypes would not produce the albino phenotype. The genotype of a pigmented individual, on the other hand, could be either +/+ or +/*a* because the possession of one dominant wild-type gene is sufficient for normal pigmentation. If there is no way to tell which genotype is appropriate, the genotype may be written as +/– or +/?, thereby indicating that either allele may be present on one chromosome of the pair.

Often the exact genotype may be inferred from a pedigree or family history, if one is available. Let us say that a case under study involves a normally pigmented man whose mother was normally pigmented but whose father was albino. The mother could have been either +/*a* or +/+, but the man under consideration must be heterozygous (+/*a*), having inherited a normal allele from his mother, accounting for his own normal appearance. The only allele for this trait that his father could give was the mutant allele.

A third way of deducing a genotype comes from knowing the frequency of the allele in the population. If the allele for albinism is very rare, one could assume with a considerable amount of confidence that a normal individual is homozygous wild type, unless there is other evidence to contraindicate this. In addition, all of these deductions assume that the gene in question is stable and did not mutate in the parental gametes.

In the study of genic transmission, there is a set mode of experimentation that is often followed. Although we will be using a human example, it

should be clear that these techniques are applied to people rarely and only with ethical difficulty. One starts with two pure-breeding parents, the P generation. The phrase *pure-breeding* means that all offspring exhibit the same phenotype (with respect to the gene under study) if the individual is mated with another of like appearance. Pure-breeding strains containing many individuals have been developed for most genetic studies. Mating within the strain produces a new generation of that strain exactly like the parental organisms. For this to happen (actually providing us with a better deductive definition of the phrase pure-breeding), a pure-breeding organism must be homozygous at the chromosomal locus in question and a pure-breeding strain must be uniformly homozygous at the locus or loci under study. If this were not so, it would be possible to produce organisms phenotypically different from the parents and the parents obviously could not be called pure-breeding.

Let us now consider the usual breeding scheme, again using albinism as our example:

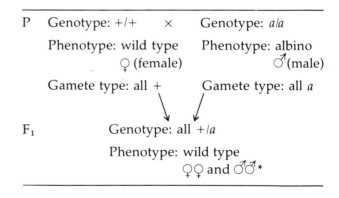

If each parent is homozygous, then all of their gametes must be uniform as well with respect to the gene for this trait. Each egg will carry one gene

---

*Double female and male symbols indicate plural number.

for the ability to be normally pigmented and each sperm will carry one gene for albinism. Every zygote produced by this union will be heterozygous, having the genotype $+/a$. But when this generation of zygotes (the first filial or F_1 generation) matures, they will all be normally pigmented. The next step is to interbreed members of the F_1 group:

F_1 Genotype: $+/a$ ×   Genotype: $+/a$

Phenotype: wild type ♀♀   Phenotype: wild type ♂♂

Gamete types: $1/2 +$ Gamete types: $1/2 +$
$1/2\ a$ $1/2\ a$

F_2 Genotypes: $1/4\ +/+ : 1/4\ +/a : 1/4\ a/+ : 1/4\ a/a$

Phenotypes: $3/4$ wild type : $1/4$ albino
♀♀ and ♂♂   ♀♀ and ♂♂

The second filial generation (F_2) as shown above represents the *expected* statistical distribution of offspring. It does *not* mean that the first three offspring *will* be normal and the fourth *will* be an albino. It does mean that each offspring has a 25 percent *chance* of being an albino and a 75 percent *chance* of being normal (25 percent chance of being homozygous normal). If we could tag each sperm and each egg, we could make more definitive predictive statements about each member of the F_2. Because we can not distinguish one egg from any other (or one sperm from any of the rest) on the basis of this gene pair, it is necessary to reason as follows. Each fertile F_1 female will produce eggs, half of which will carry the $+$ allele and half the a allele. Another way of stating this is to say that any one egg has a 50 percent chance of carrying the $+$ alternative and a 50 percent chance of carrying the a alternative. The same statements can be made relative to the sperm produced by an F_1 male. Ac-

cording to the laws of probability, the chance of two independent events occurring simultaneously is the product of the individual probabilities. If any egg can combine with any sperm with equal probability, the chance of obtaining a $+/+$ zygote is equal to the probability that any given egg will be $+$ (50 percent) times the probability that any given sperm cell will be $+$ (50 percent), or $0.50 \times 0.50 = 0.25 = 25$ percent. Similarly, the chance of any zygote being a/a is 25 percent (50 percent chance a egg multiplied by 50 percent chance a sperm). Two other possibilities exist. A $+$ egg (50 percent chance) may unite with an a sperm (50 percent chance) yielding a 25 percent chance of obtaining a heterozygote. Conversely, a heterozygote may also be formed 25 percent of the time by an a egg and a $+$ sperm. Therefore, the genotypic ratio of possibilities is 25 percent $+/+$: 50 percent $+/a$: 25 percent a/a, and the phenotypic ratio of possibilities is 75 percent wild type to 25 percent albino. These ratios are often written as 1:2:1 and 3:1, respectively. They constitute the usual genotypic and phenotypic ratios of a monohybrid cross in which one allele is clearly dominant to its alternative. The word *monohybrid* refers to the individual or to the generation of individuals that is heterozygous for the one gene pair under study.

The appearance of these ratios is clearly dependent upon the phenomenon of dominance. There are many genetic traits for which the rule of dominance does not hold, and for these characteristics the standard breeding scheme would yield modified ratios. A similar monohybrid cross is shown in Fig. 3–4, except that the trait being considered is sickle-cell anemia. There are three discernable phenotypes: normal, severely anemic, and the heterozygous condition called sickle-cell trait in which sickling does not occur unless induced by low oxygen tension, certain chemical reducing agents, or strenuous exercise. In this case, each phenotype is represented by a unique genotype resulting in a 1:2:1 ratio of the F_2 that is both genotypic and phenotypic.

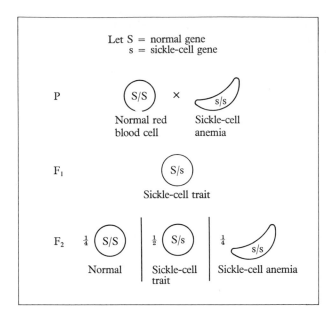

Let S = normal gene
s = sickle-cell gene

Fig. 3–4.   A monohybrid cross illustrating the F_2 generation genotypic and phenotypic ratios when complete dominance is lacking.

Fig. 3–5.   A testcross. If the unknown individual carries two dominant genes, no recessive phenotypes should appear as a result of a cross with an homozygous recessive parent. If the test organism is a heterozygote, both dominant and recessive phenotypes should appear.

ing the "test" (or genotypically unknown) organism with one that is known to be homozygous for the recessive allele. If the test organism is a heterozygote, about half of the testcross offspring should show the dominant phenotype and the remaining half should exhibit the recessive phenotype (Fig. 3–5).

On the basis of these assumptions, we then take some of the F_1 and F_2 wild-type mice and testcross them with the albino parental strain. Let us assume that the following data are collected:

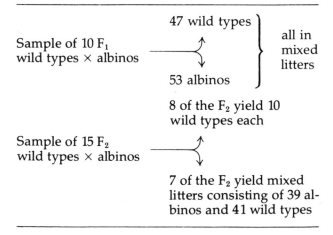

Testing the Ratio

Let us say that albinism in mice is observed as a heritable trait but the nature of transmission of this trait is unknown. We might postulate that the trait is controlled the same way in mice as in humans, namely by one set of alleles, wild-type pigmentation being dominant to the albino condition. We would then establish two pure-breeding parental lines, mate them, obtain an F_1 generation, and interbreed these animals. Let us suppose that having done this we obtain an F_1 population that is phenotypically wild type and an F_2 population that consists of two classes of mice, 41 albinos and 162 wild types. We can now test our hypothesis in two ways. The first way is by further breeding experimentation; the second method is by statistics.

If the hypothesis is correct, two genetic predictions follow: All the F_1 mice and about two-thirds of the F_2 animals should be heterozygotes. A *testcross* can be performed to determine the accuracy of these predictions. Testcrosses involve mat-

These results would appear to be fortuitous. All of the F_1 testcross animals bred as heterozygous, whereas only some of the F_2 mice followed this pattern.

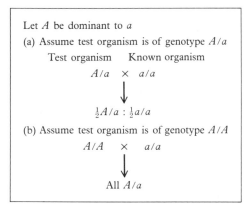

Let A be dominant to a

(a) Assume test organism is of genotype A/a

Test organism   Known organism

A/a × a/a

$\frac{1}{2}A/a : \frac{1}{2}a/a$

(b) Assume test organism is of genotype A/A

A/A × a/a

All A/a

The next step in the testing process must be statistical because, after all, the prediction that a particular egg will be fertilized by a particular sperm is a statistical statement. We cannot identify any one gamete with respect to its genes; we can only say that any one gamete has a certain *chance* of carrying a specific allele.

We wish to determine, then, if the various ratios we observed are close enough to the predicted ratios to support our hypothesis. Statistical tests cannot prove that an idea is correct. They can, however, indicate that an hypothesis is probably *incorrect* once tolerance levels for acceptance and rejection have been arbitrarily set. How are these limits set? It is important to realize that acceptable limits are man-made boundaries. If the agreed-upon limits are very narrow, that is, only very small deviations between observed and predicted ratios are allowed, accurate hypotheses will be discarded sometimes and inaccurate hypotheses will be discarded most of the time. If tolerance limits are very wide, accurate hypotheses will be saved most of the time, but some inaccurate theories will slip through as well. A balance must obviously be achieved.

One might first ask, Why are discrepancies observed at all? Why are the observed ratios not exactly the same as expected ratios if the expecta-

tions are based on the correct hypotheses? The answer is relatively straightforward. If one could test an entire, infinitely large population, there should be no discrepancies. Observations would be identical with predictions. In genetics, this would be tantamount to tagging each and every sperm and egg. If an experiment is based on only a fraction of the population, then errors of sampling can occur. The sample, however carefully chosen, may not be entirely representative of the whole population. Obviously the smaller the sample, the greater the risk of chance sampling errors.

Statisticians have constructed tables that can be used to relate sample size and magnitude of discrepancy observed to the probability that the discrepancy is caused by chance sampling errors. Within certain limitations, these tables can be used to examine not only genetic ratios, but any observed ratios that must be compared to statistically ideal ratios. It is of no consequence that these tables were not empirically derived from observations in genetics; the *probability* of 0.5 that a gamete will carry allele A as opposed to a is no different mathematically than the 0.5 probability that an honest coin, when flipped, will turn up heads.

Let us suppose that we have 10 unbiased people flipping 10 coins 10 times. Perhaps one or two of the group will get 5 heads and 5 tails. Others will get 6 heads and 4 tails, or 7 heads and 3 tails, or 4 heads and 6 tails, and so on. If each person flipped the coin 100 times, or if the number of people were increased to 100, then there would again be a distribution *centering* around 50 percent heads and 50 percent tails. The larger the group, or the more times that each coin was flipped, the more closely the results would cluster near the expected ratio. However, there would inevitably be a distribution of results. Collectively, these would average out to the ideal 50:50 ratio, but unless each person flipped until eternity, only a few people would arrive at the precise ratio individually. On this basis, one can construct statistical tables that indicate the range of values and their frequencies expected for a given sample size. Geneticists have agreed that large, repeatable discrepancies from the ideal indicate that the ideal was probably incorrect, namely, that the predicted ratios were probably based on a wrong hypothesis. Small discrepancies would be expected to occur most of the time and would, therefore, not invalidate the hypothesis (although further experimentation may point in another direction). We return to our so-far unanswered question: How much is "large" and how much is "small"? "Large" includes those discrepancies that deviate so much from the ideal as to be expected less than 5 percent of the time. "Small" includes the rest of the range of deviations that cluster about the ideal and that are expected more than 5 percent of the time.

An example may prove useful. Let us suppose that 100 coins are tossed 100 times by 100 non-biased tossing machines. We will keep score of the number of heads only; the number of tails will be 100 minus the number of heads. We could make a list of the number of heads and the frequency with which each number appeared:

Number of heads out of 100 tosses per coin	Number of coins showing particular number of heads
35 or less	0
36	1
38	1
40	1
41–59	94
60	1
62	1
63	1
64 or more	0

In about 5 percent of the tosses, fewer than 40 or more than 60 heads were observed. Results between 40 and 60 exclusive occurred about 95 percent of the time. Using agreed-upon limits, we might say that those coins that fell below 40 or over 60 probably were unbalanced or were influenced by some other extraneous factor. Although we could not be positive about it, we might reject the

50:50 ratio as applicable to these deviant coins by saying that the deviation was larger than we would accept on the basis of chance sampling errors alone. At the *5 percent level of significance*, we would say that the other 95 coins were balanced, the 50:50 ratio probably applied, and that chance sampling errors alone were responsible for their not turning up exactly 50 heads. Generally, results in genetics are not considered to nullify the hypothesis if they are above the 5 percent significance level. Results below this, or right on the borderline of 5 percent, are regarded with suspicion and some factor other than chance is sought to explain the deviation.

In order to calculate discrepancy probabilities with the least effort, the *chi-square* (x^2) method is generally used. Built into the method is the recognition that larger samples are closer to the infinitely large ideal and would be expected to deviate less than smaller samples. Let us return to the example of albinism in mice in order to explain the workings of the chi-square method. Consider first the testcross of the F_1 animals:

	Phenotype of offspring		
	Wild type	Albino	Total
Observed (o)	47	53	100
Expected (e) by hypothesis	50	50	100
Deviation (d) = o − e	3	3	
Deviation squared (d²)	9	9	
d²/e	9/50 = 0.18	9/50 = 0.18	
x^2 = sum of all (d²/e) = 0.18 + 0.18 = 0.36			

In order to determine if the deviation observed is permissible (namely, explainable at the 5 percent level of significance on the basis of chance alone), first the value of x^2 is calculated, as shown above. Note that the actual number values were used; ratios in percentages must be converted back to original number values before the x^2 is calculated. To proceed further with the test, one additional bit of information is needed. If you examine the chi-square table (Table 3–1), you will see a column that is labeled *degrees of freedom*. This figure is equal to

Table 3–1. Chart of values for chi-square (x^2).

n	P = 0.99	0.98	0.95	0.90	0.80	0.70	0.50	0.30	0.20	0.10	0.05	0.02	0.01
1	0.000157	0.00628	0.00393	0.0158	0.0642	0.148	0.455	1.074	1.642	2.706	3.841	5.412	6.635
2	0.0201	0.0404	0.103	0.211	0.446	0.713	1.386	2.408	3.219	4.605	5.991	7.824	9.210
3	0.115	0.185	0.352	0.584	1.005	1.424	2.366	3.665	4.642	6.251	7.816	9.837	11.345
4	0.297	0.429	0.711	1.064	1.649	2.195	3.357	4.878	5.989	7.779	9.488	11.668	13.277
5	0.554	0.752	1.145	1.610	2.343	3.000	4.351	6.064	7.289	9.236	11.070	13.388	15.086
6	0.872	1.134	1.635	2.204	3.070	3.828	5.348	7.321	8.558	10.645	12.592	15.033	16.812
7	1.239	1.564	2.167	2.833	3.822	4.671	6.346	8.383	9.803	12.017	14.067	16.622	18.475
8	1.646	2.032	2.733	3.490	4.594	5.527	7.344	9.524	11.030	13.362	15.507	18.168	20.090
9	2.088	2.532	3.325	4.168	5.380	6.393	8.343	10.656	12.242	14.684	16.919	19.679	21.666
10	2.558	3.059	3.940	4.865	6.179	7.267	9.342	11.781	13.442	15.987	18.307	21.161	23.209

[Abridged from Table II of Fisher and Yates: *Statistical Tables for Biological, Agricultural and Medical Research* (1953), published by Longman Group Ltd., London. (Previously published by Oliver and Boyd Ltd., Edinburgh.) By permission of the authors and publishers.]

the number of phenotypic classes minus one; in our example there were two phenotypic classes, normal and albino—therefore, one degree of freedom. As the degrees of freedom increase, so too does the acceptable level of deviation. For the non-statistician, this concept may be best explained by analogy. Let us say that we present a blindfolded person with a pair of gloves. With respect to putting on the first glove, two possibilities exist. He or she can put on the glove on the matching hand or on the wrong hand. However, once this choice is made (one degree of freedom), the second glove has nowhere to go but on the second hand. If, however, four unique gloves were presented to a person with four unique hands, the number of choices would be increased to three. Only the glove that went on the fourth hand would be open to no choice. In this case there are four possibilities (classes) and three degrees of freedom. For the two-handed person, the odds of matching glove to hand are one in two (0.5 or 50 percent); the odds of a similar match due to chance alone for the four-handed person are one in four (0.25 or 25 percent). Therefore, as the number of classes increases, tolerable deviations due to chance also increase.

In our example, we had a chi-square value of 0.36 at one degree of freedom. In the body of the table, this value of x^2 falls near a value for p (prob-ability) of 0.5. This means that 0.5 or half the time one would expect deviations as great or greater than the one we obtained. So far then, the hypothesis about the transmission of albinism in mice stands.

One can apply the test to the F_2 testcross as well:

	Compositions of testcross litters		
	All wild type (indicates *AA* test parent)	Mixed wild type and albino (indicates *Aa* test parent)	Total
Observed (o)	8	7	15
Expected (e)	5	10	15
Deviation (d)	3	3	
d^2	9	9	
d^2/e	9/5 = 1.8	9/10 = 0.9	

$x^2 = 1.8 + 0.9 = 2.7$
Degrees of freedom = 1
Probability (p) = about 0.1
Interpretation: Ten percent of the time, chance alone could account for a deviation as large as this one. At the 5 percent level of significance, the hypothesis still stands, but a wise investigator would repeat the experiment using a larger sample.

Finally, one could do a chi-square test on the distribution of offspring in the mixed-litter group. Our hypothesis would predict that the testcross parents that produced the mixed litters should be heterozygous and the litters should be comprised of about equal numbers of albino and wild-type individuals.

	Phenotypes within the mixed litters		
	Wild type	Albino	Total
Observed (o)	41	39	80
Expected (e)	40	40	80
Deviation (d)	1	1	
d^2	1	1	
d^2/e	1/40	1/40	
	= 0.025	= 0.025	

$x^2 = 0.025 + 0.025 = 0.05$
Degrees of freedom = 1
Probability (p) = about 0.7
Interpretation: In about 70 out of 100 trials, chance deviations as large or larger than this would be expected. At the 5 percent level of significance, the hypothesis stands.

Multiple Alleles

It could be assumed on the basis of logic alone that if a gene can mutate to an alternative form, further mutations of the same locus are possible and a series of alleles could result. For example, if gene A mutates to form allele A', at a different time and in another member of the population allele A could mutate to A'', and in another to A'''. Each of these alleles would be expected to specify the production of a different end product and, therefore, the population would be found to contain up to four possible proteins for that genetic locus. However, each diploid *member* of the population can carry only two representatives of the locus.

One of the classic cases of a multiple-allelic series concerns eye color in the fruit fly. The wild-type color is a particular shade of red. First to be discovered was a recessive mutant, white, and since that time many intermediates have been

found. These intermediates are distinguishable from each other and from the wild type not only visually, but also by measurements of the amount of extracted eye pigment in each pure-breeding variety (Table 3–2).

Table 3–2.   Multiple alleles in *Drosophila* eye color series (gene w).

Genotype	Name of phenotype	Relative amount of pigment
W/W	Wild type (red)	1.0000
w/w	White	0.0044
w^t/w^t	Tinged	0.0062
w^a/w^a	Apricot	0.0197
w^{bl}/w^{bl}	Blood	0.0310
w^e/w^e	Eosin	0.0324
w^{ch}/w^{ch}	Cherry	0.0410
w^{a3}/w^{a3}	Apricot-3	0.0632
w^w/w^w	Wine	0.0650
w^{co}/w^{co}	Coral	0.0798
w^{sat}/w^{sat}	Satsuma	0.1404
w^{col}/w^{col}	Colored	0.1636

[Data from D. J. Nolte (1959), *Heredity* **13**.]

In addition, measurements of pigment production have shown that within the wild-type group, differences in pigment amount exist in discrete, heritable classes. Therefore, more than one wild-type allele exists. Similarly, nonvisible chemical distinctions are found among the mutant genotypes. Alleles that are not separable visually because they share the same phenotype or phenotypic range, but are distinguishable chemically are sometimes called *isoalleles*. However, this is a semantic distinction based upon the precision of identification. Basically, isoalleles simply expand the multiple-allelic series for any particular locus.

The most widely investigated multiple alleles in man are those of the A–B–O blood group series. In 1925, it was shown that the divergence of the human population into blood groups A, B, AB, and O could be accounted for by three alleles of one gene. This gene was called the *I* gene and its possible forms are *i*, I^A, and I^B. These interact to form the four phenotypic groups as follows:

Blood group phenotype	Genotype
O	*i/i*
A	I^A/I^A or I^A/i
B	I^B/I^B or I^B/i
AB	I^A/I^B

The *i* allele is recessive to both the I^A and I^B variants; I^A and I^B do not have a dominance relationship, that is, they are codominant Hence, the heterozygote I^A/I^B contains the products of both alleles and determines blood group AB.

The particular product of gene *I* is a protein molecule (called an isoagglutinin) that appears on the surface of the red blood cell; allele I^A specifies that a particular sugar is conjugated with the protein and allele I^B specifies a slightly different sugar is conjugated to the same basic protein. However slight the difference in the sugars may be, this difference is detectable by the antibody-producing system of the body. An individual of group O carries antibodies to both the A and B substance. The antibody against A is called alpha (α) and is present in the plasma of blood types B and O. The antibody that acts against the B substance is called beta (β) and is present in the plasma of blood types A and O. The AB individual has both A and B on his red blood cells and carries neither antibody in his plasma. The interactions during transfusion of the various kinds of red blood cells and the naturally occurring plasma antibodies are shown in Fig. 3–6. The mating combinations that could produce offspring of the four types appear in Fig. 3–7.

As blood testing became more and more precise, the number of detectable alleles in the A–B–O phenotype increased. Isoalleles of this gene probably number seven or more. Blood group phenotypes distinct from the A–B–O system exist as well. Each of these phenotypic systems is also related to surface components of the red blood

cells and antibodies that can be made against them. Some of these phenotypes are listed in Table 3–3. It is quite probable that a few of these phenotypic systems are related not only to a multiple-allelic assortment at one locus, but also to interaction between two or more loci, each of which includes a series of alleles (Race and Sanger, 1962).

Table 3–3. Some of the more easily detected blood group phenotypes and the probable number of multiple alleles in each. Many of the phenotypic systems such as Duffy or Lewis are named after the person in whose blood they were first discovered.

Blood group phenotype system	Probable number of alleles
A–B–O	7
M–N–S	16
Rh	18
Kell	4
P	4
Lewis	2 or 3
Duffy	3

[Data from Race and Sanger (1962).]

LIMITATIONS AND EXTENSIONS OF INDEPENDENT ASSORTMENT

As described earlier in this chapter, Mendel found that different pairs of genes assort themselves independently during meiosis. As a consequence, when he crossed two pure-breeding varieties that differed in two gene loci, with a dominant and a recessive in each, he derived what is now known as the classical dihybrid ratio of 9:3:3:1. The results that he obtained depended upon four conditions that do not necessarily apply to any pair of randomly selected genetic traits. If any of these conditions is altered, highly significant deviations from the 9:3:3:1 phenotypic ratio would be observed. First of all, the genes considered were on different pairs of homologous chromosomes—they were

Donor		Recipient blood type A	B	AB	O
		Antibody in plasma β	α	Neither	$\alpha + \beta$
Blood type	Antigen on cells				
A	A	Antigen A + antibody β	 Antigen A + antibody α	No antibodies	Antigen A + antibody α
B	B	Antigen B + antibody β	Antigen B + antibody α	No antibodies	Antigen B + antibody β
AB	A + B	Antigens A + B + antibody  β	Antigens A + B + antibody  α	No antibodies	Antigens A + B + antibodies $\alpha + \beta$
O	Neither	No antigen	No antigen	No antibodies	No antigen

Fig. 3–6. The interactions that occur during mixing of different A–B–O blood types during transfusion. Assume that the donor contributes many cells (antigens) and little plasma (antibodies). Therefore, the relevant reaction is between donor cells and recipient plasma. If massive transfusions are given, the blood types must be matched exactly. The shaded circles represent no clumping of donor cells. The circles with clumps are those combinations in which agglutination of donor cells has occurred. If such clumping occurs within the recipient's blood vessels, vascular blockage can result.

Offspring	Possible parental combinations		
A	A×A	A×B	AB×O
	A×O	AB×AB	
	A×AB	B×AB	
B	B×B	B×A	AB×O
	B×O	AB×AB	
	B×AB	A×AB	
AB	AB×AB		
	A×B		
	A×AB		
	B×AB		
O	O×O	A×O	
	A×A	B×O	
	A×B	B×B	

Fig. 3–7. A–B–O blood groups of offspring that can be produced by different parental combinations. The large variety of parental combinations for blood group A and B offspring occur because these blood types may result from homozygosity or heterozygosity.

not "linked." Secondly, each gene affected a totally different aspect of the phenotype—there was no phenotypic interaction between genes of different loci. Many examples of two loci crosses exist in which both sets of alleles relate to the same trait. In Chapter 6 examples of this kind are given, and it will be seen that the standard ratios are modified in certain ways according to the nature of interaction between the loci. Thirdly, within each locus one of the alleles was clearly dominant to the other. Just as the phenotypic ratio for monohybrid crosses is modified when dominance is lacking between the alleles, so too is the 9:3:3:1 dihybrid ratio changed. Finally, there were only two alleles at each locus; neither of the loci included a multiple-allelic series.

Standard F_2 ratios can be constructed for any number of independently assorting genes considered together. If one were to consider a three-gene cross (loci A, B, and C, for example), one would anticipate the following results:

P	$ABC/ABC \times abc/abc$
F_1	All ABC/abc

F_2 Separate phenotypic ratios:

$3/4\ A/-$: $1/4\ a/a$ $3/4\ B/-$: $1/4\ b/b$ $3/4\ C/-$: $1/4\ c/c$

Combined phenotypic ratios:

$(3/4\ A/-)\ (3/4\ B/-)\ (3/4\ C/-)$	$=$	$27/64\ ABC/---$
$(3/4\ A/-)\ (3/4\ B/-)\ (1/4\ c/c)$	$=$	$9/64\ ABc/--c$
$(3/4\ A/-)(1/4\ b/b)\ (3/4\ C/-)$	$=$	$9/64\ AbC/-b-$
$(1/4\ a/a)\ (3/4\ B/-)\ (3/4\ C/-)$	$=$	$9/64\ aBC/a--$
$(3/4\ A/-)\ (1/4\ b/b)\ (1/4\ c/c)$	$=$	$3/64\ Abc/-bc$
$(1/4\ a/a)\ (3/4\ B/-)\ (1/4\ c/c)$	$=$	$3/64\ aBc/a-c$
$(1/4\ a/a)\ (1/4\ b/b)\ (3/4\ C/-)$	$=$	$3/64\ abC/ab-$
$(1/4\ a/a)\ (1/4\ b/b)\ (1/4\ c/c)$	$=$	$1/64\ abc/abc$

(Note: Since the presence of one dominant allele is sufficient to establish the phenotype when dealing with dominants, a shorthand method of designating a phenotype is $A/-$. The class so designated is understood to include both A/A and A/a.)

We have seen that one pair of alternative alleles may be combined to produce two phenotypic classes of offspring; that two pairs of alternate alleles combine to produce four phenotypic classes; and that three pairs result in eight phenotypic classes in the F_2. These relationships can be expressed as 2^n, where n equals the numbers of pairs of segregating alleles. Thus four pairs of segregating alleles would produce sixteen (2^4) classes, and five would produce thirty-two (2^5). In a similar manner, other relationships can be expressed as functions of n:

2^n	$=$	Number of classes of gametes formed by F_1.
2^n	$=$	Number of phenotypic classes in F_2, assuming dominance.
3^n	$=$	Number of genotypic classes in F_2.
3^n	$=$	Number of phenotypic classes in F_2, assuming lack of dominance.
4^n	$=$	Number of possible combinations of gametes to form the F_2.

QUESTIONS

1. In peas, gray seed color (G) is dominant to green (g). The following data were collected. Indicate the genotypes of each set of parents.

	Offspring	
Parents	Gray	Green
Gray × Gray	59	20
Gray × Gray	100	0
Green × Green	0	105
Gray × Green	67	60

2. In the following pedigree, the shading indicates affected individuals. Horizontal lines represent matings, squares stand for males, circles stand for females, and the vertical lines connect parents with their offspring.

Assuming the characteristic in question is caused by a single gene, is the mutant allele dominant or recessive to its normal allele? Using the letters B and b, indicate the genotype of each member of the pedigree. In cases of doubt, indicate all possible genotypes.

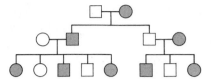

3. The human genetic disease thalassemia results in anemia. An individual homozygous for the allele T usually dies in childhood. Such homozygotes are said to have thalassemia major. Heterozygotes are mildly affected (thalassemia minor).

   With respect to this disease, what types of children and in what proportions would be expected from a woman with thalassemia minor and a normal man?

   If a child has thalassemia minor, what are all the possible genotypes and phenotypes of the parents?

4. Dark coat color in dogs is allelic to albino. Short hair is allelic to long hair. The following data are obtained:

Parents	Offspring			
	Dark, short	Dark, long	Albino, short	Albino, long
Dark, short × Dark, short	180	51	60	25
Dark, short × Dark, long	56	0	0	0
Dark, short × Dark, long	53	50	17	18

Construct an hypothesis regarding dominance within each set of alleles and linkage between them. Test your hypothesis statistically. What are the parental genotypes?

5. Four groups of a certain plant were self-pollinated. All of these plants had round, blue seeds. The progeny of each group are shown:

	Round, blue	Round, white	Square, blue	Square, white
I.	100	0	0	0
II.	68	20	0	0
III.	112	0	36	0
IV.	187	62	57	20

Construct an hypothesis regarding the inheritance of seed color and shape. Test your hypothesis. What are the genotypes of the parental groups?

6. One set of coat-color genes in guinea pigs consists of a multiple-allelic series. In the homozygous state, the phenotypes are:

$$CC \quad \text{Black}$$
$$c^k c^k \quad \text{Sepia}$$
$$c^d c^d \quad \text{Cream}$$
$$c^a c^a \quad \text{Albino}$$

Dominance is observed in the following order:

$$C \text{ dominant to all}$$
$$c^k \text{ dominant to } c^d \text{ and } c^a$$
$$c^d \text{ dominant to } c^a$$

What phenotypic ratios would be expected from the following crosses?

   a) Homozygous black × albino
   b) Homozygous black × sepia
   c) Homozygous black × cream
   d) Homozygous cream × albino
   e) Homozygous sepia × homozygous cream
   f) The offspring of (a) × the offspring of (e)

7. Assume that the Rh blood group is controlled by only two alleles, rh^- being recessive to Rh. For the M–N blood group series, assume that there are only two alleles, L^M and L^N, each of which is codominant.

   A mother of phenotype A MN rh^- has a child of phenotype B N Rh. What are the phenotypic possibilities for the father of this child?

8. In cattle, the allele R for red coat color is incompletely dominant to that for white coat color, r. The heterozygote is described as roan. Another pair of alleles that assort independently of the coat color alleles are H (hornless) and h (horned). Dominance for this set is complete.
   a) What F_1 phenotypes and genotypes would result from a $RRHH \times rrhh$ cross?
   b) What F_2 phenotypic and genotypic ratios would be expected?

4

Transmission of Genetic Information: Linkage and Recombination in Diploid Organisms

In 1905 Bateson, Saunders, and Punnett described an exception to Mendel's principle of independent assortment. They worked with two characters in sweet peas that separately followed the Mendelian predictions of unit-factor inheritance and dominance, but when combined in the same dihybrid did not show independent assortment. One character concerned flower color, purple being dominant to red, and the other concerned pollen grain shape, long being dominant to round. Instead of the F_2 assorting out in the classical 9:3:3:1 ratio that characterizes independent assortment, it appeared in the ratios shown in Fig. 4–1. Results of the chi-square test verified that the distribution of classes observed could not be ascribed to chance (Fig. 4–1). The authors proposed that these genes are "partially" linked, that is, that they tend to be transmitted more often than not in the combination in which they were inherited.

Linkage is observed between loci located on the same chromosome, and partial linkage is explained by the phenomenon of *crossing-over* between homologous chromosomes during meiosis when the chromosomes are in synapsis, thereby producing genetic combinations that were not present in either of the chromosomes prior to synapsis (Fig. 4–2). Chromosomes that show new combinations of genes are called *recombinants*, and so are the individuals that possess them. The appearance of recombinants, therefore, is interpreted as evidence that crossing-over has occurred.

Other important points concerning linkage came to light after the discovery of crossing-over and genetic recombination. For one, it was noticed that the frequency with which recombination occurred between a given pair of linked loci was constant. For example, results similar to those obtained by Bateson, et al. (Fig. 4–1) have been obtained subsequently by many investigators. Different but nonetheless constant frequencies of recombination were found to apply to other pairs of

P	Purple flowers		Red flowers	
	Long pollen grains	×	Round pollen grains	

F_1	Purple flowers			
	Long pollen grains			

	4831 (69.5%)	390 (5.6%)	393 (5.6%)	1338 (19.3%)
F_2	Purple flowers :	Purple flowers :	Red flowers :	Red flowers
	Long pollen grains	Round pollen grains	Long pollen grains	Round pollen grains

Chi-square test on results

	Purple, long	Purple, round	Red, long	Red, round
Observed (o)	4831	390	393	1,338
Expected (e)	3,911	1,303	1,304	434
Deviation (d)	920	913	911	904
d^2	846,400	833,569	829,911	817,276
d^2 expected	216	640	636	1,883

$x^2 = 3375$

Degrees of freedom = 3

p = far below 0.05

Interpretation: These results do not fit a 9:3:3:1 ratio at the 5% level of significance.

Fig. 4-1. Experimental data of sweet pea cross. (See text for discussion.) The chi-square test indicates that independent assortment of these two sets of alleles is not taking place. The hypothesis that these sets of alleles show partial linkage was then propounded.

linked loci. The question that arises is, What is the physical basis of this constancy? Another important point that came to light suggested an explanation. It was that the frequencies of recombination between pairs of linked genes were usually additive when these frequencies were small, and not additive when they were large. For example, if the genotype ABC/abc yielded recombination frequencies of 1.0 percent between loci A and B and 1.5 percent between B and C, it would probably yield a frequency of 1.0 percent + 1.5 percent = 2.5 percent between A and C. On the other hand, if the recombination frequencies between A and B and B and C were 10 percent and 15 percent, respectively, a frequency significantly less than the sum of 25 percent would be observed between A and C.

On the bases of the constancy of recombination frequencies and of the additivity of small frequencies, it was suggested that chromosomes in diploid organisms could be mapped, with recombination frequencies as the more-or-less arbitrary measure of distance between linked loci. The rationale for this suggestion is that if the crossing-over that effects recombination between any two loci is a random event, then the frequency of recombination should be a function of the linear distance between these loci. That is, a random crossover is more likely to occur between two loci that are far apart than between two that are close together. Not only are loci that are very far apart inclined to be separated rather frequently, because crossing-over could occur randomly at many points between the loci, but closely spaced loci present less distance in which a random exchange could take place. This rationale further explains why recombination values between widely separated loci are not equal to the sum of the frequencies of recombination between loci that lie between them. If two loci, A and C, are relatively far apart, then two simultaneous crossovers could take place between them. These would probably be undetected because one crossover would cancel out the effect of the other in terms of recombination (Fig. 4–2), and the relationship between the loci in question would remain unchanged—that is, no recombination. Only if the intermediate genes are included in the study can such double crossovers be detected.

THE THREE-POINT TESTCROSS

The foregoing points are neatly illustrated by a breeding technique and analysis known as the "three-point testcross." For this cross a group of individuals that are heterozygous for three linked loci, such as *ABC/abc*, are testcrossed:

$$ABC/abc \quad \times \quad abc/abc$$

For reasons that will become apparent, linked genes are indicated more conveniently by a "fraction" which positions homologous chromosomes one above the other, simulating their positions in synapsis. Thus the foregoing testcross would be represented as:

$$\frac{ABC}{a\,b\,c} \quad \times \quad \frac{abc}{abc}$$

On the assumption that the linear order of loci is that represented above and that crossing-over can occur between each of them with sufficient frequency to be detected in the numbers of off-

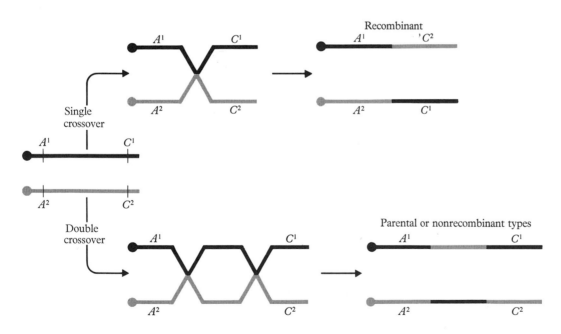

Fig. 4–2. Effect of single and double crossover events of two widely spaced loci are studied. The single exchange will lead to a different and detectable gene arrangement, whereas the product of the double exchange is identical to the parental arrangement. Assume that A^1 and A^2 differ phenotypically; C^1 and C^2 are likewise different in their actions.

spring produced by this mating, eight genotypic classes of offspring should be observed (Table 4–1). Each of these classes reveals the kind of crossover event that took place in the particular germ cells of the heterozygous parent from which the members of the class were derived. The relative frequency of each class would be determined by the crossover frequency that characterizes the segments of chromosomes involved—that is, the segment between loci A and B and the segment between B and C. The percentage of crossovers that occurred between A and B in this hypothetical example would be computed as 5.6 percent, as seen in Recombination Class II, plus 0.4 percent, as seen in Recombination Class IV, for a total of 6.0 percent. The crossover frequency between loci B and C similarly would be calculated as those in Class III (6.6 percent) plus those in Class IV (0.4 percent), or 7.0 percent.

The frequency of double crossovers is invariably much smaller than that of the singles in a three-point testcross. This is because such frequency represents the product of two simultaneous singles. One would expect that if the probability of a crossover between A and B were 6 percent and one between B and C were 7 percent, then the probability of a double crossover between A and C would be $0.06 \times 0.07 = 0.0042$, or 0.42 percent. The observed frequency (0.4 percent) is very close to this expected value. Because the frequency of double crossovers is very much smaller than that of the singles, the doubles are identifiable at a glance. In effect, a double crossover in a three-point testcross switches only the middle gene out of the original combination:

$$\frac{ABC}{abc} \rightarrow \frac{AbC}{aBc},$$

making it possible to establish instantly the true linear order of the genes involved. In the example given in Table 4–1, the linear order ABC is therefore verified.

Table 4–1. Genotypes of germ cells generated by the trihybrid $\frac{ABC}{abc}$. The frequencies of the genotypic classes are hypothetical.

Genotypic class	Frequency (percent)	Recombination class	Crossover frequency (percent)
ABC	43.7	I. Parental	
abc	43.7	(no recombination)	
Abc	2.8	II. Singles	
aBC	2.8	between A and B	5.6
ABc	3.3	III. Singles	
abC	3.3	between B and C	6.6
AbC	0.2	IV. Doubles (simul-	
aBc	0.2	taneous singles	0.4
		between A and B	
	100.0	and between B	
		and C)	

Coefficient of Coincidence—Interference. A disparity often is found between the observed frequency of double crossovers and their expected frequency. This disparity is inversely related to the magnitude of the single crossover frequencies in the region of the chromosome under study. That is, the closer together the loci involved, the greater the disparity; the farther apart, the less the disparity. This relationship would be expected if the occurrence of one chiasma tended to prevent or *interfere* with the formation of another chiasma close to it. The amount of such interference may be determined from the *coefficient of coincidence*, which is the ratio between the observed and the expected frequencies of double crossovers. The coefficient of coincidence for the chromosome region A–C in our hypothetical example (Table 4–1) is calculated as follows:

$$\frac{\text{Observed frequency of doubles}}{\text{Expected frequency of doubles}} =$$

$$\frac{0.004}{0.06 \times 0.07} = \frac{0.004}{0.0042} = 0.95$$

A coefficient of coincidence of 0.95 means that double crossovers occurred 95 percent as frequently as expected on the basis of the frequencies of the singles. If the difference is due to interference, this interference can be expressed as "one less the coefficient of coincidence," or $1.00 - 0.95 = 0.05$, or 5 percent. A coefficient of coincidence of 1.0 is interpreted as indicative of no interference, whereas a coefficient of coincidence of 0.0 expresses complete interference.

Crossover Maps. If the frequency of crossing-over is a function of the distance between loci on a chromosome, then maps that show the linear order of loci and the relative distances between them can be made using crossover frequencies as empirical units of distance, as shown in Fig. 4–3. By correlating specific mutations with structural features on chromosomes, one may go even further and construct chromosome maps that show with considerable fidelity the location of every known locus on every chromosome (Fig. 4–4).

The data of Bregger (1918) obtained from breeding experiments with corn (*Zea mays*) may be used to illustrate how crossover frequencies in a three-point testcross can be used to construct a portion of a crossover map. Bregger used three linked genes that are concerned with the development of color and texture of parts of the endosperm, which forms a conspicuous component of the seed. The gene for shrunken endosperm (*sh*) is recessive to its allele (*Sh*) for smooth endosperm; colorless (*c*) is recessive to colored (*C*); and waxy texture (*wx*) is recessive to starchy texture (*Wx*). Following is Bregger's data:

P	$\dfrac{Sh\ C\ Wx}{Sh\ C\ Wx}$	$\times$	$\dfrac{sh\ c\ wx}{sh\ c\ wx}$
F$_1$		$\dfrac{Sh\ C\ Wx}{sh\ c\ wx}$	
Testcross:	$\dfrac{Sh\ C\ Wx}{sh\ c\ wx}$	$\times$	$\dfrac{sh\ c\ wx}{sh\ c\ wx}$

Testcross Progeny

Class	Observed phenotype	Deduced genotype	Crossover type	Number (f)
1	Smooth-colored-starchy	$\dfrac{Sh\ C\ Wx}{sh\ c\ wx}$	Parental	17,959
2	Shrunken-colorless-waxy	$\dfrac{sh\ c\ wx}{sh\ c\ wx}$	Parental	17,699
3	Shrunken-colored-waxy	$\dfrac{sh\ C\ wx}{sh\ c\ wx}$	Single	509
4	Smooth-colorless-starchy	$\dfrac{Sh\ c\ Wx}{sh\ c\ wx}$	Single	524
5	Smooth-colored-waxy	$\dfrac{Sh\ C\ wx}{sh\ c\ wx}$	Single	4,455
6	Shrunken-colorless-starchy	$\dfrac{sh\ c\ Wx}{sh\ c\ wx}$	Single	4,654
7	Shrunken-colored-starchy	$\dfrac{sh\ C\ Wx}{sh\ c\ wx}$	Double	20
8	Smooth-colorless-waxy	$\dfrac{Sh\ c\ wx}{sh\ c\ wx}$	Double	12
			Total:	45,832

Immediately one recognizes that the linear order of loci is *c sh wx*, because in classes 7 and 8 of the preceding data, it is the alleles *Sh* and *sh* that have "flipped over" from the $\frac{Sh\ C\ Wx}{sh\ c\ wx}$ arrangement in the trihybrid parental genotype. On this basis, the data may be reassembled and analyzed to give the following recombination frequencies (RF):

Recombination between c and sh loci

Class	Number
3	509
4	524
7	20
8	12
	1,065

$$RF = \frac{1,065}{45,832} = 0.0232 = 2.32 \text{ percent}$$

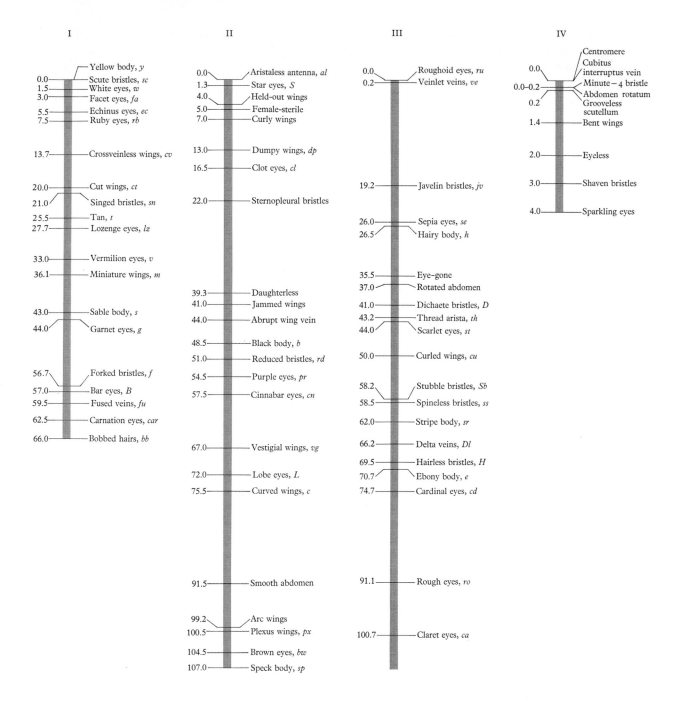

Fig. 4–3. The linkage groups (chromosomes) of *Drosophila melanogaster* showing the location of many alleles that have been mapped.

Recombination between sh and wx loci

Class	Number
5	4,455
6	4,654
7	20
8	12
	9,141

$$RF = \frac{9,141}{45,832} = 0.1994 = 19.94 \text{ percent}$$

A crossover map showing the linear arrangement and crossover units between the three loci would appear thus:

```
c        sh                              wx
|--2.32--|----------19.94----------------|
```

Interference is quite high in this region of the chromosome, because the coefficient of coincidence is 0.15. This accounts in part for the disparity between the observed frequency of doubles (0.07 percent) and the expected frequency of doubles ($0.0232 \times 0.1994 = 0.46$ percent). In any map based upon recombination data, the sum of the smaller units within a region will be more accurate than that obtained between loci that are relatively far apart. Therefore, a more accurate estimate of the distance between *sh* and *wx* can be obtained by including one or two additional loci between them. It follows that recombination frequencies observed between pairs of gene loci are approximations of linear distance, not absolute measures. This becomes more evident if the experimental work for a three-point testcross is repeated. Oftentimes the results differ slightly, probably owing to environmental influences. In addition, there appear to be regions along chromosomes that show greater or lesser rates of recombination than would be expected on the basis of distance alone. Extreme cases of this phenomenon appear in male *Drosophila* and female silkworm moths, in which crossing-over is either exceedingly rare or nonexistent.

It is possible to construct cytological maps of chromosomes by utilizing data from studies that relate phenotypic changes to chromosomal rearrangements. Figure 4–4 compares the genetic (crossover) map with the more accurate cytological map.

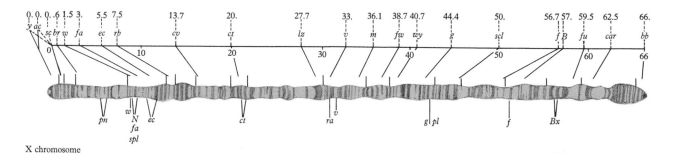

X chromosome

Fig. 4–4. A comparison of genetic and cytological maps for the X chromosome of *Drosophila melanogaster*. The maps agree with respect to gene order but differ in the relative distances between the loci. Cytological maps are generally considered to be more accurate. Cytological maps can be prepared by studying the large polytenic chromosomes in the salivary glands of *Drosophila* larvae. The banding pattern of these chromosomes is definite and microscopically visible. Even when portions of the chromosome are shifted by chromosomal mutations, the specific bands can be identified. Loci of particular genes can be determined by a detailed comparison of mutant phenotypes relative to mutant banding patterns. Genetic maps are constructed from crossover data, using recombination frequencies as relative map units. [From T. S. Painter (1934), *J. Hered.* **25**:465–476.]

CYTOLOGICAL EVIDENCE FOR CROSSING-OVER

Up to this point in the discussion, the fact of crossing-over in diploids has been a deduction based upon observations of recombination and the knowledge that pairing and chiasma formation occur during meiosis. Evidence that definitively relates recombination results to the chromosomal events that produced those results appeared in 1931 from the laboratories of Creighton and McClintock, working with indian corn (maize), and Stern, working with *Drosophila*. Crucial to these experiments was the fact that, in both cases, chance chromosomal mutations presented the investigators with particular pairs of chromosomes that had microscopically visible differences. If the members of a pair of homologues can be distinguished and if one can obtain an organism that is heterozygous for two or more loci on the same chromosome pair, then recombination can be visualized as an exchange of chromosomal segments. This type of proof is shown in Fig. 4–5.

(a)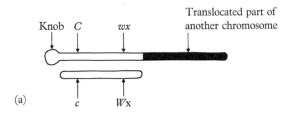

(b)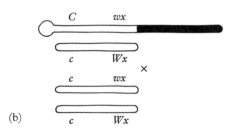

(c)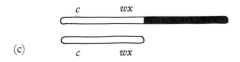

Fig. 4-5. Creighton and McClintock bred a strain of corn (maize) that was heterozygous for two pairs of genes on Chromosome 8. These genes were *C* (colored endosperm) or *c* (colorless endosperm), and *Wx* (starchy endosperm) or *wx* (waxy endosperm). In addition, one chromosome of the pair had a microscopically visible knob at one end and a translocated segment of another chromosome at the other end. This was also microscopically visible because the translocated segment rendered the total chromosome unusually long. (a) The arrangement of the alleles and chromosomal peculiarities is shown. In essence, this corn is a quadruple heterozygote, two of the traits being allelic and two being chromosomal. (b) The unusual strain was mated with a strain having no chromosomal peculiarities and which was homozygous for the *c* allele and heterozygous for the other pair of alleles. (c) Among the categories of progeny, one category was homozygous for both the *c* and *wx* alleles, had no knob, but did have the translocated segment. (d) It is shown how this must have come about by crossing-over between the *C* and *Wx* loci during gametogenesis.

(d)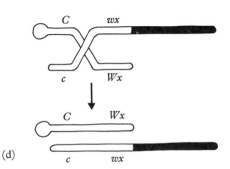

THE MECHANISMS OF CROSSING-OVER

After associating the observation of genetic recombination with the physical event of chromosomal exchange, one is prepared to ask more questions regarding the mechanisms involved.

1. When does crossing-over take place?

So far, the assumption that recombination occurs sometime during the early stages of meiosis has been implicit. Indeed, the synaptonemal complex (cf. Chapter 2), which forms during early prophase I of meiosis, appears to be concerned with the regulation of recombination; the inference can be drawn that this is the normal time for crossing-over to take place. Nevertheless, low frequencies of recombination have been associated with mitosis.

Although the occurrence of recombination other than during meiosis is probably accidental, it has been the object of considerable interest and is believed to have clinical significance. Stern (1936)

discovered evidence of mitotic crossing-over in an individual in a group of female *Drosophila* that were heterozygous at two loci on the sex chromosomes. It was significant that each chromosome of the pair carried only one of the mutant alleles (the *trans* configuration). One of the X chromosomes carried the recessive allele *y* (yellow body), and the other carried the recessive allele *sn* (singed bristles) at a different locus. The phenotypic appearance of the flies was therefore wild type for both traits. But Stern noticed that one of these flies had two unusual spots adjacent to each other on the thorax. One spot was yellow and the other had the deformed bristles characteristic of the singed mutation (Fig. 4–6). Stern proposed that the most likely explanation for the twin spots was mitotic crossing-over during development. Half of the progeny cells following the abnormal mitosis would be homozygous for the *sn* allele, and the other half would be homozygous for the *y* allele. It was suggested that the abnormality might have

Fig. 4–6.   Somatic crossing-over in *Drosophila melanogaster*. (a) The thorax of a fly heterozygous for yellow body (*y*) and singed bristles (*sn*). Note the twin spots: yellow hairs and yellow bristles in the left spot; singed hairs and bristles in the right spot. (b) Rationale for the twin spots. During development, crossing-over between chromatids occurred in one cell undergoing mitosis. One of the daughter cells thereby became homozygous for the *y* allele, whereas the other daughter cell became homozygous for the *sn* allele. All of the cells that developed from these daughter cells are part of the twin spots.

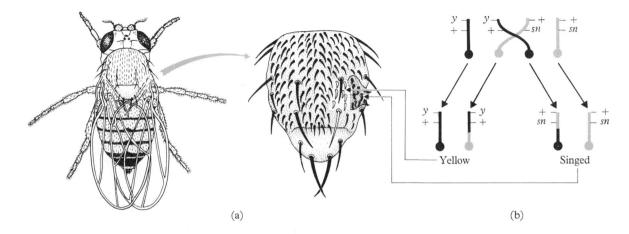

(a)   (b)

resulted from mutation rather than from crossing-over, but the fact that the phenotypically "opposite" spots were adjacent to each other and the fact that the probability of two simultaneous mutations occurring in these very loci in adjacent cells is extremely small suggest crossing-over as the more plausible explanation.

Pontecorvo (1953) showed that mitotic crossing-over can be observed readily in the fungus *Aspergillus nidulans* (Fig. 4–7). If two genetically different strains are grown together on the same plate of culture medium, fusion of some of the hyphae, which are usually haploid, will occasionally occur, giving rise to cells that are known as *heterokaryons* (Greek: different nuclei). These are cells that contain two or more nuclei of different genotype. If the strains that are grown together

differ in color, such as yellow and green (wild type), strings of differently colored conidia will be produced as each nucleus is budded off separately by mitosis from the heterokaryon. Occasionally two of the haploid nuclei will fuse to form a diploid nucleus. Conidia formed from these heterozygous nuclei will, of course, be wild type in color, but they will be distinguishable by their larger size. Measurements of DNA content confirm their diploid nature. If diploid conidia that are heterozygous for the yellow (y) allele are grown on culture medium, yellow components occasionally will be observed. Since these components (either conidia or hyphae) are products of mitosis, mitotic recombination must have occurred to produce cells that are homozygous for the yellow allele (Fig. 4–7b).

Another genus of fungus, *Neurospora crassa*,

Fig. 4–7. (a) The process of heterokaryon formation in *Aspergillus nidulans*. Occasionally a diploid conidium will show the yellow color even though the original diploid strain was a heterozygote, green (wild type) being dominant. (See text for further description.) (b) The production of yellow diploid conidia (y/y) by mitotic crossing-over is explained.

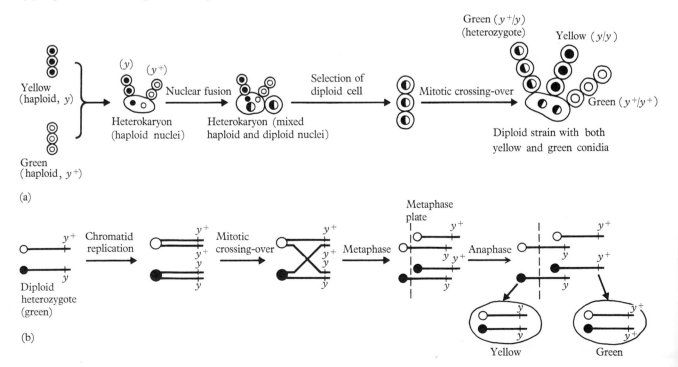

has proved to be useful in attempts to pinpoint the time of recombination in meiosis. Figure 4–8 is a diagram that shows the essential features of the reproductive cycle of *Neurospora*. The sac (ascus) that contains the meiotic products of the fusion nucleus is shaped in such a manner as to hold these products, the ascospores, in the same order in which they were produced by each meiotic and mitotic division. The individual ascospores can be picked out and grown separately in sterile culture medium. Because dominance effects do not apply in haploid cells, it is possible to determine from the mycelia they develop the exact sequence of events that occurred during the meiotic divisions that produced them.

The haploid vegetative cells of *Neurospora* contain seven chromosomes; the zygote contains seven pairs. Let us choose two linked loci on one pair of zygotic chromosomes, using a recessive mutant allele at each locus to serve as genetic "markers." If we begin with a zygote that is heterozygous for both sets of alleles and assume no crossing-over, there will be a four-to-four assortment of ascospores, as shown in Fig. 4–9(a). If crossing-over takes place in the two-strand stage—that is, before replication has occurred (premeiotically)—all of the products will be recombinants (Fig. 4–9b), whereas crossovers in the four-strand (prophase I) stage of meiosis would yield both parental (nonrecombinant) and nonparental (recombinant) progeny in equal numbers (Fig. 4–9c). Experimental observations support the latter and indicate that crossing-over occurs after the chromosomal tetrad is formed. With rare exceptions, it is a reciprocal process that involves only two of the four strands at any one particular crossover point. Although more than two strands can undergo recombination in the same cell, they do so at different points, as will be described later.

Mapping of linkage groups can be done for *Neurospora* in a manner that differs only slightly from that described for maize. Houlahan, et al. (1949) described a simple case involving crosses between an albino (*al*) strain and a wild-type strain (*al⁺*), from which they obtained the results shown

in Fig. 4–10(a). The progeny comprising Group A are the parental types; Group B includes the asci in which recombination has taken place. Note in the interpretation given in Fig. 4–10(b) that only two of the four strands underwent recombination. The recombination frequency between the *al* locus and the centromere is calculated as one-half the frequency of recombinant asci, which in this case is:

$$\frac{1}{2}\left(\frac{141}{129 + 141}\right) = 0.26, \text{ or 26 percent.}$$

The frequency of recombinant asci is multiplied by $\frac{1}{2}$ because only half of the strands in each tetrad show recombination. If the distance between two linked loci is desired, their respective distances from the centromere are calculated. These values are then subtracted if both loci are on the same side of the centromere, and added if they are on different arms of the chromosome.

The presence of ordered tetrads in *Neurospora* permits a more detailed analysis of recombination between linked loci. For example, matings of an albino of mating type A* with a wild-type strain of mating type a (*al A* x *al⁺ a*) produced the following results in terms of the kinds of asci recovered: (1) asci containing spores like the two parentals (*parental ditype*), (2) asci containing all recombinants (*nonparental ditype*), and (3) asci in which both parental and recombinant classes were present (*tetratype*). These may be diagrammed as follows:

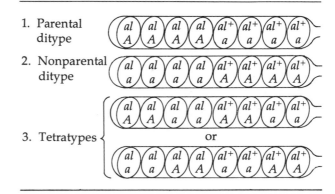

1. Parental ditype
2. Nonparental ditype
3. Tetratypes

*Neurospora, like other molds, is characterized by genetically determined "mating" types.

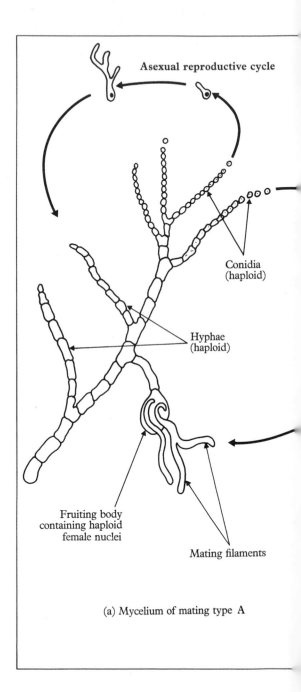

Fig. 4-8. The life cycle of *Neurospora crassa*. Two my-celia (mats of hyphae) of opposite mating types A and a are shown. All nuclei except the zygote nuclei are haploid. Both asexual and sexual reproductive cycles are indicated. The components of the latter are shown enlarged and, in this case, involve conidia from mating type A and female nuclei from mating type a. The opposite also occurs, but is not shown in detail here. (See text for additional description.)

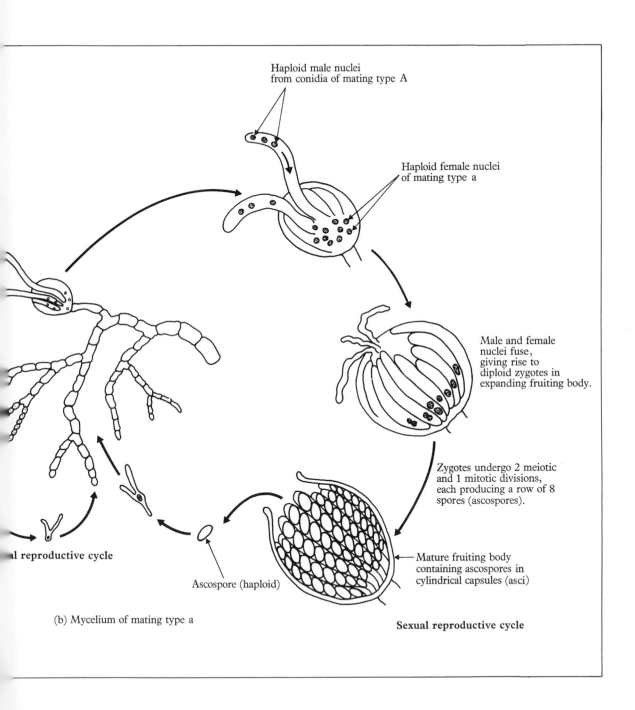

Haploid male nuclei
from conidia of mating type A

Haploid female nuclei
of mating type a

Male and female
nuclei fuse,
giving rise to
diploid zygotes in
expanding fruiting body.

Zygotes undergo 2 meiotic
and 1 mitotic divisions,
each producing a row of 8
spores (ascospores).

Mature fruiting body
containing ascospores in
cylindrical capsules (asci)

Ascospore (haploid)

l reproductive cycle

(b) Mycelium of mating type a

Sexual reproductive cycle

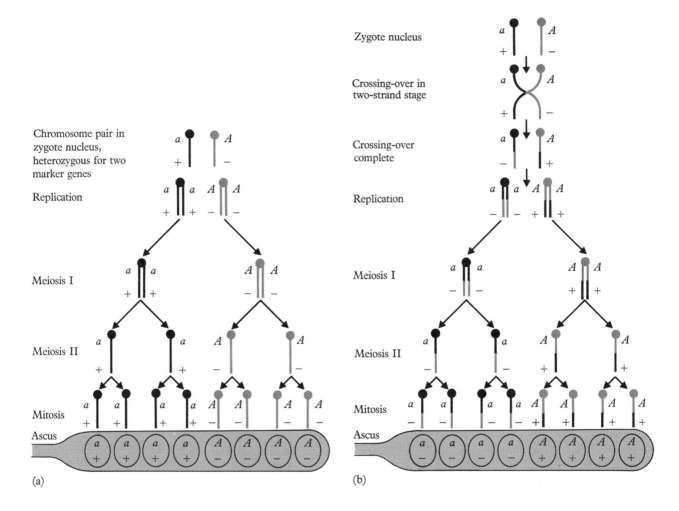

Fig. 4–9. Meiosis and ascus formation in *Neurospora crassa* using two genetic markers: + vs. − mating type and *A* (colored) vs. *a* (albino). (a) No crossing-over. A 4:4 assortment of spores would be found in the ascus. All of the chromosomes would be parental in type. (b) If crossing-over occurred in the two-strand stage, all of the products would be recombinant. This result has not been observed—therefore crossing-over is not believed to occur at the two-strand stage. (c) Crossing-over in the four-strand stage would yield a variety of products. Fifty percent of the spores in any ascus are parental in type and 50 percent are recombinant. This is, in fact, what is observed to occur.

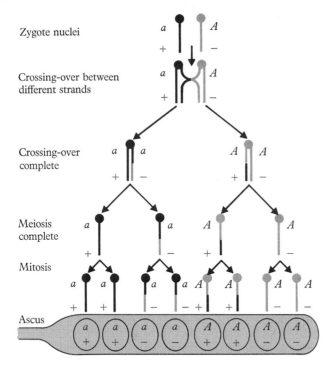

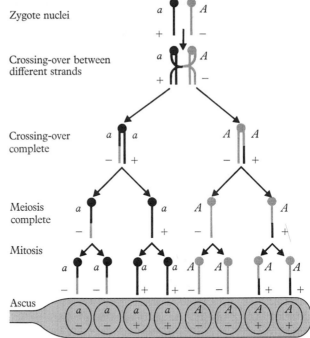

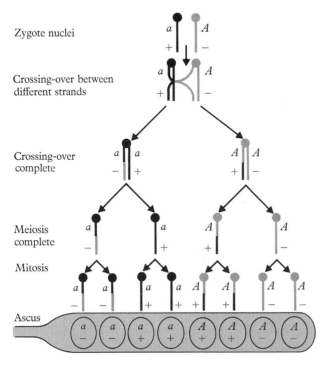

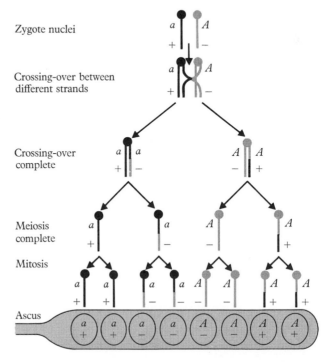

(c)

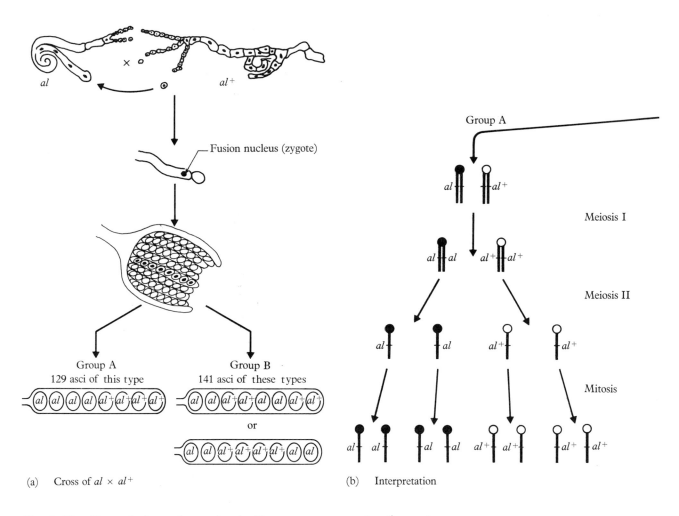

Fig. 4–10.   The technique of mapping in *Neurospora crassa,* using the centro-mere and a pair of alleles (*al* and *al*⁺) as markers. (See text for explanation.)

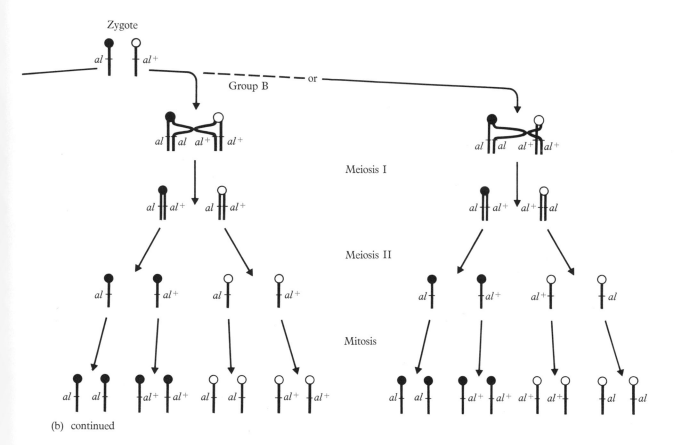

(b) continued

These results are interpreted in Fig. 4–11. Notice that the parental ditype represents tetrads in which no crossing-over (or undetected double crossing-over) has occurred between the *al* and *a* loci. The nonparental ditype represents the product of a meiosis in which crossing-over has occurred between all four chromatids, and the tetratypes result from single or odd-numbered crossovers. It is apparent from this interpretation that, although only two strands can participate in crossing-over at any given point along the chromosome, all four strands can show recombination if other points are considered.

Moreover, double crossovers and crossing-over between not only two but also three and four strands can occur at different points along the chromatids. To illustrate this point, consider the results that Houlahan and his colleagues obtained from a cross between two strains of *Neurospora* that evidenced nutritional deficiencies. One of these strains was genetically deficient in its ability to synthesize the essential chemical substance pantothenic acid, requiring supplements of it in a culture medium. The defect was shown to be due to a mutation, which was designated *pan*. The other strain carried the mutant allele *pdx*, which rendered it incapable of synthesizing another essential substance pyridoxine. The results of a cross between these two strains and their interpretation are shown in Fig. 4–12.

The recombination frequency between the *pan* locus and the centromere is:

$$\frac{1}{2}\left(\frac{17 + 13 + 1 + 1}{49}\right) = 32.7 \text{ percent.}$$

That for the *pdx* locus and the centromere is:

$$\frac{1}{2}\left(\frac{13 + 2 + 1}{49}\right) = 16.3 \text{ percent.}$$

Therefore, a map of this region of the chromosome would appear as:

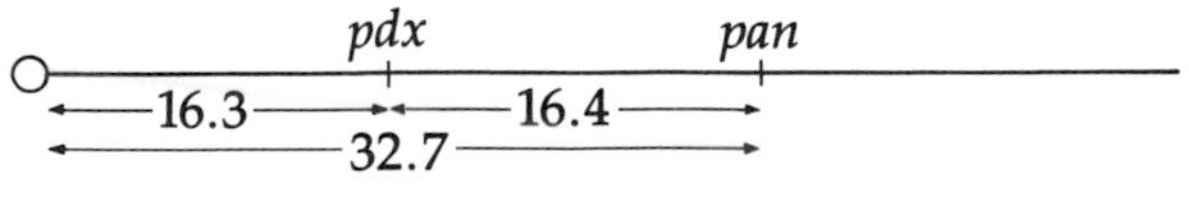

Fig. 4–11. Interpretation of the *al A* × *al⁺a* cross in ▶ *Neurospora crassa*. (See text for further description.)

2. The second question that may be asked about crossing-over is: What is its physical nature?

Two theories of crossing-over are held at the present time. The first, proposed by Belling in 1931 and later modified by Lederberg and others (cf. Chapter 10), is known as the "copy-choice" mechanism. According to this theory, homologous chromosomes first synapse and then replicate in detail, starting at some point and proceeding along their length. Because of the proximity of synapsed homologues, a newly forming chromatid may copy first one chromosome and then switch across and copy the one beside it and switch back again, as diagrammed in Fig. 4–13. One of the major weaknesses of this theory is the necessary deduction that only the two new strands being made are involved in crossing-over. Evidence from *Neurospora* clearly indicates that three- and four-strand exchanges occur, as described earlier in this chapter. Moreover, the copy-choice theory restricts crossing-over to the time of replication, which is in the S-phase of interphase, before there is any evidence of homologue synapsis. Of course, it is possible that synapsis does take place at this time, while the chromosomes are in their extended state, but this has not been demonstrated. This view may be supported, however, by the evidence of Pritchard (1955) and others that suggests that, rather than homologues pairing continuously along their entire length, small regions of close pairing alternate with longer regions of "loose" pairing (Fig. 4–14). The closely paired regions would exhibit a very high frequency of recombination, whereas the widely spaced regions would exhibit lower frequencies or no recombination at all. This concept of discontinuous synapsis could account for failure to observe synapsis during interphase, because so few chromosomes

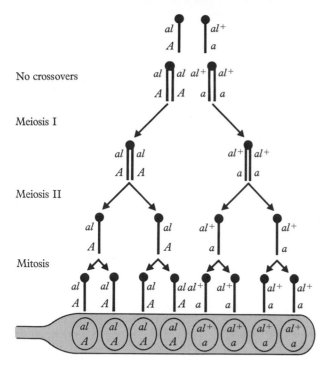

No crossovers

Meiosis I

Meiosis II

Mitosis

(a) Parental type

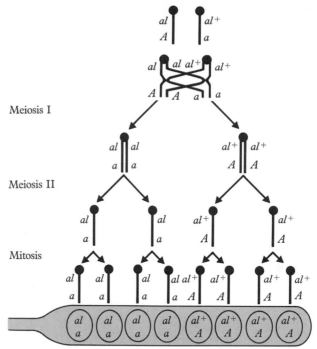

Meiosis I

Meiosis II

Mitosis

(b) Nonparental type

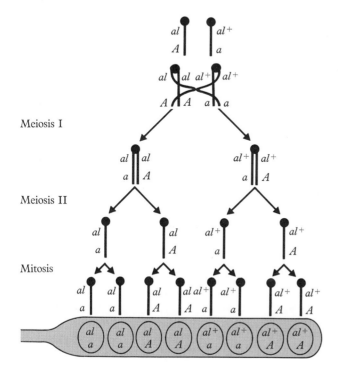

Meiosis I

Meiosis II

Mitosis

(c) Tetratype

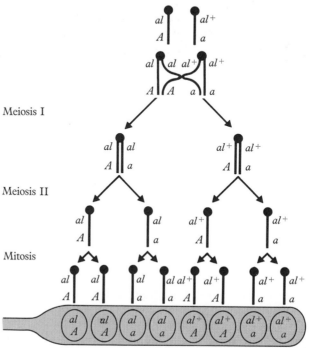

Meiosis I

Meiosis II

Mitosis

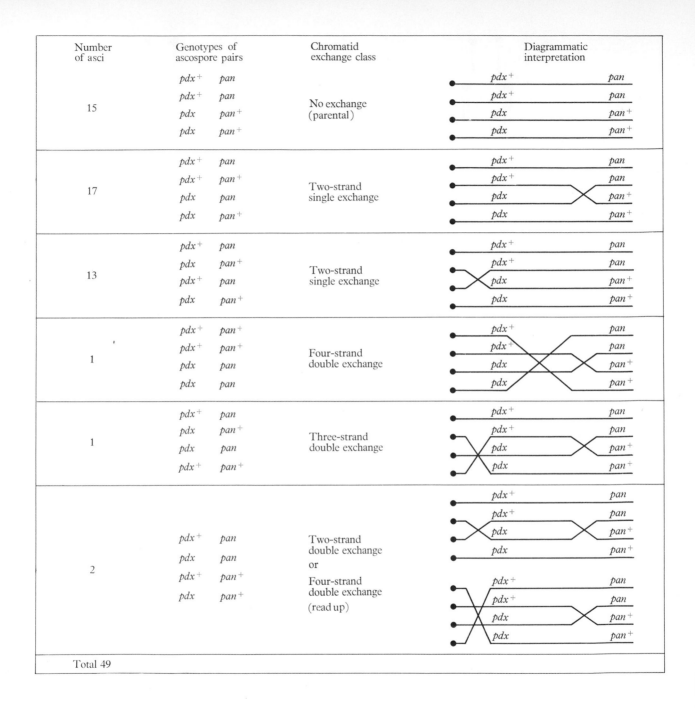

Number of asci	Genotypes of ascospore pairs		Chromatid exchange class	Diagrammatic interpretation	
15	pdx^+	pan	No exchange (parental)	pdx^+	pan
	pdx^+	pan		pdx^+	pan
	pdx	pan^+		pdx	pan^+
	pdx	pan^+		pdx	pan^+
17	pdx^+	pan	Two-strand single exchange	pdx^+	pan
	pdx^+	pan^+		pdx^+	pan
	pdx	pan		pdx	pan^+
	pdx	pan^+		pdx	pan^+
13	pdx^+	pan	Two-strand single exchange	pdx^+	pan
	pdx	pan^+		pdx^+	pan
	pdx^+	pan		pdx	pan^+
	pdx	pan^+		pdx	pan^+
1	pdx^+	pan^+	Four-strand double exchange	pdx^+	pan
	pdx^+	pan^+		pdx^+	pan
	pdx	pan		pdx	pan^+
	pdx	pan		pdx	pan^+
1	pdx^+	pan	Three-strand double exchange	pdx^+	pan
	pdx	pan^+		pdx^+	pan
	pdx	pan		pdx	pan^+
	pdx^+	pan^+		pdx	pan^+
2	pdx^+	pan	Two-strand double exchange or Four-strand double exchange (read up)	pdx^+	pan
	pdx	pan		pdx^+	pan
	pdx^+	pan^+		pdx	pan^+
	pdx	pan^+		pdx	pan^+
				pdx^+	pan
				pdx^+	pan
				pdx	pan^+
				pdx	pan^+

Total 49

Fig. 4–12. Analysis of asci recovered from the cross of $pdx^+\ pan \times pdx\ pan^+$ in *Neurospora crassa*. Only the genotypes of successive pairs of ascospores are represented, for the members of each pair, having been produced by mitosis, are genetically identical. (See text for further description.) [Data from Houlahan, et al. (1949).]

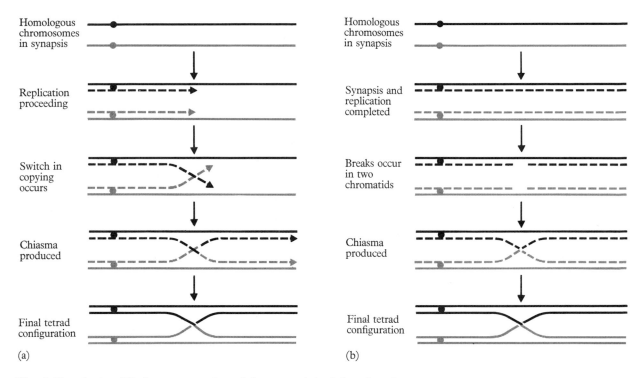

Fig. 4-13. A simplified representation of the two original theories of crossing-over: (a) the copy-choice theory, and (b) the breakage-first theory.

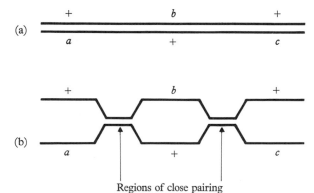

Fig. 4–14. Continuous vs. discontinuous pairing during synapsis. (a) Synapsis is classically visualized as a uniform point-for-point process with an equal, though low, frequency of crossing-over for all regions. (b) Pairing is discontinuous. The "close" regions have a high probability of exchange, whereas the separated regions have a very low probability of supporting a crossover event.

would be involved. Furthermore, this concept would explain both negative and positive interference, because single and double exchanges within the "close" regions would be expected to occur more often than one would expect on a purely random basis, and less often in the widely spaced regions.

Although there may be no satisfactory way of reconciling copy-choice with three- and four-strand crossovers, this theory could provide an explanation for a phenomenon known as *gene conversion*. Contrary to what has been stated about the reciprocal nature of segmental exchanges, sometimes nonreciprocal recombination is found. The frequency of its occurrence varies with the organism, the loci in question, and the experimental conditions. Figure 4–15 is a diagram of such an event and a copy-choice explanation is provided. Nonreciprocal recombination was termed gene conversion because it was thought that recombination *had* to be reciprocal, and hence one of the alleles must have changed or converted to its alternative after recombination.

The second theory of crossing-over is known as the "breakage-first" hypothesis, which is a modification of the so-called chiasmatype theory of Darlington. This theory holds that a necessary precondition of recombination is breakage of both participating chromatids at a point or points of ap-

proximation, which is then followed by rejoining across (Fig. 4–13), thereby effecting both genetic recombination and the chiasmata that are visible in the pachytene and diplotene stages of meiotic prophase (Fig. 2–8). This theory was firmly established by the cytological observations of Stern and of Creighton and McClintock, and has been the most widely accepted model of recombination in diploid organisms. It is possible to adapt it to the phenomenon of nonreciprocal recombination mentioned in the preceding paragraph. Evidence from certain fungi suggests that recombination and gene conversion are polarized processes; that is, that these processes preferentially occur in a

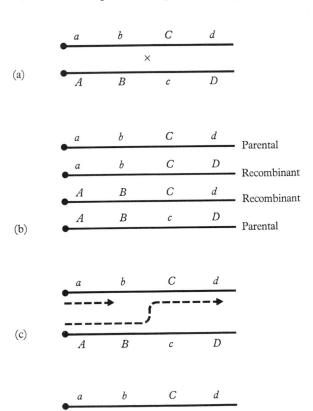

Fig. 4–15. Nonreciprocal crossing-over (gene conversion), invoking a copy-choice explanation of recombination. (a) Two fungi are crossed. (b) Tetrad analysis yields two parental strands and two recombinant strands. For gene C, however, the crossover was nonreciprocal. (c) Explanation: The replication of the lower strand advances ahead of the upper. The lower switches templates between loci B and c, thus copying first B and then C. (d) The lower strand has separated from its template, allowing the upper strand to copy C as well. It then switches templates between loci C and D, and continues along the lower strand.

fixed direction on the chromosome with reference to what is happening to the DNA in the same region. To account for this polarity, breakage-synthesis-reunion models have been proposed that would account for both reciprocal as well as nonreciprocal recombination. One of these (Whitehouse, 1969) is shown in Fig. 4–16.

3. A third question that arises relative to crossing-over and recombination is: Can their frequency be influenced?

Factors both intrinsic and extrinsic to the organism can influence recombination. Abnormal temperatures (either high or low), irradiation, and actinomycin-D are examples of some of the extrinsic factors that have been demonstrated experimentally. Even diet may be significant in this regard. The frequency of recombination can be affected from within by such considerations as maternal age, unidentified cytoplasmic factors, a single gene, a chromosome, or an entire genome (sum total of all the genes). In female *Drosophila*, the recessive allele *c3G* can inhibit crossing-over in all of the chromosomes in the homozygous condition. Chromosomal rearrangements such as deficiencies, translocations, and inversions can have at least an apparent effect upon recombination frequency.

Of greatest significance, however, is the fact that the frequency can be changed. The conditions under which the recombination frequencies are altered suggest that breakage and reunion of DNA and chromosomes are under enzymatic regulation, which argues against theories that relate crossing-over to physical stresses brought about by chromosome coiling. In fact, Clark and Margulies (1965) found mutants of *E. coli* in which the trait of recombination suppression had been correlated with defective enzymes involved in DNA breakage and repair. The consequence of this and other analyses of recombination with both prokaryotic and eukaryotic organisms has changed our concepts of genetic recombination. No longer

is recombination regarded as a relatively uncomplicated event that occurs at points along precisely paired chromosomes. Rather it is considered to comprise a series of enzymatic events including breakage, synthesis, and covalent bonding of strands of DNA spreading over a length of DNA as long as or longer than one gene (Hayes, 1968). The details and order of events within this region are still not precisely known because, as with other aspects of genetics, the deeper one delves into any process, the more complex it appears to become. It is quite possible that at least two mechanisms operate in recombination; one might function within the region undergoing recombination while another might apply to regions outside this one.

PSEUDOALLELES AND COMPLEMENTATION

In all of the examples cited in the preceding discussion on crossing-over and recombination, consideration was given only to those that evidence exchanges between whole genes; that is, to crossing-over between genes, not within them. The reason is because the vast majority of detectable cases of recombination in eukaryotic organisms have been of this kind. It is little wonder that geneticists began to look upon genes as the smallest indivisible units of inheritance. In subsequent chapters, it will be shown that in prokaryotic organisms, by virtue of the fact that they can be cultured to yield many millions of progeny in a very short time, intragenic recombination is readily observed. In these organisms it is possible to demonstrate that crossing-over can take place at any point along the DNA molecule.

Because the history of prokaryotic genetics does not include a long "classical" period, during which time concepts often become fixed and definitions somewhat rigid, the fact that a functional length of DNA is divisible by structural recombination at any point has presented no problems. Eukaryotic genetics, on the other hand, did go through such a classical period, beginning with the

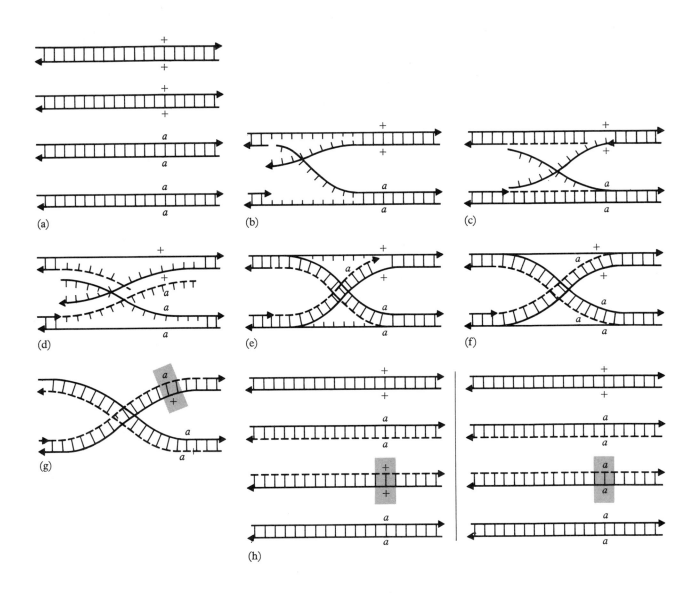

◄ Fig. 4–16. A breakage-synthesis-repair model to account for reciprocal and nonreciprocal crossing-over. (a) DNA helices from two parents, representing tetrad formation. From this point until the last drawing in this figure, only the two central helices will be considered. The arrowheads indicate the polarity of the DNA strand in each segment of DNA. (b) Breakage of two of the DNA chains. (c) Nucleotide chains are synthesized along both unbroken chains. Newly made DNA is indicated by a dashed line. (d) Dissociation of newly made DNA strands from their templates. (e) Newly synthesized DNA associates with a strand from the other molecule. (f) Crossing-over is completed. (g) The unpaired regions of the outer parental molecules are removed. Note the "heterozygous" region of DNA (shaded). This would indicate a region of faulty base pairing. (h) The original, outer parental molecules of DNA are again shown, completing the tetrad after recombination has taken place. The "heterozygous" region is repaired. If the + strand is used as a template, reciprocal recombination is observed. If the a strand is used as the template for repair, nonreciprocal recombination is observed. [From H. L. K. Whitehouse and P. J. Hastings (1965), "The analysis of genetic recombination on the polaron hybrid DNA model," *Genet. Res.* (Cambridge) **6**:27–92. By permission.]

rediscovery of Mendel's work until about the 1940's or 1950's. For eukaryotic organisms a gene was conceived to be indivisible by recombination. Therefore when data were presented in the 1940's that threatened this well-entrenched concept, the term *pseudoallele* was coined. Pseudoalleles were construed to differ from "true" alleles by being separate but adjacent components of a gene that complemented that gene's expression. True alleles, by comparison, occupy homologous segments of DNA and cannot subdivide. For example, the heterozygote

$$\frac{a}{A}$$

could not undergo the kind of recombination that would lead to the genotype

$$\frac{a\;\backslash}{/\;a}$$

True alleles conform to the classical, structural definition of genes, although if recombination within genes had not been found, they could conform to a functional definition as well. Hence pseudoalleles are adjacent portions of the same gene. Together these portions determine a trait, according to the older view, but the regions are divisible by crossing-over between the site of one part of the gene (one pseudoallele) and the other. Because they are really parts of the same gene, pseudoalleles are designated by the same basic symbol; a heterozygous organism could have the genotypes

$$\frac{a_1^+ a_2^+}{a_1 a_2}, \text{ or } \frac{a_1^+ a_2}{a_1 a_2^+}, \text{ or } \frac{a_1 a_2^+}{a_1^+ a_2}$$

for an adjacent pair of pseudoalleles (the "+" indicates the wild-type pseudoallele in each case). Hence this *pair* of pseudoalleles conforms to the definition of one gene as it is defined functionally, because a *linear* pair $a_1 a_2$ or $a_1^+ a_2$, or $a_1 a_2^+$, or $a_1^+ a_2^+$) acting together determines the function of that strand of chromosome by making a single gene product. Because crossing-over can occur between pseudoalleles, recombination can convert the genotype

$$\frac{a_1 a_2}{a_1^+ a_2^+} (cis \text{ configuration})$$

to

$$\frac{a_1 a_2^+}{a_1^+ a_2} (trans \text{ configuration}).*$$

Each of these will have a different phenotype, because in the first instance a completely wild-type gene is present ($a_1^+ a_2^+$), and in the second instance neither gene is without a mutant component.

---

*When any two genetic units are on the same member of a pair of homologous chromosomes, they are said to be in the *cis* configuration. When they are on different members of a pair, namely opposite, they are said to be in the *trans* configuration.

How is it possible to identify the presence of pseudoalleles? If the gene products could be isolated, the distinction between true and pseudoalleles could doubtless be made readily, but this is not generally possible at the present time. The only practical way in which pseudoalleles can be identified in diploid, eukaryotic organisms is by the appearance of wild-type progeny under circumstances in which they should not occur except by the highly improbable event of back-mutation. In order to illustrate what is meant by this, let us imagine a situation such as follows.

A breeder has two mutant strains that have the same phenotype. They are presumed to be different only because they arose independently in different populations. Linkage studies with other genes suggest that the mutant genes in both strains are in the same locus, or very nearly so. When the two strains are crossed, all progeny show the mutant phenotype, but when these progeny are testcrossed with either of the parental strains to yield large numbers of offspring, some of these offspring are wild type. The frequency of the wild-type offspring is significantly greater than can be explained by back-mutation (the rate of which, if it occurs, can be observed in either parental strain).

A situation such as the foregoing suggests the presence of pseudoalleles and is explainable in the following manner. The pseudoallelic mutant components are designated as a_1 and a_2, and their wild-type counterparts as a_1^+ and a_2^+ (as above).

Cross 1

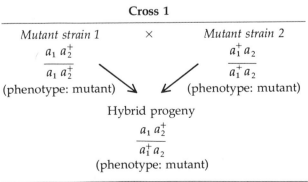

Mutant strain 1	×	Mutant strain 2
$a_1\ a_2^+$		$a_1^+\ a_2$
$\overline{a_1\ a_2^+}$		$\overline{a_1^+\ a_2}$
(phenotype: mutant)		(phenotype: mutant)

Hybrid progeny

$a_1\ a_2^+$

$\overline{a_1^+\ a_2}$

(phenotype: mutant)

Cross 2
(testcross)

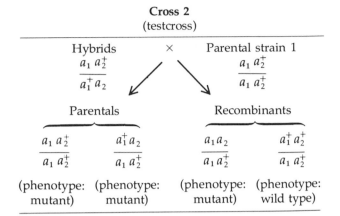

Hybrids	×	Parental strain 1
$a_1\ a_2^+$		$a_1\ a_2^+$
$\overline{a_1^+\ a_2}$		$\overline{a_1\ a_2^+}$

Parentals		Recombinants	
$a_1\ a_2^+$	$a_1^+\ a_2$	$a_1\ a_2$	$a_1^+\ a_2^+$
$\overline{a_1\ a_2^+}$	$\overline{a_1\ a_2^+}$	$\overline{a_1\ a_2}$	$\overline{a_1\ a_2^+}$
(phenotype: mutant)	(phenotype: mutant)	(phenotype: mutant)	(phenotype: wild type)

The validity of this interpretation can be tested if the organism is sufficiently prolific to provide the necessary numbers of offspring. At best, this is difficult to do with diploid organisms, and so most information on intragenic recombination is derived from studies with haploid organisms, particularly bacteria and viruses, as described in Chapter 10.

Pseudoallelism is quite readily demonstrated in certain fungi such as *Neurospora* and *Aspergillus*, which can appear as haploids, diploids, or heterokaryons, as described earlier. Recall the *pan* locus, in which the recessive mutant *pan* requires the vitamin pantothenic acid. Heterokaryons of the wild type and the mutant strain *pan₂* can grow in culture medium that lacks pantothenic acid. Case and Giles (1960) recovered 75 *pan₂* mutants that arose spontaneously. When pairs of these mutants were used to form heterokaryons, 52 of the mutants, paired in various combinations with one another, failed to grow in the absence of pantothenic acid. This indicates that the particular mutations used in the combinations are in the same gene; even more precisely, they are in the same area of the same gene. The fact that the deficiency of one strain was not compensated for or *complemented* by a corresponding normal allele in the other signifies that both strains were deficient for the same gene. This is analogous to the situation of pseudoalleles in the *trans* position. When paired in certain combinations, 23 other mutants could produce heterokaryons that grew in the absence of pantothenic acid, although growth was slower than in the wild-type organisms. For these mutants, complementation had taken place to some extent, and the defective areas of DNA were probably not areas of overlap. The complementation patterns of various arrangements produced a complex picture. For instance, mutant 1 would complement mutants 2, 3, and 5, but not 4. Mutant 2 would complement mutants 1 and 3, but not 4 or 5. On this basis, it was determined that the defec-

tive region of the pan_2 locus in mutant 1 overlapped that of mutant 4, but not that of either 2, 3, or 5. Similarly, the defective region in mutant 2 must have overlapped that of 4 and 5 (no complementation, therefore overlapping or identical regions of mutation) but not that of 1 or 3 (complementation, therefore different regions). By a stepwise sorting out and fitting together of all the data, a functional complementation map of the pan_2 locus was established. This map can be compared to a map based on the frequency of recombination in sexual crosses of one mutant strain with another (Fig. 4–17).

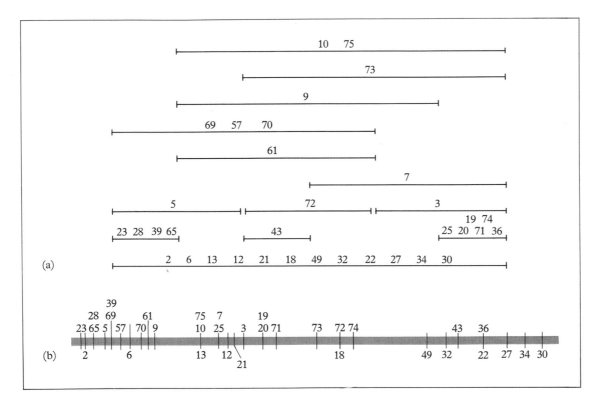

Fig. 4–17. A genetic map of the pan_2 locus in *Neurospora crassa* compared to a complementation map. (a) The complementation map based on *cis-trans* complementation. Mutations on the same line have identical complementation patterns. (b) The genetic map is based on recombination frequencies in sexual crosses. The numbers indicate the location of mutations in the various mutant strains. [From Case and Giles (1960), *Proc. Nat. Acad. Sci.* **46**:659. By permission.]

QUESTIONS

1. Linked to the A–B–O locus in man is a gene (N) causing the nail-patella syndrome that involves abnormalities of the nails and skeleton. There is a recombination frequency of 10 percent between the two loci.

A woman of blood group A who has the syndrome has a father of blood group O and no syndrome. She marries a man of blood group O who is similarly normal. List all possible phenotypes of their children and indicate which of these would be expected with a high frequency and which with a low frequency.

2. In *Drosophila*, the genes curled (*cu*) and ebony (*e*) are 20 map units apart. If females of genotype $\frac{cu\ e}{+\ +}$ were crossed with doubly recessive males, how many of each category of progeny would be expected? Assume a total number of 1000 progeny.

If the heterozygous females were of genotype $\frac{cu\ +}{+\ e}$, would the results be changed?

3. Given the following maps:

$$a \leftarrow 10 \rightarrow b \leftarrow 9 \rightarrow c \qquad d \leftarrow \!\!\!—\!\!\!—20\!\!\!—\!\!\!— \rightarrow e \leftarrow 10 \rightarrow f$$

between which set of genes would the higher coefficient of coincidence be found?

4. Given the following map based on testing for single crossovers between loci *a-b* and *b-c*,

$$a \leftarrow 4.2 \rightarrow b \leftarrow \!\!\!—\!\!\!—12.1\!\!\!—\!\!\!— \rightarrow c$$

what would be the expected frequency of double crossovers between *a* and *c*?

If the coefficient of coincidence were 0.5, what should be the actual, observed double crossover frequency?

5. Assume that genes *a* and *b* in *Drosophila* are 15 map units apart. One-thousand offspring result from the cross

$$\frac{a\ b}{A\ B} \times \frac{a\ b}{a\ b}$$

What might be a reasonable phenotypic assortment of offspring?

6. Is it possible to distinguish between two linked loci that show 50 percent recombination between them and two unlinked loci? If so, how?

7. In a particular plant, the genes *a*, *b*, and *c* are linked. In a cross between a triple heterozygote and a triple recessive homozygote, the following data were obtained. From these, determine the sequence of genes, the map distances between them, and the coefficient of coincidence.

Phenotype of offspring	Number
ABC	146
ABc	696
AbC	4
Abc	192
aBC	220
aBc	4
abC	612
abc	126

What was the arrangement of genes in the parents of the heterozygote used for the testcross?

8. In corn, the genes *an* (anther ear), *br* (brachytic), and *f* (fine stripe) are linked. Testcross data are as follows:

Progeny	Number
+ + +	88
+ + f	21
+ br +	2
+ br f	339
an + +	355
an + f	2
an br +	17
an br f	55

Determine the linkage map and the genotypes of the homozygous parents used to obtain the heterozygote for the testcross.

9. A strain of *Neurospora* nutritionally mutant for two genes *a* and *b* was crossed with a wild-type strain (+ +). The following asci were obtained. (Mitotic doubling is not indicated—only one of each pair of spores is listed.)

a b	a +	a b	a b	+ +	+ b	a +
a b	a +	a +	a +	+ b	+ +	a b
+ +	+ b	+ b	+ +	a b	a +	+ +
+ +	+ b	+ +	+ b	a +	a b	+ b
25	16	11	6	8	6	8

Are genes *a* and *b* linked? If so, construct a map showing their distances from the centromere.

10. In *Neurospora*, a cross between two strains ($+\ b$) and ($a\ +$) yielded the following tetrads. What are the map distances between each gene and the centromere?

a	b	$+$	b	$+$	$+$	a	$+$
a	b	$+$	b	$+$	$+$	a	$+$
$+$	$+$	a	$+$	a	b	$+$	b
$+$	$+$	a	$+$	a	b	$+$	b
42		40		39		42	

11. In *Drosophila*, the genotypes

$$\frac{a}{a},\ \frac{b}{b},\ \text{or}\ \frac{c}{c}$$

all produce a similar phenotype. Assuming that you have access to all the stocks of *Drosophila* that you might need, how would you distinguish whether these alleles are true alleles, pseudoalleles, or neither?

12. You are given heterozygotes

$$\frac{+\ d}{a\ +},\ \frac{b\ +}{+\ d},\ \frac{c\ +}{+\ d}$$

and homozygotes

$$\frac{a\ d}{a\ d},\ \frac{b\ d}{b\ d},\ \frac{c\ d}{c\ d}$$

for testing. Assume that all the different strains of heterozygotes have a similar phenotype. From the following testcross data, can you determine whether loci a, b, and c are really allelic?

Testcross I

$$\frac{a\ +}{+\ d}\ \times\ \frac{a\ d}{a\ d}\ \rightarrow\ \frac{+\ +}{a\ d},\ \frac{a\ d}{a\ d},\ \frac{a\ +}{a\ d},\ \frac{+\ d}{a\ d}$$

$$(25\%)\quad(25\%)\quad(25\%)\quad(25\%)$$

Testcross II

$$\frac{b\ +}{+\ d}\ \times\ \frac{b\ d}{b\ d}\ \rightarrow\ \frac{+\ +}{b\ d},\ \frac{b\ d}{b\ d},\ \frac{b\ +}{b\ d},\ \frac{+\ d}{b\ d}$$

$$(15\%)\quad(15\%)\quad(35\%)\quad(35\%)$$

Testcross III

$$\frac{c\ +}{+\ d}\ \times\ \frac{c\ d}{c\ d}\ \rightarrow\ \frac{+\ +}{c\ d},\ \frac{c\ d}{c\ d},\ \frac{c\ +}{c\ d},\ \frac{+\ d}{c\ d}$$

$$(5\%)\quad(5\%)\quad(45\%)\quad(45\%)$$

5

Sex Determination and Sex Linkage

The sexes are very readily discerned in many animals and in certain plants by distinct morphological differences such as coloration, size, distribution of hair, presence/absence of antlers, and so on. In other species, the only externally apparent morphological distinction is the external genitalia; in still others, no external morphological difference is apparent at all. In all cases of sexually reproducing organisms, however, morphologically distinguishable reproductive or "sex" cells are discernable microscopically. Because of this, sexuality is defined by this criterion, that is, as the condition wherein separate male and female cells, the spermatozoa and ova, are produced. Unfortunately this is not a very satisfactory definition from the biological standpoint, because there are organisms in which the reproductive cells, although functionally similar to sperm and eggs, show little if any morphological difference from each other. Even more extreme is the situation of some ciliate protozoans in which a process occurs that resembles sexual reproduction, inasmuch as the genes of two different cells are brought together preliminary to cell division, but fusion of the two cells does not take place to produce a zygote. Moreover, in some species of bacteria, partial mixing of genes may occur between two individual cells without either cell losing its identity or mixing any discernable amount of cytoplasm. Clearly a new definition of sex and sexual reproduction is needed.

The phenomenon that is common to all sexually reproducing organisms, as well as to those that pool genetic information without benefit of sex, is a situation of complete or partial diploidy followed by genetic recombination and haploidy. How this is accomplished varies considerably from taxon to taxon, but the end results are the same. Mechanisms that have enabled genetic recombination have provided a strong selective advantage to the species having them, and therefore such mechanisms have been of inestimable importance in biological evolution (Smith, 1971). It seems appropriate for ease of future discussion to define sexual reproduction as the biological process wherein the genes of two different individuals are included within the same cell in such a way that recombination and segregation of recombinants can occur. Variations within the process are expressed in terms of the relative duration of the diploid phase (when the two sets of genes are together) and the haploid phase (when genes exist in essentially a single set), and in terms of which phase is the "metabolic" phase (during which the major part of the organism's genome effects phenotypic expression).

Chlamydomonas. Sexuality is observed in perhaps its simplest and clearest form in certain species of the green, unicellular alga *Chlamydomonas*. Individuals in the metabolic phase are haploid and possess two flagella on the anterior end. They may reproduce asexually by mitosis, giving rise to genetically identical, haploid daughter cells. During sexual reproduction, two cells approach each other "head on" and fuse together, resulting in a single cell with two pairs of flagella and the diploid complement of genes. The flagella are very soon lost as the cell—a zygote—rounds up and develops a thick, spiny cell wall (Fig. 5–1). Prior to germination, two meiotic divisions take place, resulting in four haploid cells. Each of these may undergo one or more mitotic divisions before emerging as new individuals from the ruptured cell wall of the zygote, whereupon they grow new flagella (Bacci, 1965). By means of genetic markers, it has been shown that genetic recombination occurs during meiosis. In this kind of organism, the haploid phase is the metabolic phase and dominates the life cycle. The diploid phase is either of short duration or exists during a prolonged period of rest or metabolic inactivity. Clearly the diploid condition in this species is necessary for recombination, which appears to be its only function. The cells of the metabolic phase are all morphologically identical, with no

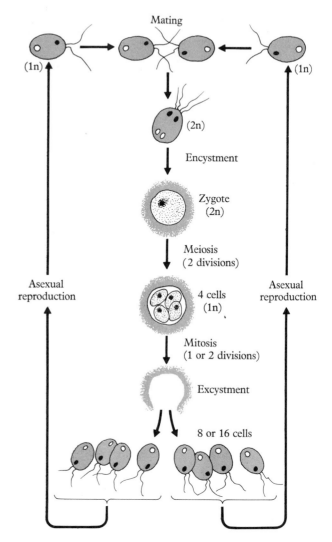

evidence whatever of sexual dimorphism (Moewus, 1936).

In other species of *Chlamydomonas*, however, a degree of sexual differentiation is expressed in the form of different "mating types," designated plus (+) and minus (−), such that mating occurs only between individuals of opposite sign. By hybridization, it has been possible to correlate the sign with one sex or the other of still another species in which sexual dimorphism does exist (Moewus,

Fig. 5–1. Diagram of the life cycle of *Chlamydomonas*. Haploid individuals meet and fuse to become a diploid zygote, which encysts. Before rupturing from the cyst wall, the zygote undergoes two meiotic divisions which result in four haploid cells. Each of these may undergo one or two mitotic divisions, producing eight or sixteen haploid cells, which emerge from the cyst as free-swimming adults. These cells may undergo a number of subsequent mitotic divisions (asexual reproduction) before another round of mating.

1936). In the latter species, the cells are either small or large, which might be considered respectively as "male" and "female" for convenience. If these are mixed with plus and minus individuals of the species characterized by mating types, it is found that the plus cells will mate with only the "male" cells and the minus cells will mate with only the "female" cells. Accordingly, plus of one species corresponds with female of the other, and minus with male. The genetic basis for difference in mating type is postulated to be two alleles, designated mt^+ and mt^-, which segregate during meiosis.

Paramecium. Several species of ciliate protozoans, including the familiar genus *Paramecium*, are also characterized by morphologically indistinguishable mating types (Sonneborn, 1937). *Paramecia* are diploid organisms that reproduce primarily asexually. As with other ciliates, they are characterized by two kinds of nuclei—two diploid micronuclei and a polyploid macronucleus, both of which divide mitotically and are represented equally in the daughter cells. However, sexuality in these species involves no loss in "parent"-cell identity through cell fusion and no meiotic division of cell cytoplasm. In this curious circumstance, individuals of different mating type effect firm contact by way of their oral grooves, and as they do the macronucleus disintegrates and the micronuclei undergo two intracellular meiotic divisions that result in eight haploid micronuclei in each cell. Seven of the micronuclei in each cell degenerate, and the eighth undergoes a mitotic division. One of the daughter nuclei is then exchanged

for one of the daughter nuclei in the other cell, so that each *Paramecium* now has two haploid micronuclei, one from each of the two individuals (Fig. 12–10). The micronuclei then fuse, restoring diploidy, but with chromosomes of different cell origin. Apparently invigorated by the experience, the two individuals then separate and undergo two mitotic divisions, regenerating new macronuclei and micronuclei from the mitotic products. The cells then undergo a variable number, usually quite large, of asexual reproductions. Hence in this species, the diploid phase is the metabolic and dominant phase. Except in terms of nuclei, haploidy really does not exist for any of the cells at any time. But since meiosis does take place and individual sets of chromosomes are segregated, recombination can occur and genetic variability is effected. If the process is visualized only in terms of micronuclei, it is seen to be exactly the same as in *Chlamydomonas*, or in mammals for that matter, differing only in the temporal relationships expressed by the duration and activities of the diploid phase as compared with the haploid phase. Presumably the basic mechanism that determines mating type is a pair of gene alleles, as in *Chlamydomonas*. The situation appears to be somewhat more complex, at least superficially, in other species of ciliates in which as many as nine mating types have been described (Elliot, 1959). Mating can occur in these animals between any two individuals of different mating type, but not between members of the same mating type.

Escherichia coli. It was only relatively recently (Lederberg and Tatum, 1946) that a sort of sexuality was observed in bacteria. This phenomenon, called "bacterial conjugation," is attributed to an episome, the F-factor, and is described in Chapters 10 and 12. It is important to note that during conjugation, the DNA molecule of the donor strain replicates, and only one of the daughter molecules is involved in migration. This is reminiscent of conjugation in *Paramecium*, but there is an important difference. Whereas in *Paramecium* an entire DNA set (haploid genome) migrates across the cytoplasmic bridge formed by the oral grooves, usually only a part of the DNA replica is transferred in *E. coli*. The reason for this appears to be that the bridge is very weak and is easily broken before completion of transfer, as was shown by Jacob and Wollman (1961). At any rate, the portion of the donor DNA that is transferred may be incorporated into one of the replicating DNA molecules of the recipient cell in which recombination may occur, whereupon the recombinant DNA will then be segregated into one of the daughter cells during the first subsequent fission. The genetic basis of sexuality in *E. coli* appears, at least as far as present knowledge goes, to be quite different from that in other kinds of organisms, for it is an episome rather than one or a group of sex genes, *per se*, that determines "sexuality," although it is conceivable that sex genes may have originated as episomal factors that became irreversibly integrated into the genome.

Neurospora. *Neurospora* and many other species of fungi have haploid nuclei, which may number one or more in a given cell. Not uncommonly, two mycelia of different individuals come in contact and some of the cells may fuse, resulting in cells called heterokaryons (Chapter 4). Experiments with mutant strains of *Neurospora* that differ with respect to certain nutritional requirements have revealed the phenomenon of "complementarity," which is discussed in Chapters 4 and 10. In essence, a cell that has two nuclei with different genomes benefits from both. If one nucleus is deficient in its ability to code for vital substance A and the other is deficient in its ability to code for vital substance B, for example, the first takes care of the requirements for substance B and the second for substance A, so that each nucleus complements the other and the cell synthesizes both substances. The formation of heterokaryons occurs quite independently of sexual reproduction in some species of fungi, whereas in others there may be a causal relationship between the two phenomena.

THE EVOLUTION OF SEXUAL REPRODUCTION

From the few foregoing examples, which were selected from many varieties of reproductive processes, some interesting speculations can be developed to account for the evolution of sex and, incidentally, of diploidy. One hypothesis that immediately arises is that bacteria represent the most primitive condition of sexuality. We can visualize that any mechanism of recombination, even if only portions of the genome are involved, would provide the organisms having such mechanisms with a selective advantage. This certainly has been proved to be the case in many pathogens, in which recombination has facilitated the very rapid derivation of antibiotic-resistant strains (Chapter 12).

Something like the situation in *Paramecium*, with some modification, might represent the next step in the evolution of sex. The important aspect here would be the exchange of a complete haplogenome between conjugating pairs, which would be followed by recombination and then segregation through cell division. The third step could be that represented by the first species of *Chlamydomonas* described earlier, the important advance in this instance being complete cell fusion with resultant zygote formation, followed by meiosis. Next we could have the genetic differentiation of mating types, which would represent an advantage because it would favor mating between genetically different individuals over genetically identical ones. From this point onward, improvements in the process would result in sexual dimorphism and acquisition of morphological and physiological characteristics that would assure outbreeding, culminating ultimately in clear-cut sexuality as we see it in all higher animals and most higher plants. The principle of the heterokaryon, that is, the selective advantage of complementation implicit in diploidy, would favor the gradual ascendancy of the diploid phase as the dominant one, with the corresponding reduction of the haploid phase to that of relatively short-lived gametes.

The ascendancy of diploidy to the dominant phase has been of very great importance in the evolution of higher forms. Not only has it provided a selective advantage by enabling heterozygosis, it has also provided a reservoir for the accumulation of genes or mutations of indifferent value which at a later time might, hypothetically, undergo a chance recombination that would represent something entirely new and highly favorable. There can be little doubt that evolution was a terribly slow process before the appearance of sexual reproduction and diploidy. The prokaryotes, in which sexual reproduction is rare and only rudimentary, have undergone very little evolutionary change since the time mitosis and meiosis were introduced. The eukaryotes, on the other hand, which are the descendants of organisms characterized by both mitosis and meiosis, have undergone a fantastic evolution since that time, giving rise to all the animals and plants that inhabit the earth today.

An alternative hypothesis for the evolution of sexuality could propose the heterokaryon as the first step. A heterokaryon would be able to survive temporary conditions of poor nutrition better than haploid cells, and then when favorable conditions returned, haploid cells could be proliferated. Fusion of haploid nuclei within the heterokaryon would represent an easy transition. This has been described in other species of mold fungi, *Aspergillus* and *Penicillium* (Pontecorvo, 1953). During DNA replication, "homologous" DNA or chromosomes might be attracted into synapsis by base pairing in the same way that DNA/RNA hybridization occurs, and thus the appropriate conditions for recombination would be established. Separation of homologues after replication would constitute the first meiotic division, and separation of replicas the second meiotic division, thereby producing four haploid nuclei all genetically different. The next stage in the evolution of sexuality could be the *Chlamydomonas* model, and so on, with bacteria and *Paramecium* representing quite distinct and possibly unrelated mechanisms.

Other hypotheses could surely be proposed, but all are speculative because there is very little hard fact to rely on at the present time. It is believed that nature has experimented with sexual reproduction many times, that many different forms have evolved quite independently of one another, and that many degenerative and/or adaptive processes have occurred. A form of sexual reproduction is found even among bacterial viruses—multiple infections by differing strains have resulted in recombinants, as described in Chapter 10. Bacterial transformation and transduction might also be considered in the same general category. The advantages of sexual reproduction and of diploidy are clear, but how and when they evolved are still matters of conjecture.

SEX DETERMINATION

The *Sexuality Theory*, advanced early in the twentieth century (Hartmann, 1931), proposes that every cell of sexually reproducing organisms is bisexual and has the full potencies of the male and female sex, becoming male or female by the preponderant development of one or the other potency. This is a useful theory because it accounts for bisexuality, hermaphroditism, sex reversal, and intersexuality, all of which have been observed throughout the animal and plant kingdoms. The Sexuality Theory does not pretend to explain how one or the other potency is developed in an individual cell or organism, however. This section will attempt to describe the most common and best understood mechanisms of sex determination encountered in nature. It will become evident that there is much still to be learned in this area of genetics.

Genetic Sex Determination

Doubtless, sex determination is basically a genetically regulated process in all sexual species. Nevertheless, a distinction may be made between situations in which sex is determined rather directly by the genotype, and those in which it is determined less directly as a consequence of factors that, in one way or another, regulate the expression of the genes that are concerned with sexual differentiation. One of the species of *Chlamydomonas* described in the preceding section appears to represent an example of straightforward genetic sex determination—cells that have the genotype *mt*⁺ differentiate under normal conditions into individuals of the plus mating type, and those with the genotype *mt*⁻ differentiate into the minus mating type. Certain strains of the genus *Mucor*, a common mold, are dioecious—that is, individuals are of either plus or minus mating type. Sexuality in these strains appears to be under direct genetic control as well.

A rather more complex genetic mechanism is found in the parasitic wasp *Habrobracon juglandis* (Whiting, 1943), in which a multiple-allelic series for sexuality has been described. The Hymenoptera, which includes bees and wasps, is one of several groups of invertebrate animals in which the males are usually haploid and the females diploid. The queen bee, for example, stores in her spermatic receptacles sperm that she has received by copulation with a single male. She ovulates haploid eggs, which she can fertilize or not as she "chooses," at the time of oviposition. Those that are fertilized become diploid and develop into females; those that are not fertilized remain haploid and develop into males. This kind of phenomenon is called *facultative parthenogenesis*. In the wasp *Habrobracon*, diploid males have been encountered. It has been found that they can be produced with regularity by inbreeding, but they have

very low viability, usually failing to hatch. By the use of genetic markers, it was determined that femaleness is the expression of heterozygosis of the sex alleles, and maleness results from homozygosis. Haploid individuals, being necessarily "homozygous," are invariably male. Assuming a tendency for outbreeding, which is very strong in wild Hymenoptera, and assuming the existence of several sex alleles, the probability of diploid homozygosis occuring by chance in nature is fairly low. If the various alleles have an equal frequency, then the chance of diploid homozygosis in freely outbreeding populations could be represented by the formula

$$\frac{1}{n+1}$$

where n is the number of allelic forms. Hence, if there are nine allelic forms, as Whiting described in random sampling, the chance of diploid homozygosis would be about ten percent for the fertilized eggs.

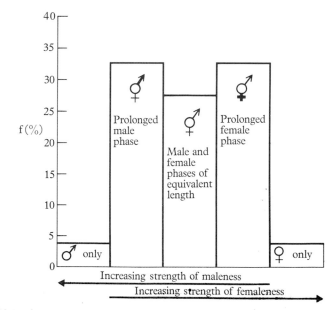

Fig. 5–2. Graphic representation of sex distribution in the marine limpets *Crepidula plana* and *Patella coerulea*. Most individuals are protandrous hermaphrodites with some variation in the relative duration of the male and female phases. Unisexual individuals occur with about equal frequency, with a combined frequency of about 7.0 percent. [Data from Bacci (1965).]

In some animals, a sort of continuous variation in sexuality is apparent. In the molluscan genera *Crepidula* and *Patella*, which are marine limpets, unbalanced hermaphroditism occurs and is expressed in a sort of bell-curve distribution of sexes (Fig. 5–2). The echiuroid worm *Bonellia viridis*, although not hermaphroditic, displays sexual indeterminacy that has much in common with unbalanced hermaphroditism. Hermaphroditism is very common among animals and plants, although it is rare in vertebrates, and so most people are not aware of its prevalence. True hermaphroditism is the condition in which an individual is functionally both male and female, either at the same time or consecutively. Species in which individuals are male during the first part of their reproductive life and female during the later part are called *protandrous* (Greek: first male) *hermaphrodites*. The reverse of this is *protogynous* (Greek:

first female) *hermaphroditism*. Some hermaphrodites have both ovaries and testes (*antheridia* in flowering plants), and others have a combined gonad (an *ovotestis*, as in most snails). In unbalanced hermaphrodite species, unisexual forms— that is, individuals that are purely and irreversibly one sex or the other—are found in populations that are otherwise predominantly hermaphroditic. Moreover, groups of individuals that are in a sense "intermediate" between the typical hermaphrodites and the unisexual individuals have been described. In *Patella coerulea*, for example, most individuals are protandrous, with male and female phases of about equal duration. Unisexual male and female individuals also occur, although rarely. But intermediate between these three types of individuals are organisms that have a prolonged male phase and correspondingly shortened female phase, on the one hand, or a fore-

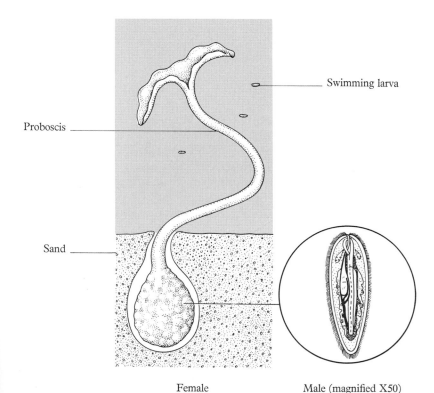

Proboscis

Swimming larva

Sand

Female   Male (magnified X50)

Fig. 5–3. *Bonellia viridis* is a burrowing marine worm that exhibits extreme sexual dimorphism. The female, shown at about actual size, is many times larger than the male, which lives in her reproductive tract.

shortened male and prolonged female phase on the other. *Crepidula plana* exhibits a similar distribution of sexuality.

Bonellia expresses its ambisexuality in a different way from *Patella* and *Crepidula*, and merits some description. A mature female *Bonellia* has a long bifurcated proboscis, which is mounted on a bulbous body about the size of a walnut. Located within her proboscis or in the ducts of her reproductive tract are one or more males, which measure between one and three millimeters in length (Fig. 5–3). Fertilization of the eggs is therefore as internal as it can be. The fertilized eggs are shed into the surrounding sea where they develop into ciliated, swimming larvae of undifferentiated sex, which ultimately settle down on the sand. For the most part, those that settle directly on the sand differentiate into females, whereas those that settle on the proboscis of an adult female differentiate into males. An aqueous filtrate of minced proboscis or reproductive tract of a mature female will induce maleness in larvae just as effectively as actual contact. However, a small number of larvae will differentiate into male or into female irrespective of whether or not they make contact with a mature female. Except for these, larvae that have begun to differentiate into one sex or the other can be induced to differentiate into the opposite sex by alteration of the environment. This confusing and rather plastic situation characteristic of *Bonellia* has been represented as analogous to unbalanced hermaphroditism, even though all mature adults are unisexual.

Bacci (1965) attributes sex determination in *Bonellia* and in unbalanced hermaphrodites to sex polygenes, as represented in Table 5–1. He postulates four sex-gene loci, each with male- and female-determining alleles. An individual that has seven or eight male alleles in its genotype would be unisexually and irreversibly male. Seven or eight female alleles would produce individuals that are unisexually and irreversibly female. Most individuals would have equal numbers of contrasting alleles and would be protandrous hermaphrodites of equal male and female duration, in the case of *Patella* and *Crepidula*; in the case of *Bonellia*, they

Table 5–1. Postulated sex determination in unbalanced hermaphrodites and *Bonellia* by sex polygenes. (See text.)

Genotypes	Percent frequency	Phenotypes in:		
		Patella coerulea	Crepidula plana	Bonellia viridis
8 dom. 0 rec.	0.4	Pure ♂	Pure ♂ (associated) with indefinite length of the male phase	Pure ♂ (Spätmännchen)
7 dom. 1 rec.	3.1			
6 dom. 2 rec.	10.9	♂ with long ♂ phase	♂ with long ♂ phase	Individuals which are easily masculinized
5 dom. 3 rec.	21.9			
4 dom. 4 rec.	27.4	♂ with equal length of ♂ and ♀ phases	♂ with equal length of ♂ and ♀ phases	Individuals which are equally easily masculinized or feminized
3 dom. 5 rec.	21.9	♀ with short ♂ phase	♀ with short ♂ phase	Individuals which are easily feminized
2 dom. 6 rec.	10.9			
1 dom. 7 rec.	3.1	Pure ♀	Pure ♀ (isolated)	Pure ♀ (larvae which will not adhere to the proboscis)
0 dom. 8 rec.	0.4			

[Reprinted with permission from Guido Bacci (1965), *Sex Determination*. Copyright © 1965, Pergamon Press Ltd.]

would be unisexual as adults but readily and equally reversible as differentiating larvae. Individuals of intermediate propensities would result from the other possible genotypes. In the case of *Bonellia*, these would fall into hypothetically two classes, one consisting of individuals which as larvae can be more readily masculinized than feminized, and the other consisting of those that can be more readily feminized than masculinized (Table 5–1).

Genes with specific effects on sexuality have been described. One of the clearest and best identified of these is the gene called *transformer* (*tra*) in *Drosophila melanogaster*. This gene was discovered as a mutant recessive in laboratory stocks in 1945 (Sturtevant), and it was found that in homozygous mutant form, it causes flies that are genotypically female to develop into sterile males. A third allele (and second mutant) with an intermediate effect on diploid females has also been described for this gene (Gaven and Fung, 1957). That a number of genes distributed throughout the genome directly influence sexuality in *Drosophila* has been known for a number of years. The *tra*+ gene, which is located on one of the autosomes (see below), is doubtless one of these.

In 1925, C. B. Bridges advanced his brilliantly conceived theory on sex determination in *Drosophila*. Called the *Theory of Genic Balance*, it proposed the existence of a number of sex-determining genes located throughout the genome, but distributed in such a way that there is a preponderance of male determiners in the "autosomes" and a preponderance of female determiners on the X chromosome. As will be described more explicitly below, many species of animals and a few species of plants have two kinds of chromosomes in relation to sexuality—the sex chromosomes, which are customarily designated X and Y, and all others, which are called *autosomes*. Female *Drosophila* have two sets of autosomes and two X chromosomes. The males have two sets of autosomes and only one X chromosome. Therefore, females have relatively more female deter-

miners than males do, and conversely, males have relatively more male determiners, and so it is the "balance" between the two kinds of determiners that tips the scales toward differentiation into one sex or the other. Bridges was able to develop this theory through the ingenious manipulation of chromosome numbers in *Drosophila*, as shown in Table 5–2. His results showed that the Y chromosome has no influence on sexuality save sperm motility, as described on page 101. This means that flies with two sets of autosomes and one X chromosome are as fully male as are flies with two sets of autosomes plus an X and a Y chromosome, except that their sperm are immotile, hence sterile. Flies with three sets of autosomes and only one X showed exaggeration of the male secondary sexual characteristics and were called "super-males." This is actually fraudulent, of course, because they are invariably sterile and lacking in normal vigor—no match for a normal male. On the other hand, "super-females," which turned out to be just as fraudulent as their male counterparts, resulted from a chromosomal formula of two sets of autosomes and three X chromosomes. Individuals with ratios of autosomes and X chromosomes intermediate between those characteristic of male

Table 5–2. The relation of the "sex index" to sex phenotype in *Drosophila melanogaster*.

Number of X chromosomes (N_X)	Number of sets of autosomes (N_A)	Sex index (N_X/N_A)	Sex phenotype
3	2	1.5	Super-female
4	4	1.0	Female (4n)
3	3	1.0	Female (3n)
2	2	1.0	Female (2n)
1	1	1.0	Female (1n)*
2	3	0.67	Intersex
1	2	0.50	Male
1	3	0.33	Super-male

[From C. B. Bridges (1925).]

*The haploid female phenotype was deduced not from an entire individual, but from the character of haploid tissues in the chromosome mosaic 3,X/6,XX.

and female developed intermediate characteristics, appropriately termed "intersex." Subsequently (Dobzhansky and Schultz, 1934), it has been shown that the X chromosome of *Drosophila* does indeed include a number of female-determining genes. In another organism, the gypsy moth (*Lymantria dispar*), races have been described that differ from one another in the "strength" (possibly the number) of their male- and female-determining genes, so that it has been possible to obtain a range of intersexual types between normal male and female (Goldschmidt, 1942; Winge, 1937). Clearly, sex determination is by genic balance in this species also.

Chromosomal Sex Determination

Specialization of sexual function would seem to be advantageous for a species in a situation of selective competition, for this would enhance reproductive efficiency. In fact, this is precisely what has happened in most higher forms of animal and plant life. Strengthening sexuality, as for any function, implies the accumulation of increased numbers of genes for differentiation of the character. In order that polygenes can have maximum effect in producing two functionally distinct phenotypes without a high frequency of intermediate types, they must be linked in some way, that is, not be subject to independent assortment. In the case of sexual differentiation, intermediate types would be intersexes, which are evolutionary deadwood. Therefore, it would be advantageous to have the bulk of determiners for one sex or the other concentrated on one member of a pair of homologous chromosomes. In order that they remain on that chromosome, some method of preventing crossing-over between the two members must be introduced, at least for the portions of the chromosome that include the sex genes (Darlington, 1931). Finally, there would be distinct selective advantage in developing a mechanism that would assure a sex ratio of fifty-fifty. The rationale for the last is as follows (Kalmus and Smith, 1960). Sexual re-

production favors a high degree of genetic diversity, which is essential for adaptation. The greater the amount of genetic diversity, the greater the potential for adaptation (cf. Chapter 14). Sexual reproduction depends upon male and female encounters and, in order to make the greatest use of genetic diversity, the variety of sexual matings must be maximum. This is achieved when there are equal numbers of males and females. It has been possible through the study of mechanisms of sex determination in lower vertebrates (fishes and amphibians) to reconstruct from species living today that each of the foregoing requirements has been met, *seriatum*, in vertebrate evolution (Ohno, 1967).

The evolution of the sex chromosomes X and Y, which have provided such selective advantages as the foregoing, although not as widespread through nature as most biology texts imply, has taken place in a number of unrelated yet evolutionarily very successful groups. For example, sex chromosomes are most common among insects, birds, and mammals, but they have been described in some crustaceans, at least one nematode, some of the lower vertebrates, and a few dioecious plants (Westergaard, 1958). The distribution of sex determiners on the sex chromosomes varies, for in some species such as *Drosophila*, the X chromosome includes a number of female determiners whereas the Y chromosome is almost devoid of sex determiners. In other species such as *Lychnis (Melandrium)* and human beings, the Y chromosome contains some if not a number of essential male-determining genes. In extreme cases, the one-sidedness has been carried to the point of total loss of the Y chromosome, as in certain insects in which the female has two X chromosomes and the male has only one, although he is diploid for the autosomes. The sex chromosome formula for such males is designated XO.

The sex that is XX is called the *isogametic* or *homogametic* sex, because all of its germ cells have equal potential with regard to sex determination. The sex that is XY or, as in the insect case cited above, XO is called the *digametic* or *heterogametic* sex, because it produces two kinds of gametes with respect to sex potential. In some crustaceans and in many insects (Lepidoptera and Chrysopidae), in some fishes, amphibians, and reptiles, and in all birds as far as is known, the males are isogametic and the females are digametic. That is, males are XX and females are XY or XO.* In all others, which include predominantly the hemipteran, dipteran, and most orthopteran insects, and probably all mammals including human beings, the males are digametic (XY or XO) and the females are isogametic (XX).

Evidence for the preponderance of female-determining genes on the X chromosome of *Drosophila* has already been mentioned (Bridges, 1925; Dobzhansky and Schultz, 1934). It is because of this that the unbalanced genomes 2A + XXY is female and 2A + X is male in *Drosophila*, where 2A refers to a diploid set of autosomes, whatever their total number may be. In *Melandrium* and in human beings, the opposite applies, although the results differ somewhat.

As shown in Table 5–3, sex in *Melandrium* is largely determined by the presence or absence of the Y chromosome. By ingenious methods that included the use of triploid plants and fragments of Y chromosomes, Westergaard (1948) showed that the differential segment of the Y chromosome (the part that does not pair with the X) consists of three regions on the basis of distribution of sex determiners (Fig. 5–4). Region I contains a gene or genes that suppress the differentiation of female flower parts; Region II includes genes essential for development of male flowering parts; and Region III contains genes essential for maintenance of the pollen mother cells. Region IV, present in both X and Y, represents the pairing segment. The differential segment of the X chromosome is Region V.

*The sex chromosomes of birds and, less frequently, of other species that show this "reversed" formula are often designated as Z (= X) and W (= Y). Thus, ZZ = male and ZW = female.

Table 5–3. The relationship between chromosome constitution and sex phenotype in *Melandrium*.

Chromosome constitution	Sex phenotype
2A + XX	♀
2A + XXX	♀
3A + XX	♀
3A + XXX	♀
4A + XX	♀
4A + XXX	♀
4A + XXXX	♀
4A + XXXXX	♀
2A + XY	♂
2A + XYY	♂
2A + XXY	♂
3A + XY	♂
3A + XXY	♂
3A + XXXY	♂
4A + XY	♂
4A + XXY	♂
4A + XXYY	♂
4A + XXXY	♂
4A + XXXYY	♂
4A + XXXXY	♂ → ♀
4A + XXXXYY	♂

[From Bacci (1965), after Westergaard (1953).]

The silkworm *Bombyx mori* (Bacci, 1965) shows a similar mechanism of sex determination, except that the Y chromosome determines femaleness, the female being the digametic sex in lepidopterans. In a race of the dioecious flowering plant *Rumex hastatulus*, the mechanism of sex determination is a combination of genic balance and strong male determiners on the Y chromosome (Smith, 1963).

The sex chromosome condition in human beings merits special consideration because of recent findings of clinical and sociological significance. The sex chromosomes of man, like those of the higher apes, are a medium-sized, metacentric X and a short, acrocentric Y (see Chapter 2). Several kinds of congenital abnormalities have been shown to be associated with departures from the normal chromosome complement in human females and males. The normal complement is 46

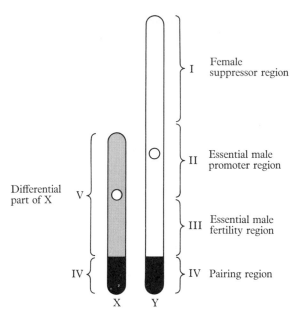

Fig. 5–4.   Analysis of the sex chromosomes of *Melandrium.* [From M. Westergaard (1948).]

Fig. 5–5.   Squamous epithelial cells from buccal scrapings of (a) a normal woman showing a single Barr body (arrow) adjacent to the nuclear membrane, and (b) a woman having the karyotype 47,XXX showing two Barr bodies (arrows). A blood neutrophil (c) from a normal woman shows a single Barr body appearing as a "drumstick" satellite (arrow). ▼

chromosomes, including two X's in females (designated 46,XX), and an X and a Y in males (designated 46,XY).

Barr Bodies and the Lyon Hypothesis.   In 1949, Barr and Bertram discovered in histological preparations of neurons of adult female cats the presence of a densely stained body adjacent to the nucleolus. It was rarely seen in males. Subsequently it was shown that this "Barr body" is one of the pair of X chromosomes that has remained condensed or pycnotic after mitosis (Fig. 5–5). After extensive studies of Barr bodies in mice, Lyon (1961) proposed that the Barr body can be either of the two X chromosomes—that is, either of paternal or maternal origin—in different cells of the same animal, and that it is genetically inactive. This proposal is known as the *Lyon Hypothesis.* She postulated further that at some time during embryonic development, the choice of which X remains pycnotic becomes fixed, so that the individual becomes a mosaic of "areas" in this respect, some areas being comprised of tissues in which one of the X chromosomes is active and others of tissues in which the other X chromosome is active. This hypothesis has been confirmed subsequently (Deys, et al., 1972). The phenotypic consequences of X inactivation in mammals will be deferred for the

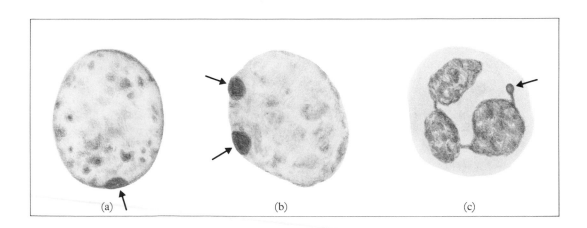

time being, because X inactivation has a different significance in the context of the role of the sex chromosomes in sex determination. The significance is that the presence of a Barr body in a mammalian cell is diagnositic of the presence of more than one X chromosome. Therefore it should be useful in indicating the sex chromosome karyotype of individuals for whom there is reason to suspect an abnormality in sex chromosome complement. The method of screening is simple and straightforward—cells lightly scraped from the inner surface of the cheek can be wiped on a microscope slide, stained with methylene blue or other chromatin dye, and examined under a microscope. In scrapings of normal XX females, a single Barr body of sex chromatin will be observed adjacent to the nuclear membrane in the majority of cells (Fig. 5–5a). Sex chromatin will be seen in very few cells of scrapings from normal XY males. Other tissues may be used as well. Blood neutrophils are often used in such diagnoses, but in them the sex chromatin appears as a satellite or "drumstick" appended to the cell's nucleus (Fig. 5–5c).

The Lyon Hypothesis also explains the fact that in most instances in which the Y chromosome is present in mammals, the resulting phenotype is male. If only one of the X chromosomes is genetically active, then regardless of the number of supernumerary X's, if a Y is present it will be able to express itself. An important modification of the Lyon Hypothesis is relevant to human chromosomes, which must be borne in mind when considering the various abnormal sex chromosome karyotypes that follow. A small section of the heteropycnotic X chromosome retains its genetic activity and does not itself become heteropycnotic, so that supernumerary X's do have some phenotypic effect, as will be shown.

1. 47,XXY. This chromosome complement in human beings has a high correlation with a congenital abnormality called Klinefelter's Syndrome. Found in males, this abnormality is characterized

by small, undeveloped testes that often fail to descend into the scrotal sac, infertility, and often abnormal development of the mammary glands. A single Barr body of sex chromatin occurs in epithelial scrapings. Patients frequently show a variety of psychological and emotional disturbances (Swanson and Stipes, 1969), and sex chromosome mosaicism is not uncommon (Spencer, et al., 1969). Telfer, et al. (1967) and Court Brown (1968), in studies on the karyotypes of criminals, found that 47,XXY is also associated with a tendency toward socially aberrant behavior. The significance of genic balance in abnormalities of this kind was suggested by the appearance of several of the characteristics of Klinefelter's Syndrome in a triploid (69,XXY) infant boy (Schindler and Mikano, 1970). Analogous abnormalities have been described in other mammals (Clough, et al., 1970) and in birds (Abdel-Hameed, 1972), so man is not alone (Fig. 5–6).

2. 45,X. Women with only one X chromosome display Turner's Syndrome, which is a congenital abnormality characterized by rudimentary ovaries, amenorrhea (failure to menstruate), and infantile development of the mammary glands and other secondary sexual characteristics. Barr bodies are rarely seen, although some women with Turner's Syndrome are mosaics, with a single X chromosome in some cells and two X's in others (45,X/46,XX) with a single Barr body, or with one X in some and three in others (45,X/47,XXX) (Bacci, 1965). The triple mosaic 45,X/46,XX/47,XXX has also been described and is also associated with Turner's Syndrome (Fraccaro, et al., 1962b). Two Barr bodies are diagnostic of the XXX cells.

3. 47,XYY. This karyotype has received considerable publicity in recent years, but to what extent the sociological implications of the anomaly are justified is still quite controversial and questionable (Hook, 1973). Several studies, most of which have been in Britain, have indicated that the XYY karyotype is associated with tendencies toward violent, antisocial behavior and with above-normal stature. The results of some other studies have been mixed, and the most that can be said at this time is that, although there is a definite association between the karyotype and its presence in individuals in mental-penal institutions, neither the nature nor the extent of this association has been determined.

The effect of the XYY karyotype on fertility in human beings is not known. In other animals with this karyotype, the supernumerary Y is discarded during meiosis, so that only normal X and Y sperms are produced and fertility is normal, whereas in still others it is associated with sterility (Cattanach, et al., 1969; Sumner, et al., 1971). In certain species of bats, the males normally have

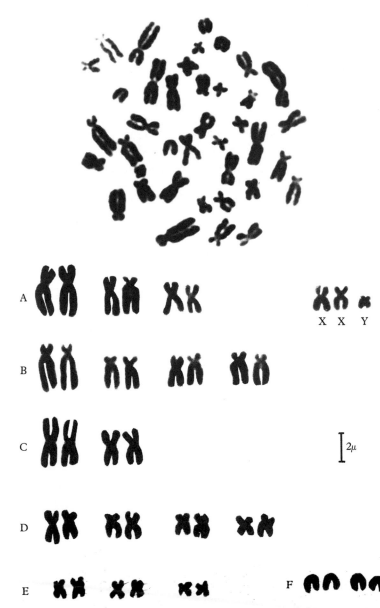

Fig. 5–6. A chromosome squash (above) and karyotype (below) of a Himalayan cat with feminine characteristics. [Courtesy of R. L. Pyle, et al. (1971), *J. Hered.* **62:** 220–222.]

two Y chromosomes, but they are morphologically different, and so the karyotype is designated XY_1Y_2 (Patton and Gardner, 1971).

4. 49,XXXXY.   A number of individuals having four X chromosomes and one Y have been described. They have been characterized in most instances by a variety of abnormalities, the principal ones being mental deficiency, dwarfism, abnormal genitalia, and unusual facial features (Joseph, et al., 1964; Fraccaro, et al., 1962a; Shapiro, et al., 1970). All have been males and their cells show typically three Barr bodies.

5. Mosaics.   A wide variety of sex-chromosome mosaics have been described in human beings. These include, in addition to those already mentioned, both double and triple mosaics. The phenotypic effects of sex chromosome mosaicism are highly variable, even among individuals with the same pattern. For example, the mosaic 46,XX/46,XY may appear as a woman, a man, or what clinicians erroneously call a true hermaphrodite (Ferguson-Smith, 1965; Fitzgerald, et al., 1970). These so-called "true" hermaphrodites have both ovarian and testicular tissue, but one or the other is always incompletely developed and nonfunctional, and sometimes both are. A really true hermaphrodite is functionally both male and female, as described earlier.

The mosaic 46,XX/47,XXY is extremely rare. In one thoroughly examined case, it was associated with Klinefelter's Syndrome (Ford, et al., 1959), and in another it was reported to be the karyotype of a prominent female athlete, although this has not been authenticated in the scientific literature to the knowledge of the authors. A woman with clinical evidence of masculinization was found to be the triple mosaic 45,XO/46,XX/47,XXy (Fraccaro, et al., 1962b). The y represents a small chromosome fragment, presumed to be a fragment of a Y.

Some insight into the causes of the variable phenotypic expression of mosaicism may be gained by considering their expression in insects.

Occasionally an insect is encountered that is part male and part female in appearance (Fig. 5–7). These individuals are called *gynandromorphs*, because they combine both feminine (gyn-) and masculine (-andro-) forms (-morphs). Examination of the karyotypes of the cells of the two parts reveals the basis of the condition. Gynandromorphs of *Drosophila*, for example, usually are XX in the female part and XO in the male part. This can also be demonstrated by use of genetic markers located

Fig. 5–7.   A bilateral gynandromorph of *Drosophila melanogaster*. This fly was one of a number of offspring resulting from a cross of a wild-type female with a white-eyed male, and thus the zygote from which it developed most probably had the genotype w^+/w (cf. p. 101). The gene marker w is recessive to wild type and is located on the X chromosome. Consequently the left half of this fly is doubtless w^+/w with two X chromosomes, and the right half is $w/-$ with only one X chromosome. [After T. H. Morgan (1911).]

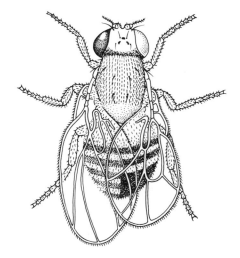

on the X chromosome, as is described below. *Drosophila* gynandromorphs have been shown to arise from XX zygotes, but in the course of mitosis one of the X chromosomes is excluded from the nucleus of a daughter cell and is lost. That cell and all the cells subsequently derived from it by mitosis have the haplo-X (XO) karyotype. Inasmuch as sexual differentiation in insects is not regulated hormonally as it is in vertebrates, the tissues whose cells are XX will develop feminine characteristics, and those that are haplo-X will develop male characteristics. If the X is lost during an early cleavage division by one of two daughter cells, the resulting gynandromorph will be partly male and partly female, as shown in Fig. 5–7. The fault may occur later in cleavage, in which case only a limited number of structures, or parts thereof, will have male characteristics.

Sexual differentiation in vertebrates, particularly in the higher vertebrates, is regulated by hormones, principally the sex hormones. The presumptive gonad has two components, an outer cortex and an inner medulla. Depending upon the genotype of the cells that constitute the presumptive gonad, either the cortex or the medulla will proliferate gonadal tissue. If the genotype is female, the cortex differentiates to become ovarian tissue; if the genotype is male, the medulla proliferates sex cords and differentiates into testes (Witschi, 1936, 1960). The hormones that the differentiating gland then produces determine in large part the sexual differentiation of the rest of the organism in a uniform way, to a great extent irrespective of the genotype of the other tissues of that organism. Hence in a sex-chromosome mosaic, if the cells of the presumptive gonad are mostly or entirely female in genotype, the cortex will differentiate and the individual will be basically female. If the presumptive gonad consists of cells that are mostly or entirely male in genotype, then the medulla will be favored and the organism will differentiate into what is basically a male individual.

Presumably what medical clinicians call a "true" hermaphrodite is the consequence of either a uniform mixture of the two genotypes in the presumptive gonads, so that they can go either way with perhaps one differentiating into an ovary and the other into a testis, or one gonad is primarily of one genotype and the other of another. In either case, both kinds of embryonic hormones, androgenic and estrogenic, are produced and sexual differentiation is highly abnormal and usually incomplete as a result. This is because the hormones not only stimulate differentiation of their corresponding sexuality, but also have an inhibitory action on differentiation of the gonad of the opposite sex. This is actually an oversimplification of a highly complex process because the hypothalamus, the pituitary gland, and perhaps to some extent the adrenal cortex are also involved in sexual differentiation in the higher vertebrates. It is not always a clearcut case of "either-or."

Often mammals that are sex-chromosome mosaics develop into *pseudohermaphrodites*, individuals that are either male or female on the basis of their gonads, but show partial differentiation of both sexes in terms of genitalia and some of the secondary sexual characteristics. A male pseudohermaphrodite, for example, typically has undescended testes, a small penis, a rudimentary vagina, no scrotal sacs, and abnormal development of the mammary glands, and is sterile. A typical female pseudohermaphrodite has underdeveloped ovaries, an infantile uterus and vagina, a rudimentary penis, little or no breast development, and amenorrhea, and is sterile. Often such unfortunate individuals can be corrected to a considerable extent in infancy by surgery and hormone therapy. The presence or absence of Barr bodies can be useful in conjunction with other information in helping the clinician decide what corrective measures should be taken.

The Origin of Sex Chromosome Mosaics.   Loss of an X chromosome during mitosis by exclusion from the nucleus at telophase was mentioned above as one means by which mosaics may arise. A mechanism that is perhaps more frequent is *nondisjunction*, which is illustrated in Fig. 9–24. For reasons still poorly understood, sister X chromatids produced during mitosis have some tendency not to separate or "disjoin" from each other as readily as do the autosomal chromatids, and occasionally both members of a pair will fail to disjoin and will move to the same pole. This results in one daughter cell having three X chromosomes and the other having only one. The one with three X's is often inviable and disintegrates, but the haplo-X is fully viable and reproduces itself as a normal cell and becomes mixed with descendants of those cells in which nondisjunction has not occurred, thereby establishing a mosaic of XX and XO cells. When this occurs during the first cleavage division, a Turner's Syndrome individual results if the XXX cell dies, and an XXX/XO mosaic is established if it survives. The later in cleavage or subsequent development nondisjunction occurs, obviously the smaller will be the population of XO cells. Consequently there is as great a variability in sexuality among sex-chromosome mosaics as the number and kinds of cells that have the abnormal sex chromosome karyotype. Figure 5–8 (Spencer, et al., 1969) illustrates a suggested mechanism for the origin of a 47,XYY/48,XXYY/49,XXXYY mosaic that was found in one individual.

In some insects such as the silkworm (*Bombyx mori*) and bees and wasps, binucleate eggs may be produced by failure of one of the polar bodies to be excluded from the egg cytoplasm (Bacci, 1965). In *Bombyx*, in which the female is digametic, one nucleus may contain an X chromosome and the other the Y. Polyspermy (penetration of the egg by more than one sperm cell) is commonplace in insects as well as in many other species, so that in such binucleate eggs a double zygote may be formed. Each sperm cell contributes an X chromosome in *Bombyx*, so that the double zygote in this instance would be a mosaic of XX and XY (Fig. 5–9). The

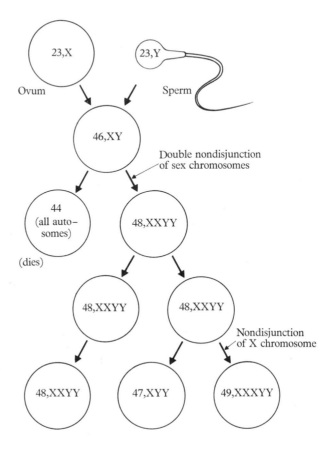

Fig. 5–8. A suggested mechanism of origin of the sexual mosaic 47,XYY/48,XXYY/49,XXXYY. [From D. A. Spencer, et al. (1969), *J. Med. Genet.* **6:**159– 165. By permission.]

two zygote nuclei become segregated into separate cells so that a sex-chromosome mosaic and consequent gynandromorph result. Similar binucleate eggs may be produced in the Hymenoptera, such as the honeybee (*Apis mellifera*) and the parasitic wasp (*Habrobracon*). If one of the nuclei fuses with a sperm cell, it becomes diploid and will develop into a female line of cells; if the other remains haploid, it will give rise to male cells and develop into a gynandromorph.

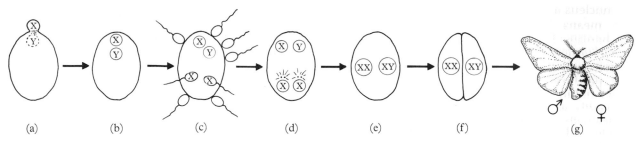

(a)   (b)   (c)   (d)   (e)   (f)   (g)

Fig. 5–9. The origin of gynandromorphism in the silkworm *Bombyx mori*. (a) A polar body fails to pinch off completely and (b) its nucleus returns to the egg, making it binucleate. (c) Egg is fertilized by two sperm cells, which (d) become male pronuclei and, led by their centrioles, migrate toward egg nuclei. (e) Fusion of male and female pronuclei occurs. (f) The two zygote nuclei of opposite sex potential are segregated into separate cells at beginning of cleavage. (g) The cells of the left half of the adult are male (XX) whereas those of the right half are female (XY).

Environmental Determination of Sex

Although in every species in which the mechanism of sex determination has been thoroughly and carefully analyzed the sex genotype proves to be the primary factor, the environmental milieu also represents a significant and, in many cases, essential vector. *Bonellia* is frequently cited as a species in which sex is determined by the presence or absence of a sex-influencing substance in the larva's environment, specifically a hormonelike substance that is produced by an adult female. As we have seen, this substance influences the expression of the genotype to a greater or lesser degree, depending upon the "strength" of that genotype. But it evidently can be influenced by other substances as well, such as the concentration of potassium and magnesium ions (Herbst, 1935, 1936). *Crepidula* and *Patella*, two molluscan species that were discussed earlier, and several other species of invertebrates are similarly influenced by the environmental milieu. Temperature, over-ripeness of the egg, and amount of stored yolk seem also to be decisive factors in some animal species, and changes in the length of daylight can decidedly influence sexuality in some dioecious plants.

As is described in Chapter 11, during embryonic development inducer substances are produced by different cells. These are genetically controlled substances that diffuse into adjacent cells and influence their differentiation. Witschi (1936) has accumulated experimental evidence showing that specific inducers are released during amphibian embryogenesis that direct the presumptive gonad to differentiate into ovary or testis, and that these inducers are determined by the genotype. Under normal circumstances, the male genotype dictates the release of an inductor that regulates the development of testes from the cells of the medulla and suppresses the development of the cortex. The female genotype has the opposite action. In neither case is the commitment irrevocable, however, for gonadal differentiation in lower vertebrates can often be reversed by administration of the sex hormones of the opposite sex (Witschi, 1960) or sometimes by transplantation. Humphrey (1945) replaced the right ovary of larval

Fig. 5–10. Predicted sex ratio of the progeny of a sex- ▶ reversed hen crossed with a normal hen.

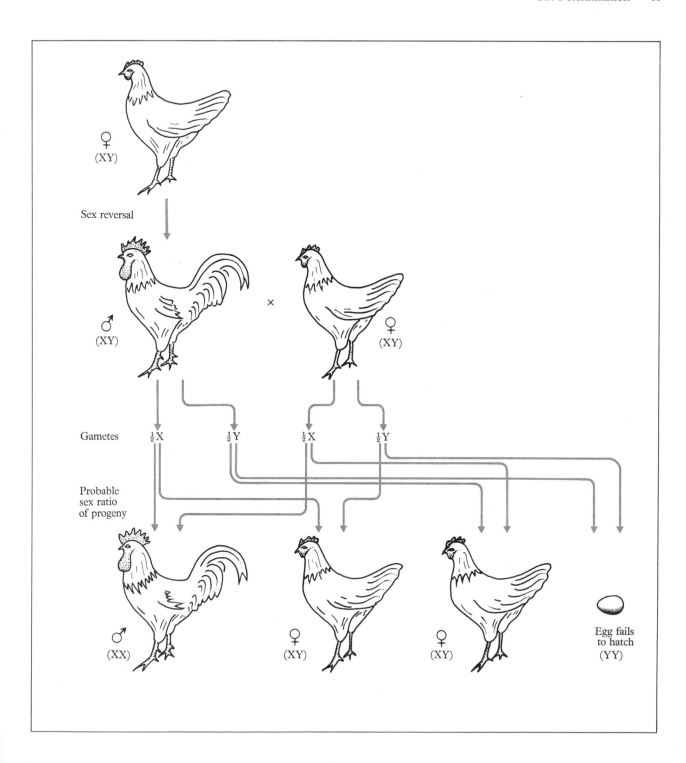

Sex reversal

♀
(XY)

♂
(XY)

×

♀
(XY)

Gametes $\frac{1}{2}$X $\frac{1}{2}$Y $\frac{1}{2}$X $\frac{1}{2}$Y

Probable
sex ratio
of progeny

♂
(XX)

♀
(XY)

♀
(XY)

Egg fails
to hatch
(YY)

female urodeles (*Ambystoma*) with testes grafted from male donors of the same age. The left ovaries in these cases transformed into functional testes and the sex of the female hosts reversed.

Sex reversal is readily effected in many species of birds as well. A classic case of sex reversal, reported by Crew (1923), occurred in a mature hen that had already demonstrated her femininity by having laid fertile eggs. Her plumage changed to that of a cock, she developed spurs, and ''he'' topped off the remarkable transformation by ''crowing lustily with a challenging note'' and siring some chicks. Chicks have two presumptive gonads, but in females only the left one develops into an ovary—the other remains undifferentiated. Postmortem examination of the transformed hen revealed that the ovary had been destroyed by tuberculosis, and the rudimentary right gonad had differentiated into a testis. Crosses of sex-reversed birds such as this with normal females produce offspring in the ratio of one male to two females, which is consistent with the XX karyotype for males and XY for females (Fig. 5–10), YY being inviable.

Sex differentiation in mammals appears to be regulated almost entirely by the sex hormones, once the presumptive gonads have been triggered toward ovaries or testes by their genotype through specific inducers. Strong evidence in support of this theory was provided many years ago (Lillie, 1916) by analysis of a type of intersex called a ''freemartin,'' which occasionally appears among cattle. Freemartins have been known to herdsmen for so many years that the origin of the name has become obscure. A freemartin is a female pseudo-hermaphrodite or intersex, that is, an undeveloped female with some development of the sexual characteristics of the male. Freemartins are invariably co-twins of a usually normal male. Lillie demonstrated that the placenta of the freemartin is fused with that of her male co-twin, with consequent mixing of their fetal blood. He concluded that the gonads of the male become hormonally functional before the ovaries of the female, and that the male hormone both suppresses further ovarian development and induces maleness in the female calf. Similar intersexes have been described in sheep, goats, and swine.

Recently Short, et al. (1969) have shown that the gonads of the freemartin are sex-reversed and that they secrete testosterone. The authors suggest that the masculinization of the freemartin's reproductive tract is owing to its own gonads, rather than to the male's, and that the sex-reversal of her ovaries is attributable not to testosterone from the male, but to a humoral inductor substance that causes retention of the medullary sex cords. This inductor substance may be derived from the male co-twin or, in the case that Short and his colleagues studied, from the presumptive gonad of the female herself, because they found extensive XX/XY mosaicism in the tissues of both co-twins.

Heritable pseudohermaphroditism has been demonstrated in laboratory rats. In one strain, there is a genetic defect in one of the enzymes in the testosterone synthesis pathway that is expressed by accumulation of one of the precursors of testosterone, *androstenedione*, and reduced levels of testosterone (Schneider and Bardin, 1970).

Sex determination in mammals appears to be a highly intricate process that is dependent upon a great amount of genetic information. The sex genes, although distributed throughout the genome, are largely concentrated in the sex chromosomes, with male determiners in the Y. As in amphibians, the primary sex determiners control the release of specific inducers in the gonadal rudiments. These inducers direct the histogenesis of the rudiments into ovaries or testes. Under the influence of pituitary hormones as well as the specific sex inductors, the gonads become functionally differentiated and synthesize androgenic or estrogenic hormones. The differentiation and development of the remaining characteristics of sexuality are then regulated mainly by these endocrine secretions.

SEX LINKAGE

Genes that are located on the sex chromosomes are said to be *sex linked*, just as genes that are located on the other chromosomes are said to be autosomally linked (Chapter 4). Sex-linked genes merit separate consideration because their phenotypic expression is inevitably correlated in some way with the sexual phenotype. This is so because crossing-over between the X and the Y is largely or totally suppressed in those species in which there is pronounced morphological and/or genetic difference between the two kinds of sex chromosomes.

In order to avoid the semantic difficulties frequently encountered in discussions of sex linkage, it would be well to make clear fine distinctions at the outset. In most species characterized by sex chromosomes, the X and the Y each consists of two functionally distinct parts or segments, the *pairing segment* and the *differential segment* (Fig. 5–4). The former is very limited in species with well-differentiated X and Y chromosomes. Genes that are located within the pairing segment would presumably be capable of recombination between the X and the Y were it not for crossover repression, which appears to be virtually complete between the X and the Y. Because of this, it is uncertain what genes or how many genes are located in the pairing segment. Most of our knowledge concerns genes located in the differential segment, and here we can clearly distinguish between X-linked genes and Y-linked genes. In mammals, most of the Y-linked genes appear to be concerned with sex determination, that is, with maleness. In *Drosophila*, a gene for sperm motility is clearly Y-linked. Because the Y chromosome always occurs singly (except in the highly abnormal situations described in the preceding section), it cannot harbor any lethal or markedly deleterious genes. Inasmuch as genes are identified through their mutant alleles and mutations are predominantly deleterious, opportunities to identify genes on the Y chromosome are severely limited. For this reason the Y chromosome is often considered, probably erroneously, to be largely genetically inert. On the contrary, X-linked genes are much more thoroughly known because the X chromosome is paired in the isogametic sex, thereby allowing X-linked mutations to persist and to accumulate throughout the species. Finally, there are two other categories of genes whose expression is to a greater or lesser extent determined by the sex of the individual, although they are not necessarily located on either the X or the Y chromosome. One of these categories is the *sex-limited* genes, and the other is the *sex-influenced* genes, which are largely autosomal.

X-Linked Genes

The first trait that was shown to be X-linked was one of the forms of white eye in *Drosophila*. T. H. Morgan discovered this mutant and found that when a white-eyed female was mated with a wild-type (red-eyed) male, all the daughters were red-eyed and all the sons were white-eyed. When he made the reciprocal cross (red-eyed females × white-eyed males), he found that all the F_1 were red-eyed, but the F_2 consisted of 2459 red-eyed females, 1011 red-eyed males, and 782 white-eyed males (Morgan, 1910). On the basis of these results, Morgan postulated (1911) that white eye is attributable to a recessive allele located on the X chromosome and that the results he obtained could be analyzed as shown below. The smaller number of white-eyed males compared to red-eyed males was probably:

P	XX	×	$X_w Y$
	Red-eyed females		White-eyed males
F_1	XX_w	×	XY
	Red-eyed females		Red-eyed males
F_2	$\frac{1}{4}$ XX : $\frac{1}{4}$ XX_w :		$\frac{1}{4}$ XY : $\frac{1}{4}$ $X_w Y$
	Red-eyed females		Red-eyed males / White-eyed males

owing to reduced viability of the former. Morgan predicted from his hypothesis that the first cross (white-eyed female × red-eyed male), if carried to the F_2, would produce progeny consisting of $1/4$ red-eyed females, $1/4$ white-eyed females, $1/4$ red-eyed males, and $1/4$ white-eyed males, as shown below. He executed the cross and the results verified the hypothesis:

P	X_wX_w	×	XY
	White-eyed females		Red-eyed males
F_1	X_wX	×	X_wY
	Red-eyed females		White-eyed males
F_2	$1/4 X_wX$: $1/4 X_wX_w$	: $1/4 XY$:	$1/4 X_wY$
	Red-eyed White-eyed females females	Red-eyed males	White-eyed males

Subsequently many other X-linked genes were discovered and described in *Drosophila*. Among the more interesting are the semi-dominant bar-eye (*B*) and a number of recessive lethal genes (*l*). The inheritance of bar-eye is diagrammed in Fig. 5–11. The use of an X-linked lethal gave interesting results because, since it is a recessive gene, a cross of a heterozygous female with a normal (necessarily) male gave progeny in the ratio of two females to one male, as diagrammed below.

$$X_lX \ \times \ XY$$
$$\downarrow$$
$$1/4\ XX \ : \ 1/4\ X_lX \ : \ 1/4\ XY \ : \ 1/4\ X_lY$$
$$\underbrace{2/3\ \text{females}} \ \ \ : \ 1/3\ \text{males} \quad \text{lethal (dies)}$$

This is especially interesting because in human beings, in which a number of X-linked lethals and sublethals have been described, the sex ratio at maturity is about fifty-fifty. It has been postulated that the Y-bearing spermatids undergo one mitotic division immediately following the second meiotic division, thereby doubling the number of Y-bearing sperm cells in comparison with the

number of X-bearing sperm cells (Shettles, 1962). Accordingly, there would be twice as many males produced as females at conception, but the mortality of XY individuals between conception and maturity by spontaneous abortion (which accounts for most of them), miscarriage, and infant mortality is such that the sex ratio at maturity is nearly equal (Shettles, 1964). On the other hand, Barlow and Vosa (1970), who distinguished X- from Y-bearing sperm by the presence of a fluorescing spot (F-body) in the latter in semen stained with quinacrine mustard, found that the two types occur with about equal frequency. If this is so, then other mechanisms must exist to account for the preponderance of male conceptions.

A partial list of X-linked mutations described in human beings is shown in Table 5–4. Among the more interesting of these are the recessive genes for red-green color-blindness, hemophilia A, and the mutation for glucose-6-phosphate dehydrogenase deficiency, which shows lack of dominance. Red-green color-blindness is one of the more widespread of the human X-linked mutants and, because of this as well as its relative harmlessness in the homozygous condition, it is found in its expected frequency among women. This frequency is very low, of course, as compared with its frequency in males, because the probability of two mutant alleles combining in the same zygote is equal to the square of the allele's frequency in the population, whereas the probability of its occurring singly, as is the condition of its expression in males, is equal to the frequency itself. For example, if red-green color-blindness occurs in five (5.0) percent of the male population, it is predictable that it will occur in twenty-five hundredths (0.25) of a percent of the female population, which in fact it does very closely. Figure 5–12 is a diagram of the typical crisscross inheritance of red-green color-blindness as it is most commonly found. The mechanism of inheritance is identical to that of white eye in *Drosophila*. Other types of color-blindness are known, and they too are attributable to X-linked mutations.

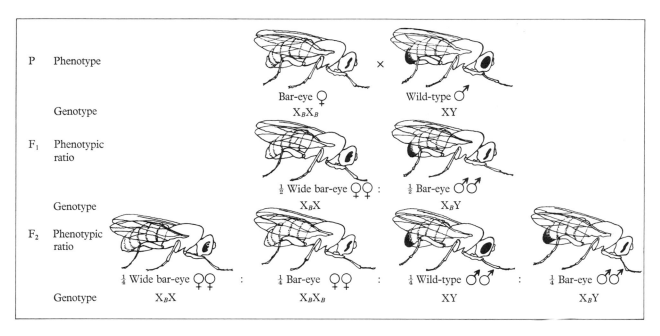

Fig. 5–11.   The inheritance of the bar-eye trait in *Drosophila melanogaster*.

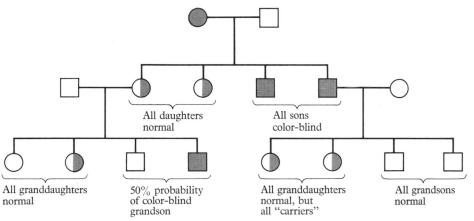

Fig. 5–12. The inheritance of red-green color-blindness in human beings. Shaded figures indicate color-blind phenotype, half-shaded figures indicate "carriers" with normal phenotype.

Table 5–4.   A catalog of X-borne mutations in man.
[From S. Ohno, *Sex Chromosomes and Sex-Linked Genes.*
Berlin: Springer-Verlag (1967).]

1. Partial color-blindness, Deutan series	33. Spinal ataxia
2. Partial color-blindness, Protan series	34. Cerebellar ataxia with extrapyramidal involvement
3. Total color-blindness	35. Spastic paraplegia
4. Glucose-6-phosphate dehydrogenase deficiency	36. Progressive bulbar paralysis
5. Xg blood group system	37. Charcot-Marie-Tooth peroneal muscular dystrophy
6. Muscular dystrophy, Duchenne type	38. Diffuse cerebral sclerosis (Pelizaeus-Merzbacher)
7. Muscular dystrophy, Becker type	39. Diffuse cerebral sclerosis (Scholz)
8. Hemophilia A	40. Hydrocephalus
9. Hemophilia B	41. Parkinsonism
10. Agammaglobulinemia	42. Ocular albinism
11. Hurler syndrome	43. External ophthalmoplegia and myopia
12. Late spondylo-epiphyseal dysplasia	44. Microphthalmia
13. Aldrich syndrome	45. Microphthalmia with digital anomalies
14. Hypophosphatemia	46. Nystagmus
15. Hypoparathyroidism	47. Megalocornea
16. Nephrogenic diabetes insipidus	48. Hypoplasia of iris with glaucoma
17. Neurohypophyseal diabetes insipidus	49. Congenital total cataract
18. Low's oculo-cerebro-renal syndrome	50. Congenital cataract with microcornea
19. Hypochromic anemia (Cooley-Rundles-Falls type)	51. Stationary night blindness with myopia
20. Angiokeratome diffusum corporis universale	52. Choroideremia
21. Dyskeratosis congenita	53. Retinitis pigmentosa
22. Dystrophia bullosa hereditaria, Typus maculatus	54. Pseudohypertrophic muscular dystrophy
23. Keratosis follicularis spinulosa cum ophiasi	55. Retinoschisis
24. Ichthyosis vulgaris	56. Pseudoglioma
25. Anhidrotis ectodermal dysplasia	57. Van den Bosch syndrome
26. Amelogenesis imperfecta, Hypomaturation type	58. Menkes syndrome
27. Amelogenesis imperfecta, Hypoplastic type	59. Albinism-deafness syndrome
28. Absence of central incisors	60. Congenital adrenal hypoplasia
29. Congenital deafness	61. Phosphoglycerokinase deficiency
30. Progressive deafness	62. Guanine phosphoribosyltransferase deficiency
31. Mental deficiency	63. Alpha-galactosidase deficiency
32. Börjeson syndrome	

Although hemophilia is a relatively rare disease thanks to a low gene frequency, it is clinically significant. Moreover, it has historical interest because of its appearance among the descendants of Queen Victoria of Britain (Fig. 5–13) and evolutionary interest because of its occurrence in dogs, in which it is also X-linked (Brinkhous and Graham, 1950). At least three types of hemophilia are known in human beings, Type A being the most common, and they are all X-linked. Since the hemophilia gene has an extremely low frequency (less than 0.01 percent), its probability of occurring homozygously in a woman is at most one in one-hundred million (0.001^2). This accounts for the fact that until fairly recently, hemophilia was unknown in women, doubtless owing to the fact that the probability of the coincidence of correct diagnosis and reporting was almost as low as the incidence in women of the disease itself.

The fact that the genes for hemophilia are

X-linked in both human beings and dogs has no Freudian or other sex-related significance as far as is known. Most likely it was a matter of chance that the ancestral chromosome that carried the genes concerned with the synthesis of thromboplastin (the essential component of the clotting mechanism that is deficient in hemophiliacs) was the one that underwent evolutionary modification to become the X chromosome. The fact that hemophilia is X-linked in dogs simply suggests that the X chromosome evolved before the evolutionary divergence of dogs and people.

The X-linked gene that is concerned with coding the synthesis of glucose-6-phosphate dehydrogenase (G-6-PD) is interesting for several reasons (cf. Chapter 14). One reason is that wherever G-6-PD deficiency has been found in mammals, it is X-linked. This observation lends support to the thought that differentiation of the X chromosome occurred in a remote mammalian ancestor—possibly even in a more remote reptilian ancestor—because the mammalian orders in which this relationship is found (marsupials, primates, lagomorphs, and artiodactyls) are known to have diverged very early in mammalian evolution (Ohno, 1967; Richardson, et al., 1971; Cooper, et al., 1971). Evidence that the gene for G-6-PD is X-linked in the horse and the ass is elegantly direct and simple, for they are readily hybridized and horse G-6-PD is electrophoretically distinct from ass G-6-PD. Hinnies are derived by crossing a stallion with a female ass, hence male hinnies receive their X chromosome from the ass and, indeed, they have the ass G-6-PD. Mules are produced by the reciprocal cross, that is, by mating a mare with a male ass. Hence male mules inherit their X chromosome from the horse and, indeed, they have the horse G-6-PD (Trujillo, et al., 1965).

Y-Linked Genes

Genes borne exclusively on the Y chromosome in mammals display what is known as *holandric transmission* (transmission from father to son in an unbroken line). Several cases of holandric inheri-

tance have been reported, but the best authenticated case in human beings is that of *hypertrichosis* of the ears. This trait appears as a conspicuous growth of hair on the outer rim of the ear. A pedigree showing its transmission through four generations is given in Fig. 5–14. This gene has high penetrance, but not 100 percent, for on occasion it may "skip" a generation. There is another gene, an autosomal dominant, that has the same phenotypic effect. Therefore, positive identification can be made only in the cases of nearly complete pedigree. Another alleged and celebrated example of holandric transmission is a type of ichthyosis—the so-called "porcupine man"—but some evidence has cast serious doubt on its being Y-linked (Penrose and Stern, 1958).

Sex-Limited and Sex-Influenced Genes

Sex-limited genes are those that normally are expressed in only one sex or the other, that is, their expression depends upon the sexual differentiation of the individual. For example, genes that regulate the size and shape of the penis, although doubtless present in both sexes, normally find expression only in the male. In contrast, genes concerned with breast development in human beings are expressed only in women. Breeders of dairy cattle recognize that the bull is just as important as the cow in a breeding program designed for high milk productivity. The role of the sex hormones in eliciting the expression of sex-limited genes has been demonstrated experimentally in mice. In this case the sex-limited gene for a serum protein variant was shown to depend upon both the proper genotype and the presence of testosterone (Passmore and Shreffler, 1971), and so this trait is limited to males. To assume that genes of these kinds are sex-linked is quite unwarranted, however.

Sex-influenced genes are those that exhibit greater penetrance or expressivity in one sex or the other. A familiar example is pattern baldness in men. The gene for this character functions as a dominant in males and as a recessive in females.

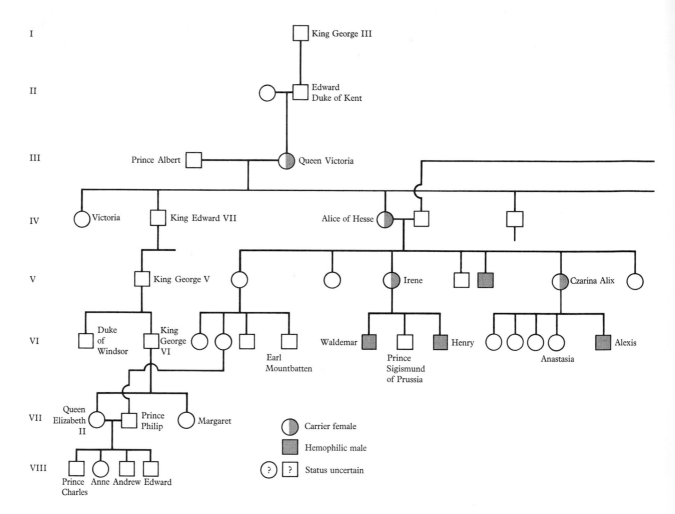

Hence, although pattern baldness will develop in a man when either homozygous or heterozygous, it will develop in women only when homozygous. The explanation for this is presumably related to concentrations of male hormone, for baldness in eunuchs is as infrequent as in women. Yet when eunuchs are treated with testosterone, a significant number of them develop pattern baldness. Similarly, women who are under male hormone therapy for estrogen-dependent malignancy frequently suffer thinning of the hair on the head and growth of hair on the upper lip.

Sheep of the Suffolk breed are hornless, whereas both males and females of the Dorset breed are horned. Among hybrids between the two breeds, the females are hornless (polled) and the males are horned, irrespective of which way the cross is made. The F_2 produced by these individuals assorts into four classes: $3/8$ polled females, $1/8$ horned females, $3/8$ horned males, and $1/8$ polled males. Obviously the gene acts as a simple dominant in males and recessive in females. Judging by the pattern of their inheritance, the genes for pattern baldness in human beings and for the polled condition in sheep clearly are not sex-linked.

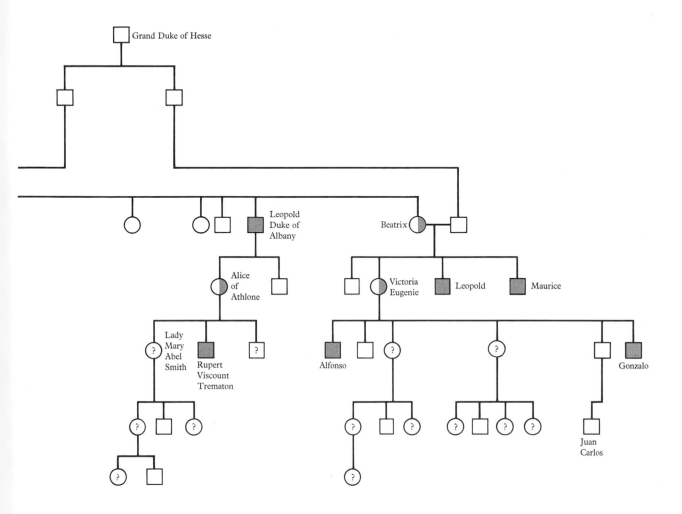

Fig. 5–13. The inheritance of hemophilia in the descendants of Victoria, Queen of England and Empress of India.

Fig. 5–14. Hypertrichosis of the ears and a pedigree showing its transmission through four generations.

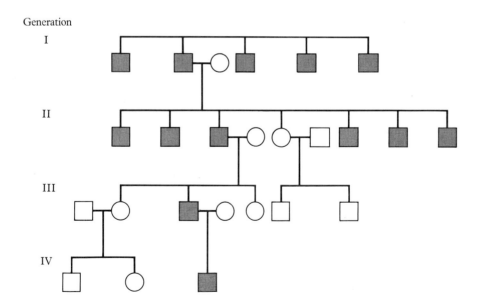

QUESTIONS

1. In some species of ciliates, as many as nine different mating types exist. Speculate on the genetic mechanism responsible for the multiple mating types. How might such an occurrence evolve?

2. What is meant by facultative parthenogenesis?

3. What is believed to be the genetic basis for the existence of unbalanced hermaphroditism in marine limpets?

4. As one ascends the evolutionary scale, the number of loci and their degree of linkage for the traits of sexual dimorphism and reproduction increases. What might have been the evolutionary pressures to cause the observed increase?

5. Contrast the mechanism of sex determination in *Drosophila* with that in human beings.

6. Reconcile the Lyon Hypothesis with the ability of females to be carriers of sex-linked recessive traits.

7. How can you account for the relationship between an abnormal sex chromosome complement and mental or behavioral abnormalities?

8. Why is it unlikely that a human being who is a mosaic with respect to the sex chromosomes will be a gynandromorph?

9. Outline the events that lead to sex-chromosome mosaicism in *Drosophila* and silkworms.

10. Why is it difficult to identify and to map Y chromosome-linked traits?

11. Can two sex-linked, color-blind parents produce a normal son or daughter? Explain.

12. What are the chances of obtaining a color-blind son from a normal woman and a normal man whose father was color-blind?

13. What are the chances of obtaining a color-blind son from a normal man and a normal woman whose father was color-blind?

14. Yellow and black coat color in cats appears to be controlled by a pair of sex-linked alleles (Y = black; y = yellow). The heterozygote is the color mixture called calico or tortoise-shell. What would be the color and genotype of the father of the following litter:

    Mother: tortoise-shell.
    Litter: 1 yellow male, 2 black males,
             2 yellow females, 3 calico females.

15. How would you account for the fact that on rare occasions a calico male cat is observed and, even more infrequently, such a male is fertile.

16. How would you distinguish a trait caused by a sex-linked recessive allele from one caused by a sex-influenced gene that is dominant in the male?

6

Genetic Interactions and Variability

The phenotype of an organism reflects the interaction of its genes and its environment. This concept accounts for the observations that different genotypes can produce similar phenotypes in the same environment and that different environments may affect the same genotype differently. The implications of this statement are manifold due to the complexity of genic functions, and the environment, both internal and external, is not only complex but is subject to great variation. Furthermore, the functions of some genes may interact with those of other genes in a number of different ways. Many characteristics that are considered as single traits represent the composite action and interaction of many sets of alleles.

There are four terms that are useful in describing certain aspects of the gene-environment-individual complex. (1) *Pleiotropy* refers to the circumstance in which one gene has multiple phenotypic effects. That this should be so is consistent with the metabolic role of genes. One gene may specify enzyme X, but X may participate in a pathway that bifurcates and produces two or more end products, each of which may act in different and seemingly unrelated ways (Fig. 6–1). (2) *Phenocopy* refers to an environmental mimic of gene action. A striking example of a phenocopy results from maintaining a pregnant mouse at reduced atmospheric pressure—the offspring produced often lack regions of the urogenital system in a pattern identical to that attributed to the expression of the mutant gene *Danforth shorttail*. Certain human abnormalities, such as retinal or skeletal defects caused by drugs or disease, may resemble similar hereditary disturbances. The well-known cases of thalidomide-induced limb defects and of blindness resulting from German measles during pregnancy are examples of these. (3) *Penetrance* is a statistical concept of the regularity with which a gene is expressed. For example, if a given gene is expressed phenotypically in only 75 percent of the individuals having it in a genetic

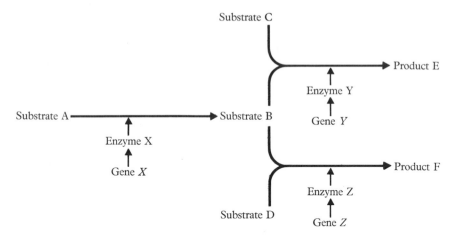

Fig. 6–1.   The relationship between genes and the enzymatically regulated steps of two divergent "metabolic pathways." In this example, enzyme X catalyzes the conversion of substrate A to substrate B. Some of substrate B may react or combine with substrate C under the influence of enzyme Y to form product E. At the same time, other molecules of substrate B may combine with substrate D in the presence of enzyme Z to form an unrelated product F.

combination favorable to its expression (namely, homozygous for a recessive; homozygous or heterozygous for a dominant), the penetrance of that gene is said to be 75 percent. To what extent reduced penetrance is attributable to the modifying action of other genes or to undetected alterations of environmental influences is not always known, but such factors do represent possible explanations for the phenomenon. The *blue sclera* anomaly in human beings is attributable to a dominant gene with a penetrance of about 90 percent. It is expressed by a blue discoloration of the white (sclera) of the eye and by brittleness of the bones. (4) *Variable expressivity* refers to the situation in which the degree of gene expression is variable. A well-documented example of variable expressivity in human beings is juvenile cataract of the eyes (Lutman and Neel, 1945) due to a dominant gene whose expression may vary from a slight milkiness of the lens to dense opacity. As with partial penetrance, variable expressivity may be explained by the modifying action of other genes and/or by environmental influences.

Individually and collectively, these terms are useful in a number of ways. First, they apply to measurable or quantifiable characteristics and, therefore, are important in evaluating certain phenotypes. Secondly, they suggest approaches that often lead to understanding how particular genes exert their effects. If a phenocopy results by experimentally blocking an enzyme, then perhaps the gene that is being copied works in the same way. Similarly, if a particular pattern of penetrance or expressivity is discovered, this can often lead to an understanding of how genes interact with the environment. Finally, comparing two groups of individuals with respect to the data obtained often gives clues to how much of a particular phenotype is genetic in origin and how much is determined by environmental effects.

GENES AND ENVIRONMENT

The complexity of development is such that it may never be possible to know all of the steps involved in the production of any one trait. For some rather straightforward characteristics, especially those caused by fully penetrant genes, a fairly detailed understanding is conceivable. But the more subtle aspects of phenotype, such as those that relate to behavior, may never be completely understood. For these traits, it would be enough to know how great a role, if any, the genes play. Unravelling the complexities of the environmental component would be far simpler if it were known how the basis of the trait is established. It is known that temperature, light, moisture, pre- and postnatal nutrition, maternal environment, age, sex, and many other parameters can influence the phenotype, but first the genetic component must be known. There is no reason to believe that behavioral problems will disappear when their genetic and environmental aspects are sorted out, but at least the problems themselves will be more rationally definable.

Logical analysis leads to the belief that differences between two organisms with identical genotypes should be attributable to environmental differences, although the converse that similar phenotypes produced in varying environments have a basically genetic origin does not necessarily follow. Studies that attempt to partition the phenotype into genetic and environmental components are more easily performed with nonhuman organisms, but ironically, it is precisely in the area of genetic and environmental action in human variation that we are most interested.

Studies on Human Twins

A start has been made in heredity/environment analysis in humans by comparing identical (monozygotic) twins, fraternal (nonidentical or dizygotic) twins, and nontwin siblings reared apart and together. Identical twins (Fig. 6–2) are not only genetically alike but are also subject to virtually the same intrauterine environment. Fraternal twins, arising from separate zygotes, are probably no more alike than nontwin siblings, but they shared the same intrauterine environment at the same time. Before any comparisons are made, it should be conceded that (1) the numbers of such comparisons are small, (2) identical twins and other siblings raised together may be subjected to many subtle differences of environment, (3) twins who know that they are being reared apart may be subject to all sorts of emotional stresses, and (4) in some cases it is difficult to distinguish mono- from dizygotic twins without extensive testing, including skin grafting.

One of the most interesting studies on twins is concerned with the event of twinning itself. The frequencies of monozygotic and dizygotic twinning (per one-thousand births) are given in Table 6–1. Evidently monozygotic twinning is an accidental occurrence that is not related to genotype in human beings as it is in the nine-banded armadillo, and so the frequency is relatively uniform in different populations. Dizygotic twinning rates are not as uniform, which implies that there may be a genetic basis for multiple ovulation that is reflected by the different gene pools of different groups of people.

The data presented in Table 6–2 are given either as measurable differences or as percent concordance within the group studied. *Concordance* refers to the situation in which both members of a pair show the trait or do not show it. Pairs that are not concordant are said to be *discordant*, that is, the members of the pair are phenotypically dissimilar. Discordance within a group is 100 percent minus the percent concordance. In data such as these in Table 6–2, two considerations contribute to the entries in the column labeled "Comments." First, concordance between identical twins varies with different traits. High degrees of concordance may be attributable to the twins' genetic identity, to the similarity of their environments (assuming that

Fig. 6–2. Joyce and Janet are twins who *appear* to be identical. There were three sets of fraternal twins in their immediate family, but no known incidence of identical twinning.

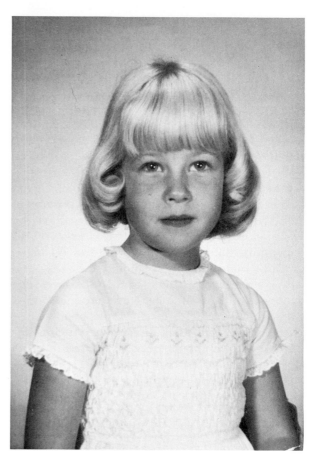

Table 6–1. A comparison between the incidence of monozygotic and dizygotic twins in samples of four populations. The figures are numbers per 1000 births.

Population	Monozygotic twins	Dizygotic twins
African negro (one study only)	?	20.00
U.S. white	3.85	7.44
U.S. negro	4.08	10.01
Japanese	4.25	2.72

[From *Heredity, Evolution, and Society* by I. Michael Lerner. San Francisco: W. H. Freeman and Company. Copyright © 1968.]

Table 6–2.   An analysis of genetic concordance of
monozygotic and dizygotic twins with respect to a variety
of traits and conditions.

Trait	Identical twins		Fraternal twins (same sex)		Comments
	No. pairs	% concordance	No. pairs	% concordance	
Alcohol drinking	34	100	43	86	These traits are probably of
Coffee drinking	34	94	43	79	environmental origin
Smoking	34	91	43	65	
Measles	189	95	146	87	
Eye color	256	99.6	194	28	Basically genetic, but with
Feeblemindedness	217	94	260	47	intrauterine environment of importance
Mongolism	18	89	60	7	Accidental (?) chromosomal event
Hair color	215	89	156	22	These traits are basically
Rickets	60	88	74	22	genetic with differential
Schizophrenia	395	80	989	13	expressivity
Diabetes mellitus	62	84	70	37	
Age when sitting up begins	63	82	59	76	Dependent on basic
Handedness	343	79	319	77	developmental genes
Manic-depressive psychosis	62	77	165	19	Genetic, partial pene-
Epilepsy, idiopathic	61	72	197	15	trance, or expressivity
Blood pressure	62	63	80	36	Basically environmental
Age when walking begins	136	68	128	31	but with possible genetic
Criminality	143	68	142	28	limitations
Pulse rate	84	56	67	34	These traits show minimal
Club foot	40	32	134	3	genetic involvement
Paralytic polio	14	36	33	6	
Stomach cancer	11	27	24	4	
Mammary cancer	18	6	37	3	
Uterine cancer	16	6	21	0	

[Adapted from M. W. Strickberger (1968), *Genetics*. New York:
Macmillan. Copyright © 1968 by Monroe W. Strickberger.]

they were reared together), or to a combination of both factors. Low degrees of concordance almost certainly reflect the manifestations of the environment, especially with reference to accidental events and traits that are expressed in the adult period when environmental similarities are probably at their lowest. The second consideration is the difference between concordance for monozygotic and dizygotic twins, because hypothetically these two categories of twins differ only in the degree of their genetic similarity. Therefore, if concordance is high for both groups, the probability is that the extra genetic similarity of the identical twins is not very contributory, and the basis of the trait is the fact that both members of either kind of twin pair share similar environments in addition to being genetic siblings. Much information resulting from testing for personality types and for intelligence exists, but it is omitted from our consideration due to subjectivity in the testing procedures and lack of agreement among those who have interpreted the data. In addition, it is well to remember that studies on twins consider the terms "genetic" and "environmental" as collectives. No amount of interpretation can answer such questions as, How many or which genes are acting?, or, Which aspects of the environment are responsible for a particular phenotype?

Interaction Between Genes—Epistasis

Interaction between the products of gene function must take place in order that growth and development and, by consequence, adult functions may be coordinated and integrated processes. There are probably no genes that say, "Make hazel eyes," or "Fashion one human stomach," but rather there are genes that are involved in the determination of each of the relevant cells and tissues, the relationships between the tissues, the rate of growth, the form of growth, and the cessation of growth. Furthermore, the involvement of genes is doubtless enzymatic, and enzymes (cf. Chapter 8) most often operate in long chains of

reactions or "pathways" that result in one of many essential end products. Therefore, many genetic pathways are needed to produce all the essential substances for hazel eyes or a stomach, and they must be coordinated. It has become increasingly clear that genes act developmentally; there are many metabolic pathways that function in the context of metabolic events that have occurred elsewhere either in time or locality within the organism. For example, the pituitary gland and the gonads may commence their differentiation relatively independently of each other, but eventually their subsequent development is profoundly affected by the mutual interaction that occurs between them.

We tend to think of hair color as a phenotype that is developed in a relatively straightforward manner. Figure 6–3 shows a simplified pathway that leads to the production of coat color in guinea pigs. As complex as this pathway may be, there is little doubt that almost all traits involve at least this degree of complexity. It is evident from this figure that no one gene is responsible for coat color, although the mutant form of one gene, albino, may nullify the effect of the rest of the coat-color genes. In general, genes act together to modify, complement, inhibit, or partially suppress the activity of other genes in the same and related pathways. These activities may occur at the level of the gene itself, or they may involve the feedback control of the operative enzymes. In this respect, interaction between sets of alleles can be thought of as an extension of the relationships of dominance within one set of alleles.

The list of observable genetic interactions is practically infinite. Figure 6–4 shows a variety of phenotypic ratios from dihybrid crosses. Within each set of alleles there may be degrees of dominance or there may be codominance. Between sets of alleles there may be no apparent relationship. On the other hand, they may act to produce a new phenotype, or one or both sets may mask the appearance of the other in a variety of ways, a condition known as *epistasis*. This may be defined as the

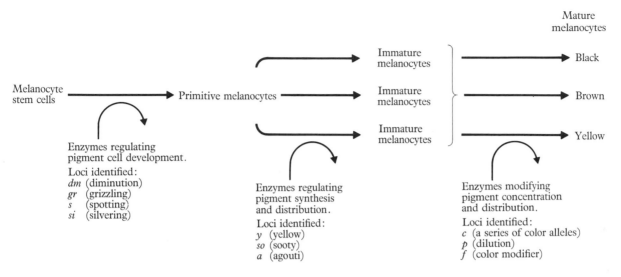

Fig. 6–3. Diagram showing the relationship of known genes to the development of coat color in guinea pigs. [Modified from S. Wright, "Genic interaction," *Methodology in Mammalian Genetics*, W. J. Burdette (ed.). San Francisco: Holden-Day (1963).]

Fig. 6–4. *Overleaf.* Some phenotypic ratios of progeny of dihybrid crosses under different conditions of genic interaction. Phenotypes are represented as abbreviations of the allele or alleles expressed. The numbers in parentheses refer to squares in the checkerboard (below) that are included in each phenotypic class. Note the genotypic patterns in the checkerboard. Classic examples are shown.

$AaBb \times AaBb$

Gametes	AB	Ab	aB	ab
AB	1 $AABB$	2 $AABb$	3 $AaBB$	4 $AaBb$
Ab	5 $AABb$	6 $AAbb$	7 $AaBb$	8 $Aabb$
aB	9 $AaBB$	10 $AaBb$	11 $aaBB$	12 $aaBb$
ab	13 $AaBb$	14 $Aabb$	15 $aaBb$	16 $aabb$

Gene relation	Phenotypic ratios and checkerboard reference
I. Gene pairs affecting different traits.	
A dominant to a; B dominant to b.	$\frac{9}{16} AB$: $\frac{3}{16} Ab$: $\frac{3}{16} aB$: $\frac{1}{16} ab$ (1–4, 5, 7, 9, 10, 13) (6, 8, 14) (11, 12, 15) (16)
A dominant to a; B codominant with b.	$\frac{6}{16} ABb$: $\frac{3}{16} AB$: $\frac{3}{16} Ab$: $\frac{2}{16} aBb$: $\frac{1}{16} aB$: $\frac{1}{16} ab$ (2, 4, 5, 7, 10, 13) (1, 3, 9) (6, 8, 14) (12, 15) (11) (16)
A and a codominant; B and b codominant.	$\frac{4}{16} AaBb$: $\frac{2}{16} ABb$: $\frac{2}{16} AaB$: $\frac{2}{16} Aab$: $\frac{2}{16} aBb$: $\frac{1}{16} AB$: $\frac{1}{16} Ab$: $\frac{1}{16} aB$: $\frac{1}{16} ab$ (4, 7, 10, 13) (2, 5) (3, 9) (8, 14) (12, 15) (1) (6) (11) (16)
II. Gene pairs affecting the same trait.	
A dominant to a; B dominant to b. A interacts with B, producing new phenotype. $aabb$ produces a fourth phenotype.	$\frac{9}{16} AB$: $\frac{3}{16} A$: $\frac{3}{16} B$: $\frac{1}{16} ab$ (1–4, 5, 7, 9, 10, 13) (6, 8, 14) (11, 12, 15) (16)
A dominant to a; B dominant to b; A epistatic to B.	$\frac{12}{16} A$: $\frac{3}{16} B$: $\frac{1}{16} ab$ (1–10, 13, 14) (11, 12, 15) (16)
A dominant to a; B dominant to b; aa epistatic to B.	$\frac{9}{16} AB$: $\frac{3}{16} B$: $\frac{4}{16} b$ (1–5, 7, 9, 10, 13) (6, 8, 14) (11, 12, 15, 16)
A dominant to a; aa epistatic to B; B dominant to b; bb epistatic to A. A and B same expression.	$\frac{9}{16} A$ or B : $\frac{7}{16}$ non-A or B (1–4, 5, 7, 9, 10, 13) (6, 8, 11, 12, 14–16)
A dominant to a; B dominant to b; A epistatic to B; bb epistatic to aa. A and bb same expression.	$\frac{13}{16} A$ or b : $\frac{3}{16} B$ (1–7, 9, 10, 13) (8, 14, 16) (11, 12, 15)
A dominant to a; B dominant to b; A epistatic to b; B epistatic to a. A and B same expression.	$\frac{15}{16} A$ or B : $\frac{1}{16}$ non-A or B (1–15) (16)

Examples

Garden pea, seed color and shape

$\frac{9}{16}$ Yellow-round : $\frac{3}{16}$ Yellow-wrinkled : $\frac{3}{16}$ Green-round : $\frac{1}{16}$ Green-wrinkled

Cattle, horns and color

$\frac{6}{16}$ Polled-roan : $\frac{3}{16}$ Polled-red : $\frac{3}{16}$ Polled-white : $\frac{2}{16}$ Horned-roan : $\frac{1}{16}$ Horned-red : $\frac{1}{16}$ Horned-white

Human blood types: M, N, MN; Sickle-cell anemia and trait

$\frac{4}{16}$ MN : $\frac{2}{16}$ M : $\frac{2}{16}$ MN : $\frac{2}{16}$ MN : $\frac{2}{16}$ N : $\frac{1}{16}$ M : $\frac{1}{16}$ M : $\frac{1}{16}$ N : $\frac{1}{16}$ N
+   +   +   +   +   +   +   +   +
Trait   Trait   Normal   Anemia   Trait   Normal   Anemia   Normal   Anemia

Domestic fowl, comb shape

$\frac{9}{16}$ Walnut : $\frac{3}{16}$ Rose : $\frac{3}{16}$ Pea : $\frac{1}{16}$ Single

Onions, color

$\frac{12}{16}$ Red or purple : $\frac{3}{16}$ Yellow : $\frac{1}{16}$ White

Mice, coat color

$\frac{9}{16}$ Agouti : $\frac{3}{16}$ Black : $\frac{4}{16}$ Albino

Sweet pea, color

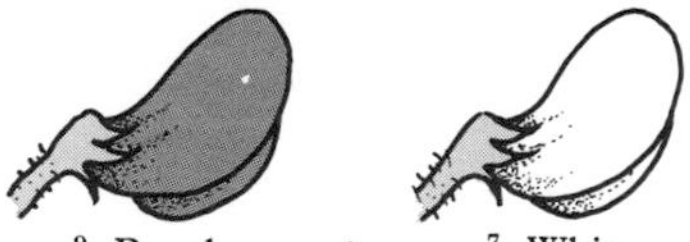

$\frac{9}{16}$ Purple : $\frac{7}{16}$ White

Domestic fowl, color (B) and color inhibition (A)

$\frac{13}{16}$ White : $\frac{3}{16}$ Colored

Shepherd's purse, seed capsule

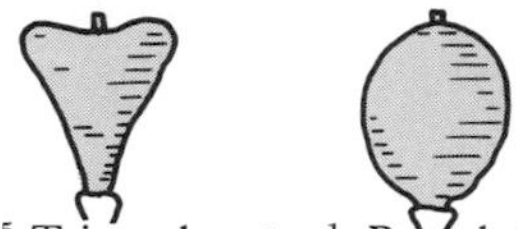

$\frac{15}{16}$ Triangular : $\frac{1}{16}$ Rounded

situation in which the action of one gene or complex of genes masks or inhibits the expression of other nonallelic genes.

Examples of epistasis are numerous. Mutations of genes operative in the development of color, such as hair color, provide instances of this. When homozygous, the allele for albinism effectively masks the expression of all other genes for color. Three genes that have been shown to be involved in feather coloration in domestic fowl, for example, are C, O, and i. Their respective alleles (c, o, and I) show epistasis in such ways that cc is epistatic to O, oo is epistatic to C, and I is epistatic to both C and O. The three breeds of pure white fowl,

white silky ($CCooii$), white wyandotte ($ccOOii$), and white leghorn ($CCOOII$), show the epistatic action of each of these alleles. An analysis of these breeds is given in Table 6–3.

In its extreme form, epistasis can account for the death of the organism by the action of so-called *lethal* genes. These are genes that so depress the vitality of the individual that death ensues. Although lethal alleles may be mutations of an otherwise undramatic set of genes, such as those that determine coat color or stature, their action may be *pleiotropic* and cause serious developmental errors with which the organism cannot cope. For example, a mouse homozygous for the dominant

Table 6–3.   Epistasis among three sets of alleles in domestic fowl.

Series I crosses			Series IV crosses		
P	White silky × Rhode Island red		P	White silky × White wyandotte	
	$CCooii$ $CCOOii$			$CCooii$ $ccOOii$	
F_1	All colored		F_1	All colored	
	$CCOoii$			$CcOoii$	
F_2	3 colored 1 white		F_2	9 colored 7 white	
	$3/4$ CCO-ii : $1/4$ $CCooii$			$9/16$ C-O-ii : $3/16$ C-$ooii$: $3/16$ ccO-ii : $1/16$ $ccooii$	
Series II crosses			**Series V crosses**		
P	White wyandotte × Rhode Island red		P	White silky × White leghorn	
	$ccOOii$ $CCOOii$			$CCooii$ $CCOOII$	
F_1	All colored		F_1	All white	
	$CcOOii$			$CCOoIi$	
F_2	3 colored 1 white		F_2	13 white 3 colored	
	$3/4$ C-$OOii$: $1/4$ $ccOOii$			$9/16$ CCO-I- : $3/16$ $CCooI$- : $1/16$ $CCooii$: $3/16$ CCO-ii	
Series III crosses			**Series VI crosses**		
P	White leghorn × Rhode Island red		P	White wyandotte × White leghorn	
	$CCOOII$ $CCOOii$			$ccOOii$ $CCOOII$	
F_1	All white		F_1	All white	
	$CCOOIi$			$CcOOIi$	
F_2	3 white 1 colored		F_2	13 white 3 colored	
	$3/4$ $CCOOI$- : $1/4$ $CCOOii$			$9/16$ C-OOI- : $3/16$ $ccOOI$- : $1/16$ $ccOOii$: $3/16$ C-$OOii$	

allele Y (yellow coat color) dies as an embryo, whereas the heterozygote differs from the wild type by being yellow (and alive). There is a qualitative difference between life and yellowness, to be sure.

Yy	$\times$	Yy	
¼ YY	:	½ Yy	:	¼ yy
(dies)		Yellow		Agouti

The conditions for lethality may differ for other lethal alleles, and lethal genes may differ in their penetrance. In the case of infantile amaurotic idiocy (Tay-Sachs disease) in man, the recessive allele must be homozygous to be lethal. On the other hand, homozygotes and heterozygotes for the dominant gene *Epiloia* in man usually die from multiple skin lesions and brain damage accruing from the expression of this gene.

A curious phenomenon known as a *balanced lethal system* is encountered when two linked sets of lethal genes are positioned in a specific way. For example, the dominant allele beaded (*Bd*) in *Drosophila* affects wing shape when heterozygous and acts as a lethal when homozygous. Closely linked but not allelic to beaded is the recessive mutant gene *l*, which is lethal when homozygous. If crossing-over does not occur, two mating flies with the genotype $\frac{Bd\ l^+}{Bd^+\ l}$ will produce only heterozygous living offspring.

$$\frac{Bd\ l^+}{Bd^+\ l} \times \frac{Bd\ l^+}{Bd^+\ l}$$

$\dfrac{Bd\ l^+}{Bd\ l^+}$	$\dfrac{Bd\ l^+}{Bd^+\ l}$	$\dfrac{Bd^+\ l}{Bd\ l^+}$	$\dfrac{Bd^+\ l}{Bd^+\ l}$
Dies (*Bd/Bd*)	Viable		Dies (*l/l*)

Interaction Between Genes—Polygenes

In 1934, Sewall Wright reported the results of a cross between two inbred strains of guinea pigs that differed in the number of toes on the hind feet. Strain 2 had the normal number of three, whereas strain D had an extra toe and was, therefore, polydactylous. The F_1 of this cross were all wild type and the F_2 consisted of about 75 percent wild type to 25 percent polydactylous. When F_1 heterozygotes were testcrossed to the four-toed stock, approximately equal numbers of three- and four-toed offspring resulted. All of these data suggested that the trait was dependent on a single pair of alleles, that for the wild-type condition being dominant. Wright then mated the three-toed testcross progeny (who were presumably heterozygotes and genotypically identical to their F_1 parent) to members of the original four-toed strain. He should again have obtained a 1:1 ratio because this cross was theoretically no different from an ordinary testcross. However, instead of the expected 1:1 ratio, the offspring assorted into a ratio of 77 percent four-toed to 23 percent normal. The hypothesis that this trait was monogenic, that is, under the control of one set of alleles, was not confirmed.

Wright then proposed that polydactyly of this sort was regulated by four sets of identical, codominant alleles, operating in cumulative fashion with a physiologic threshold of activity. Although phenotypic variation was discontinuous with respect to toe number, genotypic variation was continuous. Let us say that in each of the four sets of alleles Aa, Bb, Cc, and Dd, the lower-case letters contribute nothing toward polydactyly but the upper-case ones do. Let us also say that the threshold of activity is five upper-case genes, that is, any genotype including less than five upper-case alleles will be three-toed; five or more upper-case alleles in any combination will result in four toes. The wild-type parental stock had from zero to two of the polydactylous alleles and was, therefore, well below the threshold. The parental four-toed stock was well above the threshold, most individuals having seven or eight of the needed alleles. Members of the wild-type F_1 generation were still below the threshold, though closer to it than the wild-type parents. The F_2 and testcross progeny distributed themselves above and below the five-allele threshold, as shown in Fig. 6–5.

Analytical difficulties are also encountered with the many traits that show continuous variation. For such characteristics as stature or skin color in human beings, or seed weight in plants, the number of phenotypic classes is limited by the discriminatory powers of our measuring tools. There are no obvious discontinuities as in toe number, no present/absent alternatives, but instead a distribution of phenotypes, often in accordance with a normal distribution curve. Equally

important are observations that these very characteristics are ones that are most influenced by environmental change. For example, western travelers to Japan are often startled to find tall, strapping youths conforming neither to the travelers' preconceptions of Japanese as short people nor to the stature of their own parents. Many studies have shown a correlation between the increase in stature and medical and dietary changes. Certainly the genes that function in determining height could not have undergone such radical changes in one generation.

Fig. 6–5.   Results of crosses between two strains of guinea pigs. (See text for description.) Graphed at the right is the hypothetical distribution of the alleles contributing toward polydactyly in each case. [Adapted from S. Wright (1934), and Stern (1960).]

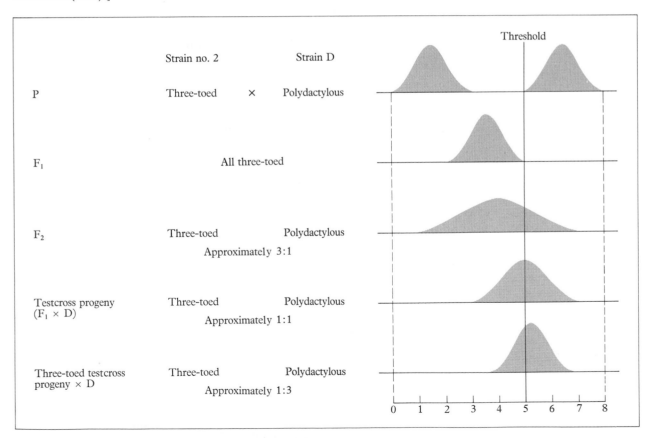

Early geneticists asked, "How is it possible for a pair of genes to produce continuous phenotypic variability?" Many of them have concluded that some traits are totally the result of environmental modification of gene function. Others reasoned differently and their ideas have become incorporated into the modern theory of polygenes. The *Theory of Polygenic Inheritance* holds that most continuous and many discontinuous traits are effected by many sets of alleles. In fact, those traits that are monogenic, such as albinism or sickle-cell anemia, represent a small and special class of characteristics that is of minor significance in determining the total phenotype of members of a species. Each set of alleles in a polygenic group segregates in a normal Mendelian manner and functions normally in the biochemical sense. Therefore, the principles of heredity that have been considered thus far in this book apply equally well to polygenes. However, such segregation and biochemistry are extremely difficult, if not impossible, to analyze by the usual genetic techniques, because the allelic sets interact in an additive or multiplicative manner. The activity of a single allele or allelic pair is thereby obscured by the activity of a multiplicity of allelic pairs, each exerting a small, and oftentimes identical, but cumulative effect on the phenotype. Because of such activity, balanced allelic substitutions can occur without changing the phenotype. Hence these alleles can only be studied statistically as a polygenic group of genes within the population. Furthermore, the effect of any one or a few polygenic alleles is usually no greater than the phenotypic variability caused by environmental activity alone. Therefore, identical phenotypes can be exhibited by a wide variety of genotypes interacting with a wide variety of environmental forces. In addition to these special complexities inherent in the nature of polygenes, one also encounters pleiotropy, differential penetrance, and variable expressivity.

In order to assess the significance of studies relating to polygenic traits, it is important to realize that the kinds of questions asked are, by necessity, different from those encountered previously in this book. Because it is assumed that polygenes are basically no different from "orthodox" genes, issues of biochemistry and gene action are relegated to the background. Because it is not possible to isolate the effects of any one set of polygenic alleles from those of other genes (and in some cases from environmental action as well), the mode of inquiry revolves not around gene action *per se*, but rather around statistical analyses of variation between organisms. This is done in order to answer questions of causality, many of which are important in our daily lives. According to Falconer (1963), these questions fall basically into four groups, all of which bear directly on the degree to which such traits can be modified either by a deliberate change in the environment or by deliberately imposed genetic selection of certain members of the group as parents of the next generation. Polygenic studies speak not only to the mathematical theoretician, but also to the plant or animal breeder, the eugenicist, the ecologist, and the evolutionist.

1. Are there genetic determinants for a trait that shows variability?

Variability will be apparent in many aspects of phenotype within any population of organisms. For instance, constancy will be shown regarding body plan—one heart, two kidneys, four limbs, etc.—in any given group of human males and females, barring genetic or developmental rarities. Variability will be noticed for such things as height, weight, skin color, eye color, hair color and form, nose length and shape, fertility, and perhaps certain aspects of personality and special abilities. In point of fact, every trait above and beyond those basically associated with the species will differ within the group.

Most biologists assume that these variable traits and their degrees of variability have some basis in genetic diversity. In other words, they conclude that genes play a role in determining the trait of height *per se*, and that variation in height results from an interaction of a polygenic series with environmental forces. However, it is important to realize that this point of view is not a universal one. Many individuals believe that although genes may contribute to the raw materials of development, it is totally in the power of the environment to shape these raw materials. As an example, one need think only of Western law which, except for individuals classified as legally insane, assumes no genetic predisposition and therefore no diminution of responsibility for any social offenses. Whether or not this assumption is valid has not yet been determined. Similarly, other schools of thought are philosophically based on genetic equality and depend upon environmental manipulation to control variability. Social workers and Marxists are two prominent, though different, examples of this idea. Polygenes thus confront us with the old nature-nurture dichotomy, the confrontation resolving itself as a balance of genetic and environmental forces. The confirmed geneticist says that genes control the appearance of the trait *and* its basic variability, the environment acting as a "fine-tuning" agent. The environmentalists credit genes with the physical framework necessary for the trait, namely, basic anatomy and physiology, but credit the environment with the development of all variability in expression of the potentials of structure and function.

Although the question of the degrees of importance of human nature and nurture has been with us for millenia, we seem able only to push the issues to more and more sophisticated planes. The questions have never been rationally answered. That this should be so is really not surprising, although it may be frustrating for confirmed believers in either the genetic or environmental camp. The probability is high that the answers will not be forthcoming for two very simple reasons. On the one hand, data from organisms other than man are difficult to extend to man, as well as being unacceptablle to those who consider man as a totally different order of being. Secondly, the classic

tests for partitioning phenotypic variability into genetic and environmental components cannot be applied to man without both sharp conflicts between ethical systems and technical breakthroughs to simplify the issues. Therefore, although the implications of polygenes may offer interesting food for thought about many fascinating aspects of human existence and variability, we must confine our discussion to mice, or corn, or flies — individuals with no moral scruples about brother-sister matings and the like, and for which the environment is more amenable to control.

It is self-evident that if one wishes to partition variation into genetic and environmental components, one must keep one of these two variables constant in order to measure the effect of the other. The simplest way to do this is by placing two different strains of the same species in the same environment, or in environments as nearly similar as possible. If the mean value for a variable trait differs between the strains, polygenic differences are held responsible for this variation.

2. What is the degree of genetic determination?

Once some degree of genetic involvement has been established for a trait showing phenotypic variation, the next question is, How much? The methodology will be briefly discussed here; the interested reader should consult any of the excellent books on the subject included in the bibliography for details of experimentation and their statistical ramifications.

The basis for partitioning variability can be expressed by the formula

$$V_G + V_E = V_P,$$

where V_G is variation due to genetic causes, V_E is variation due to environmental causes, and V_P is the resulting variation in phenotype. It should be noted that a third source of variation exists as the feedback or interaction of genes with the environment to produce differences in expressivity or penetrance. For the sake of simplicity, this factor will be ignored.

If one wishes numerical values for V_G and V_E, it is not sufficient simply to place different genetic strains in the "same" environment, because the degree of genetic similarity between the strains is unknown, as are subtleties in the environment itself. To eliminate one of the variables completely it is necessary to create its total uniformity. Of the two, genetic uniformity is far easier to achieve, because one needs only to create a highly inbred strain. Such a pure-breeding strain will be uniformly homozygous for most of its alleles, and all variation is assumed to be the result of environmental activity. For monecious organisms, homozygosity is most easily accomplished by selfing (self-fertilizing) for ten generations (Fig. 6–6). For dioecious organisms, homozygosity is assumed to be nearly complete after twenty generations of brother-sister (full-sibling) matings (Fig. 6–6). In practice, these procedures are modified slightly because the very peculiarity of being almost totally homozygous can itself lead to peculiarities in

phenotypic variability. The following example, though hypothetical, will serve as an introduction to methodology.

Let us begin by measuring flower length in a sample of a wild population of a particular plant species. For convenience, let us specify that this species is normally cross-fertilizing, but if grown under controlled conditions, it can be selfed. The measurements of flower length are arranged in a frequency distribution, as shown below.

Flower length (to nearest mm)	No. of plants	Flower length (to nearest mm)	No. of plants
14	1	34	15
15	1	35	13
16	1	36	11
17	1	37	10
18	3	38	8
19	4	39	8
20	4	40	7
21	5	41	6
22	5	42	5
23	6	43	5
24	6	44	5
25	7	45	4
26	7	46	3
27	8	47	3
28	10	48	3
29	11	49	2
30	11	50	2
31	12	51	1
32	14	52	1
33	14		

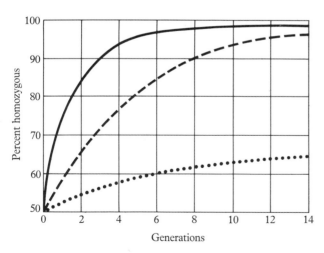

Fig. 6-6. The relationship between percent genetic homozygosity and generations of breeding by self-fertilization (solid line), brother-sister matings (dashed line), and first-cousin matings (dotted line). [Adapted from S. Wright (1921), "Systems of matings," *Genetics* **6**:111–178.]

Then the mean or average value for flower length is calculated. To derive the mean value ($\overline{X}$), all of the values ($X_1, X_2, X_3, \ldots, X_n$) are summed and divided by the number (N) of measurements:

$$\overline{X} = \frac{X_1 + X_2 + X_3 + \ldots + X_n}{N}$$

Because the original observations are arranged by class frequency (f), the calculation of the mean is simplified. One need only multiply the class value in mm (X) by the number or frequency of plants in that class (f), sum the products (Σ), and divide by N. This is usually written:

$$\overline{X} = \sum \frac{f(X)}{N}$$

X (class in mm)	f	f(X)	X	f	f(X)
14	1	14	37	10	370
15	1	15	38	8	304
16	1	16	39	8	312
17	1	17	40	7	280
18	3	54	41	6	246
19	4	76	42	5	210
20	4	80	43	5	215
21	5	105	44	5	220
22	5	110	45	4	180
23	6	138	46	3	138
24	6	144	47	3	141
25	7	175	48	3	144
26	7	182	49	2	98
27	8	216	50	2	100
28	10	280	51	1	51
29	11	319	52	1	52
30	11	330			
31	13	372		243	7975
32	14	448			
33	14	462	$\overline{X} = \dfrac{7975}{243}$		
34	15	510			
35	13	455	$= 32.82$ mm		
36	11	396			

The next specification to be met is a description of the variation that this population shows relative to the mean. As seen in Fig. 6–7, many different distributions can have the same mean, but each has a different shape when a graph is made of the values and frequency for each value. What is important to characterize is the spread or shape of the distribution curve. Many biological

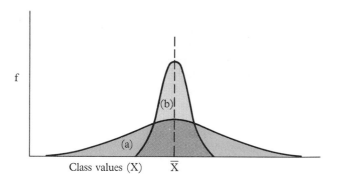

f

Class values (X) $\overline{X}$

distribution curves are of the so-called *normal* or *bell-shaped* type, that is, the mean coincides with the greatest frequency (or close to it) and the frequency curve falls away symmetrically on either side of the mean to extend indefinitely in both directions. If variation from the mean is small, the normal curve will have a narrow base; larger variations from the mean will produce a curve with a wide base. It should be noted that there are several mathematical descriptions of a normal curve just as there are for circles and ovals. Of greater concern at the moment is the fact that there are many different shapes that can be called normal curves, just as there are circles of many different diameters.

There are two terms, variance and standard deviation, that are used to describe the degree of variation from the mean. *Variance*, designated as s^2 for a sample within a population,* is defined as the sum (Σ) of the squared differences between each measurement and the mean, divided by one less than the total number of observations:

$$s^2 = \frac{\Sigma f(X - \overline{X})^2}{N - 1}$$

Standard deviation (s) is the square root of variance:

$$s = \sqrt{\frac{\Sigma f(X - \overline{X})^2}{N - 1}}$$

*Variance of a whole population, as opposed to a sample within a population, is designated as σ^2.

◀ **Fig. 6–7.** The shapes of representative normal distribution curves. Curve (a) denotes a high degree of variability about the mean ($\overline{X}$), and curve (b) denotes a fairly uniform sample of values with relatively little variation. In both cases, the area under the curve is approximately equal on either side of the mean, which is characteristic of a normal distribution.

Because of the method of calculation, variance is in square units, whereas standard deviation is in the same units as the mean. In addition, the calculation of standard deviation is such that 95 percent of the population lies between $1.96s$ below the mean and $1.96s$ above the mean, and 99 percent of the population lies between $-3s$ and $+3s$ of the mean. Therefore, the magnitude of s is a measure of the spread of the curve and the variability of the population. A large value for s is characteristic of a wide curve (high variability); a small value for s reflects a narrow curve (low variability). (See Fig. 6–7.)

While s is an index of spread to be used with the mean in defining a measurable and variable trait, it has mathematical limitations. (For derivation and rationale, the student should consult a book on statistics.) Although its units are different from the mean, variance is useful because it can be partitioned into components that reflect various causative factors, and this is an obvious advantage in nature-nurture studies. Therefore s is used as a defining term, while s^2 can be used analytically to probe the mechanisms underlying variability.

The calculations for variance and standard deviation for our flower population appear below, with 33 mm as the mean value:

X	f	f(X)	$(X-\overline{X})$	$(X-\overline{X})^2$	$f(X-\overline{X})^2$	X	f	f(X)	$(X-\overline{X})$	$(X-\overline{X})^2$	$f(X-\overline{X})^2$
14	1	14	−19	361	361	40	7	280	7	49	343
15	1	15	−18	324	324	41	6	246	8	64	384
16	1	16	−17	289	289	42	5	210	9	81	405
17	1	17	−16	256	256	43	5	215	10	100	500
18	3	54	−15	225	675	44	5	220	11	121	605
19	4	79	−14	196	784	45	4	180	12	144	576
20	4	80	−13	169	676	46	3	138	13	169	507
21	5	105	−12	144	620	47	3	141	14	196	588
22	5	110	−11	121	605	48	3	144	15	225	675
23	6	138	−10	100	600	49	2	98	16	256	512
24	6	144	− 9	81	486	50	2	100	17	280	588
25	7	175	− 8	64	448	51	1	51	18	324	324
26	7	182	− 7	49	343	52	1	52	19	361	361
27	8	216	− 6	36	288		243	7975			14,524
28	10	280	− 5	25	250						
29	11	319	− 4	16	176						
30	11	330	− 3	9	99						
31	12	372	− 2	4	48						
32	14	448	− 1	1	14						
33	**14**	**462**	**0**	**0**	**0**						
34	15	510	1	1	15						
35	13	455	2	4	52						
36	11	396	3	9	99						
37	10	370	4	16	160						
38	8	304	5	25	200						
39	8	312	6	36	288						

$$s^2 = \frac{\Sigma f(X - \overline{X})^2}{N-1}$$

$$s^2 = \frac{14{,}524}{242}$$

$$s^2 = 60.02$$

$$s = \sqrt{60.02}$$

$$s = 7.75$$

With a standard deviation of 7.75, we can say that 95 percent of our flowers measure somewhere between about 18 mm and 48 mm, which is a 30 mm range (2 × 1.96s). This represents a considerable spread of values that reflects the high degree of variability characteristic of wild populations.

To achieve genetic uniformity (homozygosity), a plant from either end of the spectrum is chosen and propagated by selfing. This procedure is repeated for ten generations so that any variation in flower length can be thought to result from modifications of the environment. We then take 50 seeds from each of the two cultivated lines and measure the flower length in the plants that result. Our observations are as follows:

Strain A

X	f	f(X)	$(X-\overline{X})$	$(X-\overline{X})^2$	$f(X-\overline{X})^2$
18	1	18	−3	9	9
19	4	76	−2	4	16
20	8	160	−1	1	8
21	23	483	0	0	0
22	8	176	1	1	8
23	5	115	2	4	20
24	1	24	3	9	9
	50	1052			70

Strain B

X	f	f(X)	$(X-\overline{X})$	$(X-\overline{X})^2$	$f(X-\overline{X})^2$
36	1	36	−3	9	9
37	3	111	−2	4	12
38	9	342	−1	1	9
39	24	936	0	0	0
40	10	400	1	1	10
41	2	82	2	4	8
42	1	42	3	9	9
	50	1949			57

Strain A	Strain B
$\overline{X} = \dfrac{1052}{50} = 21$ mm	$\overline{X} = \dfrac{1949}{50} = 39$ mm
$s^2 = \dfrac{70}{49} = 1.43$ mm^2	$s^2 = \dfrac{57}{49} = 1.16$ mm^2
$s = 1.20$ mm	$s = 1.08$ mm

In each of these we see much reduced variability. In the case of strain A, 95 percent of the flowers measure between 18.65 mm and 23.35 mm, a spread of only 4.7 mm. Strain B shows even less variability.

We now cross strain A with strain B and measure the flowers produced by the F_1 and F_2 generations:

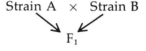

Strain A × Strain B
F_1

mm	f	f(X)	$(X-\overline{X})$	$(X-\overline{X})^2$	$f(X-\overline{X})^2$
27	1	27	−3	9	9
28	8	224	−2	4	32
29	6	174	−1	1	6
30	20	600	0	0	0
31	8	248	1	1	8
32	6	192	2	4	24
33	1	33	3	9	9
	50	1498			88

$$\overline{X} = \frac{1498}{50} = 30 \text{ mm}$$

$$s^2 = \frac{88}{49} = 1.80 \text{ mm}^2$$

$$s = 1.34 \text{ mm}$$

$$F_1 \times F_1$$

$$F_2$$

mm	f	f(X)	(X−X̄)	(X−X̄)²	f(X−X̄)²
24	1	24	−6	36	36
25	2	50	−5	25	50
26	4	104	−4	16	64
27	6	162	−3	9	54
28	10	280	−2	4	40
29	15	435	−1	1	15
30	23	690	0	0	0
31	18	558	1	1	18
32	8	256	2	4	32
33	6	198	3	9	54
34	5	170	4	16	80
35	1	35	5	25	25
36	1	36	6	36	36
	100	2998			504

$$\bar{X} = \frac{2998}{100} = 30 \text{ mm}$$

$$s^2 = \frac{504}{99} = 5.10 \text{ mm}^2$$

$$s = 2.26 \text{ mm}$$

In order to compare variance in each of the populations, the data are summarized as follows:

Group	Mean (X̄)	Variance (s²)	Cause of variation
Wild type	33 mm	60.02 mm²	Genes, environment
Strain A	21	1.43	Environment
Strain B	39	1.16	Environment
F_1 (A × B)	30	1.80	Environment
F_2 (F_1 × F_1)	30	5.10	Genes, environment

If strains A and B are both homozygous, then the F_1 will be genetically uniform, though heterozygous.

The variance that is apparent will be due therefore to environmental causes. However, the F_2 generation will show genetic as well as environmental variance because the F_1 heterozygotes will produce many different gene combinations in their gametes. The genetic component of variance in the F_2 will be:

$$5.10 \text{ mm}^2 = F_2 \text{ variance, genetic and environmental}$$
$$-1.80 \text{ mm}^2 = F_1 \text{ variance, environmental}$$
$$\overline{3.30 \text{ mm}^2 = F_2 \text{ variance, genetic}}$$

The percent of genetic determination of flower length in the F_2 will be:

$$\frac{3.30 \text{ mm}^2}{5.10 \text{ mm}^2} = 64.7 \text{ percent}$$

3. What predictions can be derived from studies of variance?

Once phenotypic variance has been partitioned into its genetic and environmental components, further analysis is possible. Genetic variance is most commonly subdivided into additive and nonadditive portions. Additive genetic variance (V_A) results from the independent effect of each allele in a polygenic series. Nonadditive genetic variance (V_D) is produced by dominance between alleles as well as by epistasis between sets of alleles. According to complex statistical methods developed by Mather (cf. Falconer, 1963), phenotypic variance in different generations theoretically assorts as follows, where E represents the environmental component:

$$\text{Variance in parental line 1} = E$$
$$\text{Variance in parental line 2} = E$$
$$\text{Variance in } F_1 = E$$
$$\text{Variance in } F_1 = \tfrac{1}{2} V_A + \tfrac{1}{4} V_D + E$$
$$\text{Variance in backcross} = \tfrac{1}{4} V_A + \tfrac{1}{4} V_D + E$$

Variance in backcross applies regardless of whether the hybrids (F_1) are backcrossed to parental line 1 or 2.

Although such analyses may be interesting for purely intellectual and/or theoretical reasons, they have been put to very practical uses as well. Much of the recent success in averting famines of major proportions in many areas of the world owes itself to the production of high-yield, hardy grain and livestock. These new strains have been created by selective breeding procedures, in which predictability is a virtue of the highest order. If one wishes to develop a high-yield strain of rice, for example, variance analyses are imperative. One must choose parental lines that show a low degree of environmental variability in order to assure that the new strain can be grown successfully in many kinds of land and by many kinds of farmers. Conversely, the greater the proportion of genetic variability, the greater the predictability of the final product. Furthermore, the difficulty of predicting how genes will assort during gamete production makes it mandatory that in vital activities such as these, the greatest component of variance must be not only genetic but also additive, in which each gene can be considered in relative theoretical isolation.

The term *heritability* (h^2) has been coined to describe the ratio of additive genetic variance to the total phenotypic variance:

$$h^2 = \frac{V_A}{V_P}$$

In some respects, this choice of terminology is unfortunate because it is conceptually suggestive of inheritance *per se*. For instance, if a trait is totally controlled by one gene, and if there is no variability in its appearance, the value for h^2 will be 0 in spite of the obvious genetic nature of the phenotype. It becomes important to realize that heritability refers not to any trait in and of itself, but rather to the inherited nature of its *variability* in the population for which heritability is estimated. Therefore, heritability estimates between populations may differ. Values for h^2 can range from 1 (all variation of a characteristic in a population is caused by additive polygenes) to 0 (all variation in a population is caused by the environment). This

aspect of heritability is better illustrated by the more empirical definition:

$$h^2 = \frac{R}{S},$$

where S is the selection differential (the difference between the mean value of the selected organisms and the mean of all measured organisms out of which they were selected), and R is the response obtained (the difference between the mean of the nonselected population and the mean of the offspring of selected parents).

In essence, estimates of heritability provide a tool to predict the degree of resemblance between the total phenotype of successive generations if the mating system is controllable and/or known.

4. How many genes are there in a polygenic series?

For most quantitative characteristics, the question of how many genes are at work is difficult to answer. Statistical methods have been developed to provide estimates, but because of the simplifying assumptions that must be made, most estimates are generally low. The difficulty stems from not knowing the effect of each gene in the series. Therefore, one must assume that each gene in a polygenic series either affects the trait or does not (no partial effectiveness), and that the effect of any one allele is equal to that of any other allele in the set.

Table 6–4 shows the frequency distributions of phenotypes produced by increasing sets of alleles. These distributions assume no environmental effects. In these situations, the number of genes involved can be deduced from the frequency of recovery of the extreme types. However, once a trait involves more than five sets of alleles, recovery of parental extremes becomes difficult. The following formula has been devised for rough estimates of the number (n) of operative pairs of alleles in a polygenic series:

$$n = \left(\frac{1}{8}\right)\left(\frac{R^2}{V_{F_2} - V_{F_1}}\right)$$

Table 6–4. The relationship between the number of alleles operative in the expression of a trait of continuous variation and the expected frequency of extreme values for the trait in the F_2. Assuming equal contribution of the alleles governing a trait, one may approximate their number by observing the frequency of the extremes.

Number of genes	Number of alleles	Fraction of F_2 as extreme as either parent (n = number of genes)
1	2	$\frac{1}{4} = \left(\frac{1}{2}\right)^2$
2	4	$\frac{1}{16} = \left(\frac{1}{2}\right)^4$
3	6	$\frac{1}{64} = \left(\frac{1}{2}\right)^6$
4	8	$\frac{1}{256} = \left(\frac{1}{2}\right)^8$
5	10	$\frac{1}{1024} = \left(\frac{1}{2}\right)^{10}$
10	20	$\frac{1}{1,048,576} = \left(\frac{1}{2}\right)^{20}$
1n	2n	$\left(\frac{1}{2}\right)^{2n}$

R is the difference between the mean values of two different inbred strains. V_{F_2} and V_{F_1} are the variances for the F_2 and F_1 generations produced by a cross of the parental strains. For the example of flower length, the calculation is as follows:

$$n = \left(\frac{1}{8}\right)\left(\frac{(39 - 21)^2}{5.10 - 1.80}\right) = \left(\frac{1}{8}\right)\left(\frac{324}{3.30}\right) = \left(\frac{1}{8}\right)(98.2)$$

n = about 12 genes

It is stressed that this value is at best a rough estimate because the formula assumes the following conditions:

1. All of the contributing alleles are in one parental strain; all of the noncontributing alternatives are in the other parental strain.

2. The parental strains are homozygous.

3. The allelic sets show no dominance.

4. None of the genes in this series is linked to any others in the series.

5. Each set of alleles has an equal and additive effect on the trait.

If in fact these assumptions do not apply, the estimate will be mildly or extremely low. More sophisticated models can be constructed to take dominance and linkage into consideration, but they are beyond the scope here. Comfort can or cannot be taken from the realization that the real world of genes is almost always more complicated than even the most painstaking model situations devisable.

QUESTIONS

1. The basic body color of *Drosophila* is light brown, but T. H. Morgan discovered a mutation that is expressed as yellow body when homozygous. The F_2 of crosses between yellow-body flies and wild type assorts in a typical Mendelian 3:1 ratio. Many years later, Rappoport discovered that if the larvae of wild-type flies are raised in medium containing silver nitrate, they develop the yellow-body phenotype as adults and are indistinguishable from the mutant strain that Morgan described. If the progeny of these adults are reared in silver-free medium, they develop into wild-type adults. How can these observations be explained?

2. Persons having hereditary glandular deficiencies may often lead normal lives if they follow a regimen of "replacement therapy." For example, diabetics may take insulin. Would you expect such replacement therapy to eliminate, modify, or have no effect upon the chance of transmitting the genetic defect to the individual's children? Explain.

3. In the following pedigree, the circles represent females and the squares represent males; Roman numerals designate generations; parents are joined by a horizontal *marriage line*; offspring are connected by short vertical lines to a horizontal *sibship line* below their parents.

Individuals phenotypically affected by a hypothetical dominant gene *M* are shaded. Assuming that gene *M* is very rare, what are the probable genotypes of all the individuals in generation II of this pedigree? Explain.

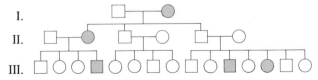

4. The degree of concordance of the congenital abnormality clubfoot is about 32 percent in monozygotic twins and about 3 percent in dizygotic twins. Chronic alcoholism, on the other hand, shows concordance of nearly 100 percent and about 85 percent in monozygotic and dizygotic twins, respectively. What are the relative influences of heredity and environment for these two traits? What additional information, if any, is needed to give you more confidence in your answer?

5. If it can be demonstrated that the potential for intellectual development has a firm, hereditary basis, would it necessarily follow that different populations of individuals would have different (a) IQ means? (b) IQ ranges? Explain.

6. Speculate on reasons accounting for the observation that some traits show little variation in nature whereas others show wide variation (for example, brain size as compared to artistic ability).

7. Both sexes of Dorset sheep have horns, but neither sex of Hampshire sheep is horned. If the two breeds are crossed, the resulting ram lambs are horned and the ewe lambs are hornless (polled), irrespective of the sex of the horned parent. How might this be explained as the expression of an environmental influence? (Hint: Sub-

stances circulating in the blood and body fluids constitute part of the environment of the body—the "internal environment.")

8. A new breed of sheep, Polled Dorsets, has been established in which neither sex is horned. Occasionally though, "scurs" (rudimentary horns) appear in the lambs of this breed, more frequently in rams than in ewes. What does this suggest regarding the penetrance and expressivity of the genetic component and the role of the environmental component in horn development in sheep?

9. Landauer described a mutant in domestic fowl that was characterized by abnormally short legs. He called them Creeper fowl. When Creepers are crossed, the chicks that hatch consist of both Creeper and normal in the ratio of two to one, respectively. No pure-breeding Creepers have been found. Explain. How might you test your hypothesis?

10. Measurements were made on the ear lengths of four inbred varieties of corn to be used for hybrid seed production. The following data were obtained:

Variety	Mean length (cm)	Variance (cm²)
1	12.5	6.25
2	14.0	4.41
3	20.3	10.25
4	21.0	38.44

What is the standard deviation of each variety? How do these strains compare in genetic uniformity?

11. When varieties 1 and 2 above were crossed, the mean ear length and variance of the progeny were 22.2 cm and 21.94 cm². Explain.

7

The Molecular Theory of Inheritance

In the early 1870's, a German chemist named Friedrich Miescher demonstrated that cell nuclei contain the unique chemical substance that is known today as nucleoprotein. He succeeded in isolating from the nuclei of fresh fish sperm the substance that later became known as DNA. The importance of his findings was immediately appreciated, for it was known that the nucleus is essential both for survival in most kinds of cells and for reproduction in all cells. Of course his discovery took place many years before chromosomes and their genetic significance were discovered, and it was not until 1970 that sufficient information existed that enabled scientists to construct a synthetic gene (Maugh, 1973). However, when the Chromosome Theory of Inheritance was finally accepted and it was learned that the chromosomes are the major repository of nucleoprotein, chemists and embryologists were stimulated to direct their attention more closely to nucleoprotein chemistry.

The nucleoproteins were shown to be large molecules that are formed by the conjugation of proteins with nucleic acids. Proteins were known to be complex molecules as varied as life itself and to be the very essence of living matter. The nucleic acids were demonstrated to be complex molecules that are composed of subunits of sugar, a variety of nitrogenous bases, and phosphoric acid. The bases were identified as purines—predominantly adenine and guanine—and pyrimidines—predominantly thymine, cytosine, and uracil. Therefore, the variety of nucleic acids that could be formed appeared to be limited by these five bases. It was soon realized that these components of the nucleic acids are assembled in molecular units called *nucleotides*, which consist of one five-carbon sugar (pentose), one phosphate group, and one of the five kinds of bases. Therefore, the kinds of nucleic acids that are possible is a function not only of the sequence of the five kinds of bases, but also the number of nucleotides that combine to comprise the nucleic acid molecule.

Proteins nevertheless offered much more promise of being ideal molecules for gene structure. Not only do proteins display a large range of molecular weights, they also show great variety in chemical properties, some being basic, some neutral, and some acidic to varying degrees. But one of the most impressive properties of proteins is their specificity. This property is expressed to a remarkable extent by catalytic proteins—enzymes—in terms of the reaction which they will or will not catalyze, and by antibody proteins in terms of the molecules with which they will react. For example, the enzyme lactase will split the molecule lactose ($C_{12}H_{22}O_{11}$), but it will not affect the similar molecules maltose ($C_{12}H_{22}O_{11}$) or sucrose ($C_{12}H_{22}O_{11}$). Antibody discrimination is even more striking. For example, antibodies against Type A blood in human beings are unreactive against Type B blood, although the differences between the two kinds of red blood cells is too subtle to be detected by electrophoresis or any other routine analytical technique. If the differences between species and even between individuals within a species are to be attributed to genes, surely the proteins gave promise of providing the enormous amount of variety and versatility implied in gene structure. And so until the definitive work of Avery, MacLeod, and McCarty (1944), genes were generally assumed to be proteins.

DNA AS THE GENETIC MATERIAL

The studies that led to the elucidation of the chemical composition of genes were prompted by a fundamental discovery by Griffith in 1928. Griffith had been working with the bacterial organism *Diplococcus pneumoniae,* the pathogenic microorganism implicated in bacterial pneumonia. In order to understand Griffith's work as well as that which follows in this and subsequent chapters, it is helpful to be familiar with certain techniques of

bacterial culture and with the life cycle of a representative bacterium such as *Escherichia coli.*

Culturing Bacteria

Most bacteria, such as *E. coli,* can be grown conveniently either in a liquid nutrient medium (broth) or on the surface of a gelatinous nutrient medium (agar plate). A nutrient medium is a concoction that contains the substances required by the organism for normal growth and reproduction. The nutrients required by *E. coli* are often provided in the form of a sugar as a source of carbon and hydrogen, and inorganic salts as a source of nitrogen, phosphorus, sulfur, calcium, potassium, magnesium, and iron. These ingredients are dissolved in water in sufficiently dilute concentrations to avoid harmful osmotic effects, and buffered to an optimal pH. Bacteria may be introduced or "inoculated" into a broth prepared in this way, and their rate of growth will be predictable for any given temperature. Optimal growth is obtained for most kinds of bacteria geneticists work with at 37°C, which is the average human-body temperature. This temperature is not a coincidence—most of the bacteria used in experimental work are normal parasites or symbionts of mammals, and so they have evolved to accommodate the temperature requirements of their host.

When a bacterial cell grows, it undergoes binary fission to produce two cells, each of which then grows to the size that the parent cell was. These cells then divide (again by binary fission) and the cycle continues (Clark, 1968). The rate of growth can be expressed in terms of the number of cells produced per unit time, and it is graphically represented in the form of a growth curve (Fig. 7–1). The rate of growth of *E. coli* is such that under optimum nutritional conditions, the number of individuals will at least double every twenty minutes at an incubation temperature of 37°C. At this rate, if nutrient requirements are met, a single bacterium will have progeny in excess of 10^{21} individuals within 24 hours. The number of indi-

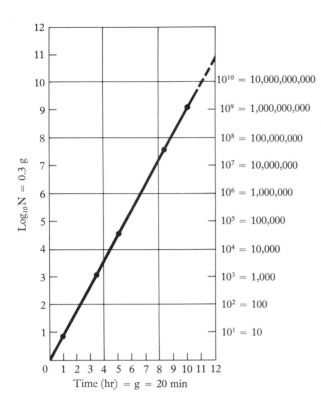

Fig. 7–1. A plot of the growth rate of *E. coli* under optimal nutritional and environmental conditions at 37°C. Since bacterial cells reproduce by binary fission, the population doubles each succeeding generation. Hence the number of cells (N) after any given number of generations (g) can be expressed as $N = 2^g$. Since growth rates are exponential, they are more conveniently converted to logarithmic functions. Thus $N = 2^g$ can be converted to $\log_{10}N = g \log 2$. Since $\log_{10}2 = 0.3$, $\log_{10}N = 0.3$ g.

viduals in a broth culture at a given time can be estimated by optical density (absorbance), by direct counting, or by plating out a sample of the culture.

By the first of these methods, the concentration of cells in a culture is estimated by measuring by means of a photocell the amount of scatter or *absorbance* of light passed through a culture sample. The greater the absorbance, the greater the number of cells per unit volume of culture. Calibration curves can be developed that relate the number of cells in a culture to the *optical density* of the culture. The *direct count method* involves the use of a special microscope slide that is constructed with a counting chamber in its center. The counting chamber has a grid of 400 squares etched over a 1-mm² area depressed in such a way that when a cover glass is mounted over it, there is a space of

0.02 mm between the counting chamber surface and the lower surface of the cover glass. Hence, the volume over each square equals 0.02 mm × 0.0025 mm², or 0.000,05 mm³ (0.000,000,05 ml). After a drop of bacterial suspension is placed in the chamber and covered with the cover glass, direct count is made with the aid of a microscope. In practice, the bacterial sample is diluted with an isotonic diluent which serves to facilitate counting by reducing the numbers of organisims in the visual field. Correction for the dilution is subsequently made.

Counting by means of the *plating out method* involves first agitating a culture in order to assure as uniform distribution of cells as possible, and then transferring a 0.1-ml sample of the culture sequentially through a series of known volumes of isosmotic diluent, as shown in Fig. 7–2. Then 0.1

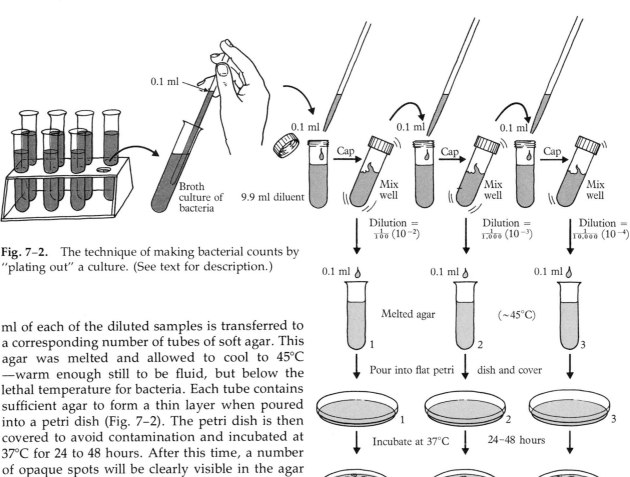

Fig. 7–2.   The technique of making bacterial counts by "plating out" a culture. (See text for description.)

ml of each of the diluted samples is transferred to a corresponding number of tubes of soft agar. This agar was melted and allowed to cool to 45°C —warm enough still to be fluid, but below the lethal temperature for bacteria. Each tube contains sufficient agar to form a thin layer when poured into a petri dish (Fig. 7–2). The petri dish is then covered to avoid contamination and incubated at 37°C for 24 to 48 hours. After this time, a number of opaque spots will be clearly visible in the agar (Fig. 7–2). Each spot is a colony of bacteria, and each colony represents the progeny of either a single bacterium or a cluster of bacteria, each of which constitutes what is known as a "viable unit." From the number of colonies, the number of viable units in the original broth culture at the time the sample was taken can easily be calculated, as described in Fig. 7–2.

The colonies that bacteria form on an agar plate have a distinctive morphology and color that are constant and characteristic of the species or strain of that microorganism. The appearance of an atypical colony often indicates a genetic change which may be extremely useful to the geneticist.

Number of viable units per ml^3
in original broth culture = number
of colonies
$\times\ 10^2$ (plate 1)
$\times\ 10^3$ (plate 2)
$\times\ 10^4$ (plate 3)

Bacterial Transformation

Griffith (1928) worked with different virulent strains of *Diplococcus pneumoniae* which may be designated as Types I, II, and III. They differed from one another in the serological characteristics of the polysaccharide capsule that these organisms secrete about themselves. He also used avirulent strains of *D. pneumoniae* which differed from their virulent counterparts in their inability to cause disease, inability to form a capsule, and in colony morphology. These characteristics are related —the lack of a protecting capsule renders the bacterium susceptible to the defense mechanisms of the host and is also responsible for the appearance of the colony. The avirulent strains produce rough-surfaced (R) colonies on solid medium, whereas the colonies of the virulent strains are smooth-surfaced (S) (Fig. 7–3). He inoculated mice with a small number of live R cells together with a quantity of Type II–S cells that had been killed by heat. A number of mice died of pneumonia and he was able to isolate live Type II–S organisms from their blood. When he repeated the experiment using heat-killed cells of Type III-S, he obtained the same effect, except that the live organisms recovered were of Type III-S. Griffith systematically varied the type of heat-killed S strains and live R strains, and he regularly obtained transformation from the avirulent R to the virulent S strain of the type that was used (Table 7–1). Moreover, the transformed bacteria were propagated in culture and bred true—that is, they continued to produce virulent S individuals of the corresponding strain. Therefore, the change was a genetic change. It is obvious that the substance implicated in the transformation, the so-called "transforming principle,"

Table 7–1. Results of some of Griffith's experiments. (a) If mice are inoculated with any of the avirulent R strains of *D. pneumoniae,* the bacteria are destroyed and the mice survive. (b) If mice are injected with virulent S strains of *D. pneumoniae,* they die of pneumonia in a few days, and live organisms of the injected S strain can be recovered from the blood and propagated in culture tubes. (c) Mice injected with heat-killed suspensions of S-strain bacteria develop no symptoms of pneumonia and remain healthy. (d) If mice are inoculated simultaneously with live R-strain *D. pneumoniae* and heat-killed bacteria of any S strain, a significant number of them will die of pneumonia, and live organisms of the corresponding S strain can be recovered from their blood. The bacteria so recovered can be cultured in nutrient broth and will be found to breed true.

Inoculum	Effect	Bacteria recovered
(a) Live R	All mice survive	None
(b) Live II–S	Most mice die	Live II–S
Live III–S	Most mice die	Live III–S
(c) Dead II–S or III–S	All mice survive	None
(d) Live R + Dead II–S	Some mice die	Live II–S
Live R + Dead III–S	Some mice die	Live III–S

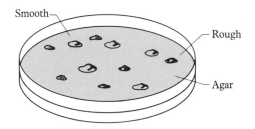

Fig. 7–3. The appearance of "smooth" and "rough" colonies of *D. pneumoniae* on an agar plate.

was heat stable and capable of being absorbed in some way by the living bacteria. Proteins are generally permanently altered by heat, and so for the first time serious doubt was cast on the hypothesis that genes are proteins.

Later investigations showed that bacterial transformation can be effected *in vitro* (in culture tubes). In essence, this is done by mixing live R strains with heat-killed S organisms, incubating them, and then killing off the R individuals with R-antiserum in order to separate out the small number of transformed individuals from the enormously large number of bacteria that have not been transformed. (This is essentially what happens in the living host.) When this is done, pure cultures of virulent S organisms are obtained. These experiments were followed by experiments in which the heat-killed S organisms were lysed (burst or broken down) and then filtered in order to obtain cell-free extracts from them. When such cell-free extracts were mixed with and incubated with live R cultures, live S individuals were obtained, demonstrating that the transforming principle is not cellular.

The next logical procedure was to take various fractions of the cell-free extract and identify which chemical substance constituted the active substance. This was done by Avery, MacLeod, and McCarty in 1944. They demonstrated conclusively that the transforming principle is DNA. The crucial experiments of their work are represented graphically in Fig. 7–4. In order to rule out any possible role of protein, they treated a control sample of DNA extracted from III–S bacteria with the enzyme *deoxyribonuclease* (DNase), which specifically hydrolyzes DNA without affecting protein. The DNase-treated extract did not transform the R-strain organisms.

Fig. 7–4. Avery, McLeod, and McCarty effected transformation of *D. pneumoniae* by the use of DNA extracted from heat-killed virulent S strains.

Viral Life Cycles and Transduction

The second set of evidence pointing to the genetic role of DNA emerged from a study of viral life cycles. Although this topic is discussed at greater length in Chapter 10, some description is necessary at this point in order to appreciate what follows. Many viruses consist only of nucleic acid and protein. The nucleic acid in some viruses is RNA, whereas in others it is DNA. The most useful viruses for genetic studies have been some of the viruses that infect bacteria. These are known as

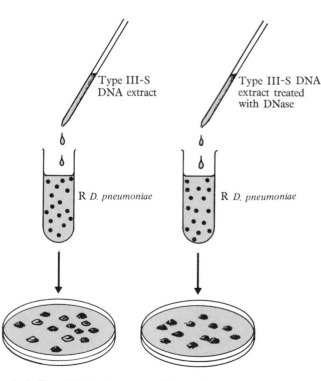

Type III-S
DNA extract

Type III-S DNA
extract treated
with DNase

R *D. pneumoniae*

R *D. pneumoniae*

Both Type III-S and
rough colonies produced

Only rough
colonies produced

bacteriophage, or more simply, *phage*. The structural components of bacteriophage were shown by Brenner and his collaborators (1959) to consist of DNA and protein molecules comprising a "head," a contractile "sheath," a tubular "core," and a series of tail filaments (Fig. 7–5). Phages have been given such cryptic and arbitrary names as T1, T2, T3, T4, T5, T6, P1, λ, etc. The most useful ones have been the "T-even" (T2, T4, and T6) and λ phages.

The life cycle of a phage particle will be treated as representative of virus reproduction. A T2 phage particle, for example, becomes attached to the surface of a bacterial cell, in this case a cell of *E. coli*, by its tail filaments, and a hole is made in the bacterial cell wall. The sheath contracts and the DNA molecule of the phage particle is literally injected via the core into the bacterial cell through the hole, leaving most if not all the protein components of the particle outside. Hershey and Chase (1952) demonstrated this by differentially labeling T2 protein with a radioactive isotope of sulfur (^{35}S) and its DNA with a radioactive isotope of phosphorus (^{32}P). Most proteins, including phage proteins, contain sulfur but little or no phosphorus, whereas DNA contains phosphorus but no sulfur. Hershey and Chase showed that the ^{32}P entered the bacteria, whereas all but a minute fraction of the ^{35}S remained on the outside. They removed the viral protein carcasses from the bacteria by applying fluid shearing forces with a blendor and washing. Subsequently the bacteria burst (lysed), liberating a new generation of phage particles complete with new, unlabeled protein components. Their experiments demonstrated that DNA is the only chemical link between phage generations; therefore it constitutes the genes of the phage particles.

The events that occur between the moment that the phage DNA enters a host cell and the moment phage progeny are released are well, but not completely, understood. In most cases, as soon as the phage DNA enters the host cell, it mobilizes that cell's replicative enzymes, other proteins, nucleic acids, and their precursors to

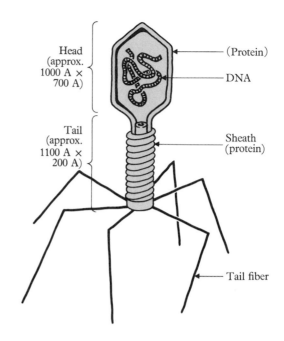

Head (approx. 1000 A × 700 A) — (Protein)

DNA

Tail (approx. 1100 A × 200 A)

Sheath (protein)

Tail fiber

Fig. 7–5.   Diagram of the structure of a T-bacteriophage particle, a DNA virus that parasitizes *E. coli*.

synthesize multiple replicas of itself. These DNA replicas then code the synthesis of new protein coats which form around each of the phage DNA molecules. In this way, new particles numbering from a hundred to a thousand or more arise within the host cell. Also coded by the viral DNA, the enzyme lysozyme is synthesized. This enzyme lyses the host cell, thereby releasing the fully formed phage progeny from the cell, which are then ready to infect other cells (Fig. 7–6).

With certain phages (called *temperate* phages) under special circumstances, a different kind of viral life cycle may be evident. The infecting viral DNA may become inserted into the DNA molecule of the host cell, where it may remain integrated for an indeterminate period of time and number of cell generations. In this condition it is known as *prophage*, and it replicates synchronously with the

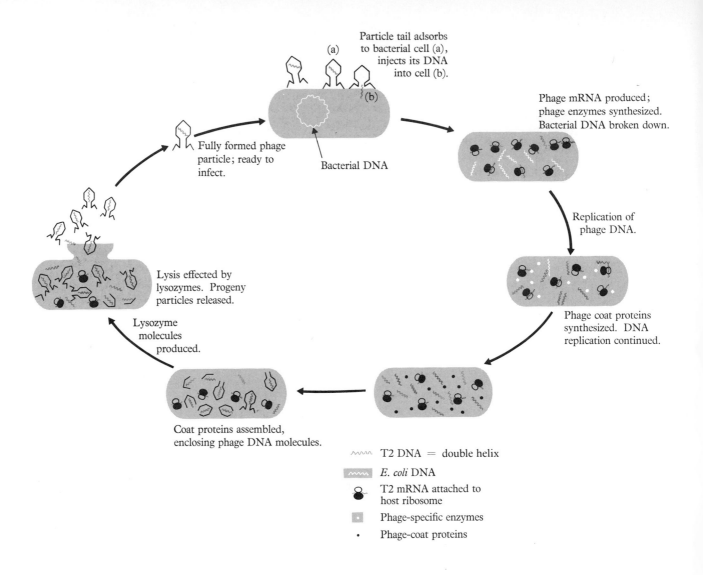

Particle tail adsorbs to bacterial cell (a), injects its DNA into cell (b).

(a)

(b)

Bacterial DNA

Fully formed phage particle; ready to infect.

Phage mRNA produced; phage enzymes synthesized. Bacterial DNA broken down.

Replication of phage DNA.

Lysis effected by lysozymes. Progeny particles released.

Phage coat proteins synthesized. DNA replication continued.

Lysozyme molecules produced.

Coat proteins assembled, enclosing phage DNA molecules.

〰〰 T2 DNA = double helix

〰〰 *E. coli* DNA

T2 mRNA attached to host ribosome

Phage-specific enzymes

Phage-coat proteins

Fig. 7–6. Diagram of the life cycle of the DNA bacteriophage T2. (See text for description.)

host cell's DNA during the latter's cell cycle. In this way, the viral DNA is integrated with the DNA of all the host cell's progeny as well. Under the influence of an appropriate stimulus, the integrated viral DNA may be released, whereupon it may replicate itself independently of the host DNA and reproduce a new generation of viral particles, which may then lyse the host cell. Because cells

that have an integrated virus can at any moment become lysed by the virus they harbor, they are known as *lysogenic* cells. Figure 7–7 describes the life cycle of phage λ, which is a temperate phage. Ultraviolet radiation is usually used as the stimulus to bring about reversal of phage DNA integration.

When prophage becomes released or excised,

it occasionally carries with it a short segment of the host cell's DNA and incorporates this DNA segment in some of the new phage particles in place of its own DNA. When these particles reinfect a new cell, they introduce the bacterial DNA segment into the new host. Virus particles that do this are deficient in their own DNA and are incapable of reproducing themselves. But the segment of bacterial DNA that they have introduced may become inserted, probably by crossing-over during DNA replication, into the host cell's DNA molecule and thus express itself in that cell's progeny and breed true thereafter. The transfer of DNA from one cell to another by means of an intermediate vector is known as *transduction* (Zinder and Lederberg, 1952) and is described at greater length in Chapter 10.

The foregoing demonstrations of bacterial transformation, viral infection, and transduction all implicate DNA as the genetic material in the bacterial organisms and viruses used. All evidence available so far justifies the sweeping generalization that, with the exception of the RNA viruses which will be discussed later, the genetic material of all cells is DNA.

NUCLEIC ACID AND THE STRUCTURE OF DNA

The two kinds of nucleic acid, DNA and RNA, were found to differ with respect to their pentose residues and their pyrimidine bases. The pentose of RNA is d-ribose ($C_5H_{10}O_5$), and that of DNA is deoxyribose ($C_5H_{10}O_4$). The RNA includes the pyrimidines cytosine and uracil for the most part, whereas DNA uses the pyrimidines cytosine and thymine exclusively (Fig. 7–8). It was also found that DNA, as a polymer of repeating nucleotides, can be very large, attaining molecular weights in excess of 500,000 (now known to exceed many times this value). Even though DNA could form such large molecules, it was still difficult to understand how DNA could satisfy the two essential

requirements of gene function—(a) reproduce itself (replicate), and (b) provide the structural and/or chemical variability consistent with the variety of genetic traits already observed in nature.

In 1947, the biochemist Chargaff analyzed DNA from a number of different species of organisms. Contrary to what had been believed up to that time, he demonstrated that the four nucleotides are not present in equal amounts and, what is more, that their ratios vary from one species to another. However in later work, Chargaff showed that the amount of adenine is always equal to that of thymine, and the amount of guanine is always equal to that of cytosine. Accordingly, the adenine-plus-thymine to guanine-plus-cytosine ratio varies from species to species (Table 7–2).

Table 7–2. Comparison of the base ratios of several species. The variations between the ratios of adenine-to-thymine (A:T) and guanine-to-cytosine (G:C) are very small in all species in which DNA is double-stranded. This suggested the hypothesis of complementary base pairing. The ratios of the frequencies of A+T pairs to G+C pairs (A+T:G+C) vary from species to species and, in the case of closely related species, may be used as a means of refining the measure of relationship.

Species	Adenine	Guanine	Thymine	Cytosine	A+T:G+C
T2 Phage	32.6	18.2	32.6	16.6	65:35
Sea urchin	32.8	17.7	32.1	17.4	65:35
Yeast	31.3	18.7	32.9	17.1	64:36
Human	31.0	19.1	31.5	18.4	63:37
Salmon	29.7	20.8	29.1	20.4	59:41
X174 phage (single-stranded)	24.7	24.1	32.7	18.5	57:43
Rat	28.6	21.4	28.4	21.7	57:43
Cattle	28.7	22.2	27.2	22.0	56:44
E. coli	26.0	24.9	23.9	25.2	50:50
Mycobacter	15.1	34.9	14.6	35.4	30:70

[Data after Chargaff (1947).]

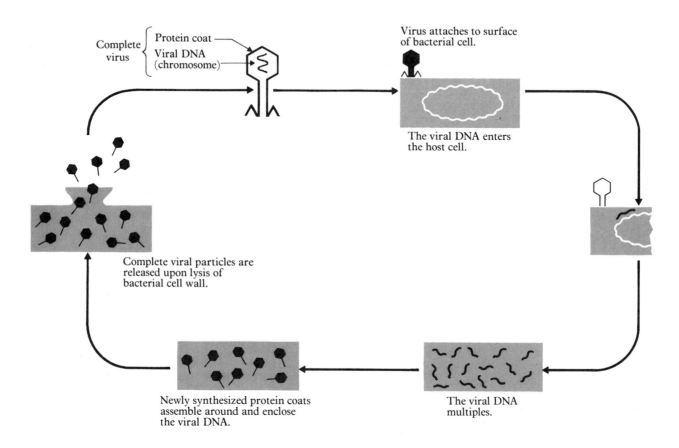

Fig. 7–7.   Diagram of the life cycle of the temperate bacteriophage **λ**. (See text for description.)

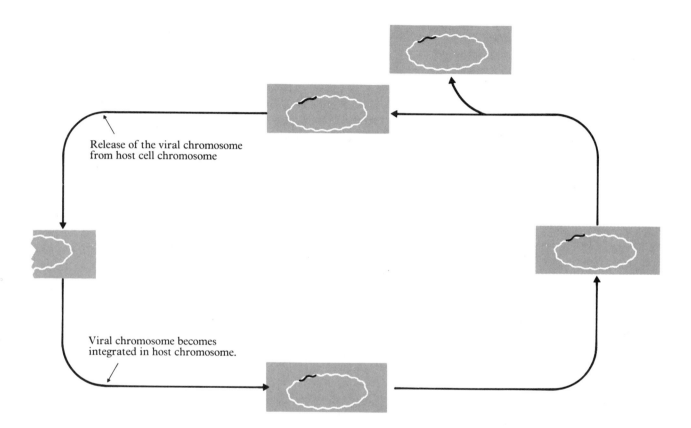

Release of the viral chromosome
from host cell chromosome

Viral chromosome becomes
integrated in host chromosome.

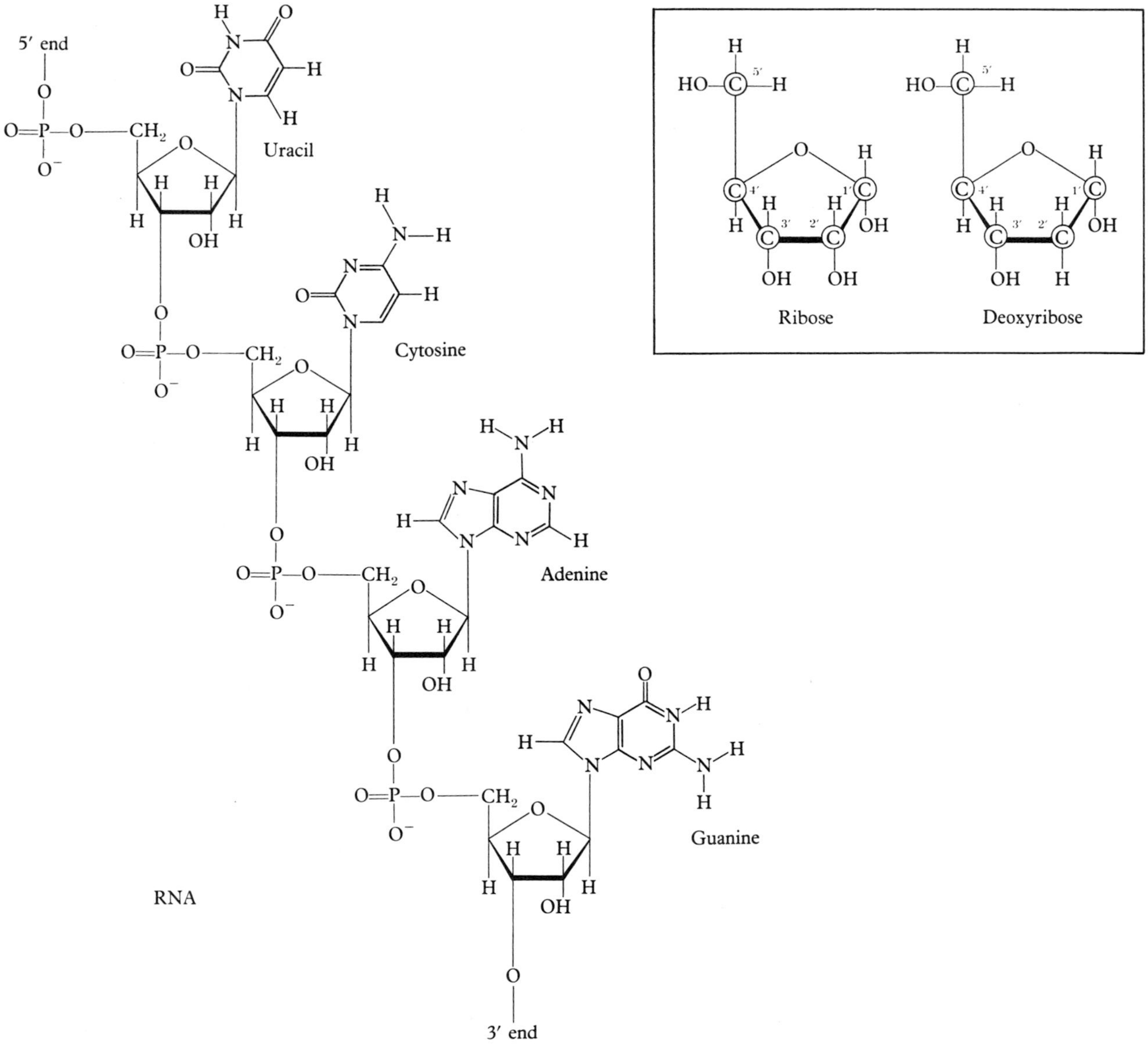

Fig. 7–8.   Comparison of the molecular structure and composition of RNA and DNA. Inset: Detailed structural formulas of the two pentoses, ribose and deoxyribose. Note that the pentose-phosphate backbones of the two DNA strands have opposite orientation (antiparallel), as indicated by the number of the carbon of the pentose to which the phosphate groupings are bonded (designated either 3' or 5', as shown in the inset).

DNA

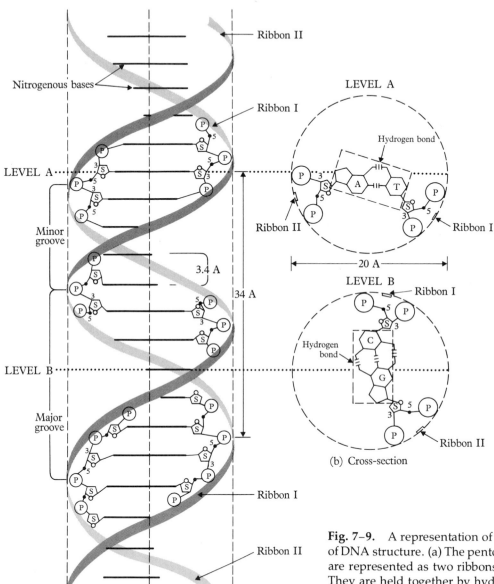

(a) Side view

(b) Cross-section

Fig. 7–9. A representation of the Watson-Crick model of DNA structure. (a) The pentose-phosphate backbones are represented as two ribbons arranged in a helix. (b) They are held together by hydrogen bonding between their bases, all of which project inward toward the central axis. The helical configuration reflects the shape and spatial (three-dimensional) attitude of the bases and their covalent bonds. Constant features are the number of nucleotide pairs per gyre (ten), dimensions of each gyre (20 A diameter, 34 A length) and the presence of major and minor grooves (spacings between the ribbons). [From W. Etkin (1974), *Bio. Science* **23**:652–653. With permission of the author.]

By logical yet ingenious extension of the fundamental findings of chemists and X-ray crystallographers, Watson and Crick in 1953 presented their theory of the molecular structure of DNA. The model that they proposed, the so-called "double helix," lent itself to meeting the two foregoing requirements of gene function. They proposed that a molecule of DNA consists of two polynucleotide strands that lie side-by-side in a helical configuration, and that these strands are bound to each other via their bases, which are arranged in the complementary pairs adenine-thymine and guanine-cytosine, as implied in Chargaff's discoveries. This model is represented in Fig. 7–9. The "backbone" of each strand consists of alternating phosphoric acid and sugar units chemically bound to each other by strong diester bonds. This configuration positions the bases as lateral projections off the sugars, with which they are covalently bonded. The two component strands of DNA present their bases toward each other, the purine adenine connected to the pyrimidine thymine by two hydrogen bonds, and the purine guanine connected to the pyrimidine cytosine by three hydrogen bonds. The phosphodiester bonds give each backbone a right-handed helical twist which is further stabilized by the cross-bonding between complementary strands by the base-pair bonds. Hydrogen bonds are weak as compared to covalent bonds, however, and are easily broken. They arise between a covalently bonded hydrogen atom with some positive charge and a covalently bonded atom with negative charge (Fig. 7–10).

The concept of base pairing is one of the essential features of the Watson-Crick model, for this

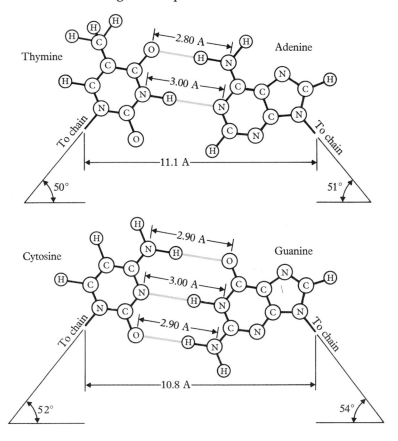

Fig. 7–10. The molecular basis of base pairing in DNA. The tendency for the purine adenine to form hydrogen bonds is matched exactly in terms of number (two) and position of such bonds by the pyrimidine thymine. Similarly, the purine guanine is matched in these respects by the pyrimidine cytosine, with which three hydrogen bonds are formed. The constant configurational characteristics of the DNA double helix are attributable in part to the dimensions and bonding angles indicated. [From J. D. Watson (1970), *Molecular Biology of the Gene* (2nd edition). Menlo Park, Calif.: Benjamin. By permission.]

provides the basis for the first requirement of gene function—replicability. Watson and Crick proposed that a molecule of DNA replicates by "unzipping" along the weak hydrogen bonds between base pairs, whereupon each single strand then serves as a template for the synthesis of a new complementary strand, as shown in Fig. 7–11. The explanation for the specificity of A-T and G-C pairing resides in the stability and the location of the amino (NH_2) and keto (CO) configurations in these bases, and the fidelity of action of a specific enzyme, DNA polymerase.

The requirement of structural variability is met by the Watson-Crick model in the sequence of bases that characterizes one or the other strand of the molecule. For example, if the four bases are represented by their initial letters A, G, T, and C, one of the strands might begin with the base sequence -C-A-T-T-A-G-A-T-T-. . ., whereas the comparable strand in another DNA molecule might start with the sequence -T-A-C-C-A-T-A-G-G-. . ., and another with the less euphonious sequence -A-G-G-G-C-T-T-C-G-. . . . In other words, the variety that can be derived by varying the order or sequence of bases in a molecule as long as DNA can be very great indeed. Watson (1970) calculated that for a single gene with the typical length of about 1500 nucleotides, the number of permutations of the four bases is 4^{1500}, more than satisfying the requirement for gene variety.

The Watson-Crick model of DNA has been tested in several ways. It predicts a cross-sectional diameter of about 20 A. Electron micrography of single DNA molecules confirms this. The model also predicts that DNA serves as its own template. This was confirmed by the experiments of Kornberg (1960) in which DNA was synthesized *in vitro* in a system that included tree nucleoside triphosphates and "primer" DNA serving as the template. The primer DNA was varied in terms of its ratio of A-T to G-C pairs. In each experiment, the DNA synthesized had the same base ratio as the primer DNA used (Table 7–3), and like the primer it had a double-helical structure.

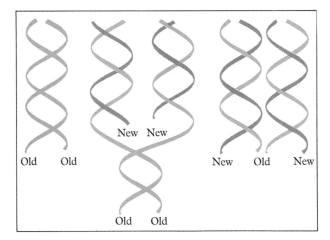

Fig. 7–11. Diagram of the mechanism by which DNA replication can take place according to the original Watson-Crick hypothesis.

Table 7–3. Relation of the base ratios of various "primer" DNA's to the base ratio of DNA derived from them by *in vitro* DNA synthesis. The A-T copolymer was a synthetic DNA composed of alternating A- and T-deoxyribonucleotides:

$$- A - T - A - T - T - T - A - - -$$
$$- T - A - T - A - T - A - T - - -$$

	Primer DNA	Product DNA	Source of primer DNA
(a) A+T:G+C	0.49	0.48	M. phlei
	0.97	1.02	E. coli
	1.25	1.29	Calf thymus
	1.92	1.90	Phage T2
	–	>40	A-T copolymer
(b) A+G:T+C	1.01	0.99	M. phlei
	0.98	1.01	E. coli
	1.05	1.02	Calf thymus
	0.98	1.02	Phage T2
	1.00	1.03	A-T copolymer

[Data from Kornberg (1960).]

A third prediction based on the Watson-Crick model is that in each replica, one strand is derived intact from the parent molecule and the other is newly synthesized. This is to say that one strand is conserved (*semiconservative replication*) in each of the new double-stranded helices formed. Experimental evidence of semiconservative replication was provided by Meselson and Stahl (1958) by labeling the DNA of *E. coli* with the heavy isotope of nitrogen (^{15}N), and then following this label through a number of DNA replications. Their procedure and interpretation of results are described in Fig. 7–12.

To conceive of a molecular structure that can serve as its own blueprint for replication, and to demonstrate that the predictions inherent in the concept are in fact realized, are quite different matters from the question of how the process of replication actually takes place. What initiates the unzipping of component strands? At which end of the molecule do unzipping and/or replication begin, or may it begin at more than one site? How can strands that are twisted about each other replicate without becoming inseparably intertwined?

At first, the answers to the questions regarding how DNA replicates seemed to be relatively easy and straightforward. Kornberg (1960) presumably had identified the enzyme (DNA polymerase) in *E. coli* which clearly functions by bonding triphosphonucleosides together in sequence to form a polynucleotide chain or strand. He found that the direction of polymerization was very precise, however, which posed a problem. Examination of the structure of the DNA segment in Figs. 7–8 and 7–13 reveals that the two strands run antiparallel to each other. That is, the phosphodiester bonds on the upper part of each sugar molecule in the left strand are at the 5' position, and those on the lower part are at the 3' position. Thus the left-hand strand might be said to have a 5'-to-3' polarity from top to bottom. The right-hand strand has the opposite polarity, reading 3' to 5' from top to bottom. The reason for the reversal is due to the molecular configuration of the component nu-

cleotides. In order to form hydrogen bonds between bases, the bases must be aligned according to this configuration. Kornberg (1969) found that DNA polymerase assembles polynucleotides in the 5'-to-3' direction only. DNA polymerase that assembles in the opposite direction has not been found.

One of the problems presented by the absence of a 3'-to-5' polymerase is that, assuming unzipping of the DNA strands in Fig. 7–13 starts from the top, only the right-hand strand could replicate until the parent strands were completely unzipped. This conclusion conflicts with observations that show simultaneous replication of both strands (Cairns and Davern, 1967). The apparent answer to this problem came when it was observed that short polynucleotide chains are recoverable during replication, which suggests that these can be inserted along corresponding sections of the parent DNA strands as the latter become sufficiently separated (Okazaki, et al., 1968; Segino and Okazaki, 1972) (Fig. 7–14).

At one time, the replication of DNA might have been as simple and straightforward as the foregoing descriptions imply. But evolution rarely preserves or even leads to simplicity—on the contrary, the more safeguards a cell has against replicative and other kinds of biochemical errors, the greater evolutionary advantage it has. All cells existing today have survived at least a billion years of natural selection. Therefore, we should not be surprised to find that today's cells have an intricately constructed and delicately tuned machinery of specific enzymes that not only minimize error, but even correct errors induced by such uncontrollable external forces as radiant energy. The picture of DNA replication that is emerging is not a simple one at all. For one thing, DNA appears to be circular in many species, and may be circular in all. When the double helix begins to replicate, one or both of the strands breaks at a point or points, the location of which is precisely regulated by a "nicking" enzyme, an *endonuclease*. As replication proceeds, the parent strands must unwind, but

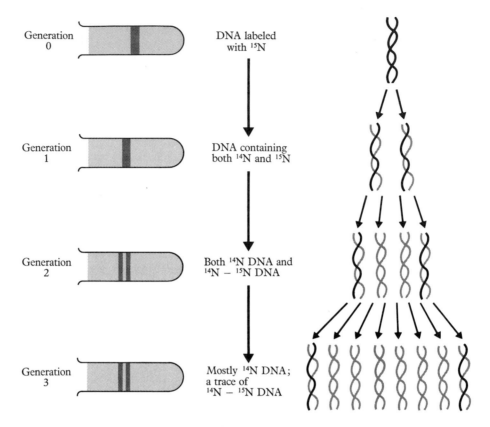

Fig. 7–12.   Interpretation of the results of the Meselson–Stahl experiments with labeled DNA in *E. coli.* Bacteria were cultured in a medium containing the heavy isotope of nitrogen ^{15}N for a sufficient number of generations to label all DNA with the isotope. DNA was extracted from the bacteria, suspended in a solution of cesium chloride, and then centrifuged in an ultracentrifuge for a prolonged period of time (up to 2 days). When spun in this manner, the cesium chloride tends to become more concentrated centrifugally, forming a gradient of concentration (and therefore of density) from one end of the tube to the other. The DNA accumulates in the zone of density that corresponds with its own density, and this position is represented as a dark band in a photograph of the preparation taken by ultraviolet light. This is because DNA absorbs ultraviolet to a much greater extent than the solution of cesium chloride does. Accordingly, the position of this band can be used as a measure of the density of the DNA molecules used.

This figure shows the position of the bands of DNA obtained from ^{15}N-labeled bacteria after each of several successive generations cultured on the normal isotope of nitrogen (^{14}N), as well as an interpretation of these results. The solid lines represent ^{15}N-rich DNA; the shaded lines represent DNA synthesized in ^{14}N medium.

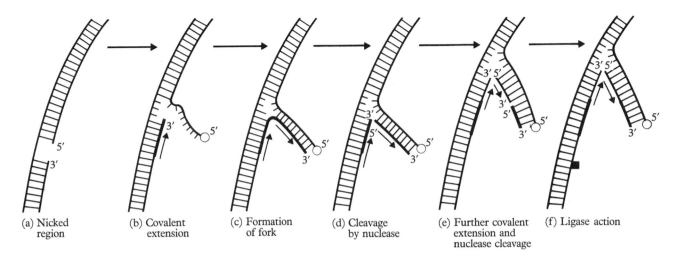

(a) Nicked region

(b) Covalent extension

(c) Formation of fork

(d) Cleavage by nuclease

(e) Further covalent extension and nuclease cleavage

(f) Ligase action

Fig. 7–13.   A proposed scheme for unidirectional replication of DNA. [From Kornberg (1969), *Science* **163**:1417. Copyright © 1969 by the American Association for the Advancement of Science. By permission.]

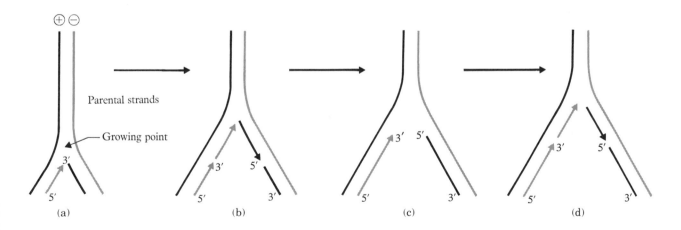

(a)

(b)

(c)

(d)

Fig. 7–14.   A scheme for DNA replication simultaneously along both strands. (a) Partial unraveling of parental helix followed by 5′-to-3′ synthesis of short fragments on both + and − parental strands. (b) Enzymatic joining of short fragments to progeny + and − strands. (c) Further unwinding of unreplicated portions of parental helix. (d) Still newer short fragments form along recently opened-up single-stranded templates.

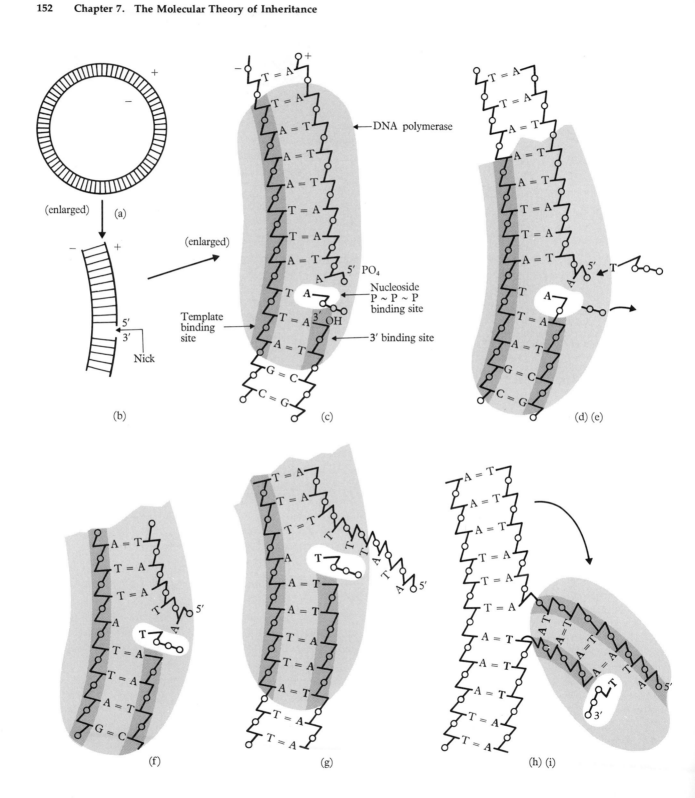

(a)

(enlarged)

(b)

Nick

(enlarged)

DNA polymerase

5' PO₄

Nucleoside
P ~ P ~ P
binding site

Template
binding
site

3' OH

3' binding site

(c)

(d) (e)

(f)

(g)

(h) (i)

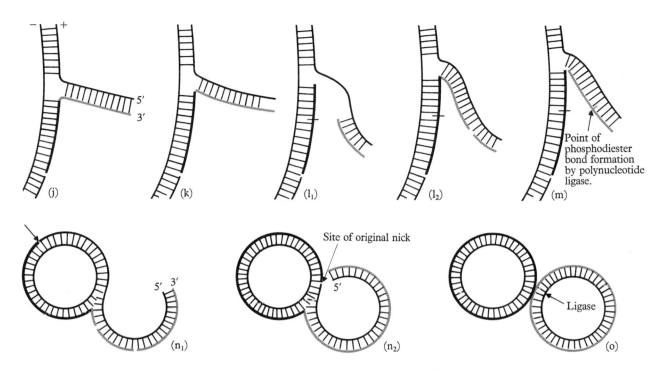

Fig. 7–15. A "rolling circle" model for DNA replication. (a) The DNA starts as a circular, double-helical molecule, one strand of which may be designated + and the other −. (b) An endonuclease breaks or "nicks" the phosphodiester bond at a specific location in one of the strands—let us say the + strand—exposing a 3'-hydroxyl group on one broken end and a 5'-phosphate on the other. (c) At the level of the nick, one molecule of DNA polymerase provides binding sites for (1) the − strand, (2) the 3' end of the broken strand, and (3) a nucleoside triphosphate. A carrier molecule that "recognizes" the proper nucleoside may be involved in this step. (d) The nucleoside triphosphate becomes linked by hydrogen bonds with the complementary base on the − strand as this base breaks its hydrogen bonds with its former partner on the + strand. (e) Simultaneously with step (d), a phosphodiester bond forms between the 3' end of the broken strand and the new nucleotide, with the release of two phosphates as inorganic pyrophosphate. (f) The new 3' end of the + strand shifts down into the 3' binding site on the DNA polymerase molecule, accompanied by a compensatory shift of the intact − strand relative to its binding site. (g) The process is repeated with successive nucleoside triphosphates being inserted and added onto the 3' end.

(h) The 5' end, which is now single-stranded, becomes associated with DNA polymerase (whether it is the same molecule already in use or another is conjectural) occupying the intact-strand (template) binding site. (i) The 3' end of the growing strand recurves sharply and binds to the 3' binding site on the new (or repositioned) DNA polymerase. (j) A polynucleotide fragment is synthesized in the 3'-to-5' direction using the unpaired + strand as a template. The newly synthesized DNA, therefore, becomes partly a copy of the − strand and partly of the + strand. (k) The phosphodiester bond between these two parts of the new DNA strand is cleaved by an endonuclease. (l) The entire process is repeated, proceeding from the new nick until two polynucleotide fragments are synthesized in position along the parental + strand. (m) The enzyme polynucleotide ligase connects adjacent ends of the fragments by a phosphodiester bond. (n) The process continues in the foregoing manner until replication has proceeded all the way around the circle back to the site of the initial nick (arrow). (o) At some point, an endonuclease cuts the phosphodiester bond between the old and new + strands, and polynucleotide ligase joins the free ends of the two double helices derived in order to establish two complete and separate molecules of circular DNA.

how this is accomplished is unknown. One of the first mechanisms proposed postulated a molecular "swivel" that would unwind the strands during replication. In recent years, a number of new enzymes and other proteins concerned with DNA synthesis have been discovered (Moses and Richardson, 1970; Salas and Green, 1970; Richardson, 1969a), but none appears to qualify as the swivel.

A replication scheme that appears to be consistent with much of what is known about DNA synthesis in prokaryotic organisms is a modification of what is known as the "rolling circle" model. Since it was first expounded by Gilbert and Dressler in 1968, this model has undergone several modifications in light of new discoveries. Because the model is largely speculative, it will doubtless undergo additional modification by the time this book is in print. The essential steps in the rolling circle model are outlined and diagrammed in Fig.

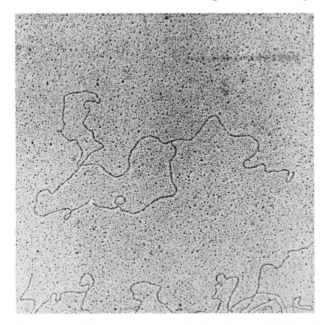

Fig. 7–16.   Electron micrograph of replicating DNA of the vegetative bacteriophage λ, showing a distinct "tail." [J. A. Kiger and R. L. Sinsheimer (1971), *Proc. Nat. Acad. Sci.* (U.S.) **68**:112–115. By permission.]

7–15. It is consistent with electron autoradiomicrographs of replicating DNA (Cairns, 1964) to the extent that it provides for the replication of both strands as they separate, and it is consistent with the unidirectional (5' to 3') nature of DNA polymerase activity (Kornberg, 1969). Electron micrographs often fail to show a loose end or "tail" during replication, as this model requires for one of the strands, yet sometimes they do (Knippers, et al., 1969; Dressler, 1970; Kiger and Sinsheim, 1971) (Fig. 7–16). There is evidence that bacterial DNA becomes attached to the cell membrane during replication (Ryter, 1968; Rosenberg and Cavalieri, 1968), and it has been postulated that it is the tail that is involved in this attachment (Kornberg, 1969). If this is so, then cell-free isolates of replicating DNA, in which an attachment would necessarily be severed as a consequence of the techniques of preparation, might secondarily attach the tail to some other part of the replicating circle. Alternatively, the membrane attachment point may conceal the postulated swivel mechanisms. Other proposals for DNA replication have been set forth (Knippers, et al., 1969; Haskell and Davern, 1969; Dressler, 1970), but they all embrace principles similar to those observed in the rolling circle model. In addition to endonucleases, ligases, and DNA polymerases, there is evidence that RNA is directly involved in at least the initiation of DNA replication (Lark, 1972). Moreover, Brewin (1972) postulated that a kind of RNA functions as an adaptor molecule that uses a triplet of internal bases as a template for the synthesis of a complementary trinucleotide of DNA. When the latter has been synthesized, it is transferred to the 5' end of the growing DNA strand, proper alignment being assured by base-pair bonding between the same internal RNA triplet that served as the template and the complementary triplet of the parent strand of DNA (Fig. 7–17).

Finally, it should be noted that the role of DNA polymerase is not as clear as it was when these models of replication were proposed. A mutant form of *E. coli* has been isolated and described

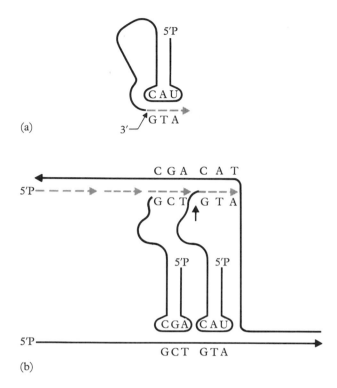

(a)

(b)

Fig. 7–17. Postulated role of an adaptor molecule of RNA in DNA synthesis. (a) A new trinucleotide of DNA is synthesized at the 3' end of the RNA adaptor molecule by base pairing with an internal template. (b) The charged adaptor then becomes positioned on the parental DNA by base pairing along one strand with coincident base pairing of the new DNA trinucleotide with the other strand. The uncharged adaptor is then released and extension of the nascent DNA chain proceeds in the 5'-to-3' direction. Newly synthesized DNA is indicated by dotted lines; double-helical regions are indicated by close parallel lines. [N. Brewin (1972), *Nature New Biology* **236**: 100. By permission.]

polymerases have been found and described in *in vitro* systems and one of them, DNA polymerase III, is concerned with DNA replication *per se* (cf. Kornberg and Gefter, 1972).

In summary, the broad outlines of the semiconservative nature of DNA replication are known. We know that it is a process involving many different enzymes and complementary base pairing. However, the details of how the various enzymes behave in relation to the DNA double helix is still a matter of active investigation.

Single-Stranded DNA

The DNA of some viruses has been found to be single-stranded. This discovery was based on the observation that their complementary bases did not occur in equimolar amounts—that is, the amount of adenine was not the same as the amount of thymine, and the amounts of guanine and cytosine were not equal either. As with other prokaryotic organisms, the DNA of these viruses exists as a circle when the organism is not actively infecting. As soon as the single-stranded DNA enters a host cell, it assumes the circular configuration and functions as a template for the synthesis of a complementary strand. It thus becomes double-stranded, but this state is maintained only during replication. While replicating, only the complementary strand serves as a template, thereby polymerizing single-stranded replicas of

(DeLucia and Cairns, 1969) which has less than one percent of the normal level of DNA polymerase, but DNA replication appears to be unaffected. Moreover, the mutant strain is highly sensitive to ultraviolet radiation. Because DNA was shown to have a marked capacity to repair itself after radiation damage, this finding implies (and subsequent work has shown) that the function of the DNA polymerase with which so much *in vitro* work had been done may be concerned with the repair of DNA rather than with DNA replication *per se*. The sensitivity of DNA to the damaging effects of ionizing radiation has been known for many years, but it is only recently that an understanding of how living cells have been able to cope with this ever-present destructive force has been achieved. A mechanism of DNA repair is perhaps as essential as the mechanism of replication itself, and certainly the two processes share several common features. Other DNA the original infecting strand. How this is regulated is not known. Possibly the viruses code the synthesis of their own DNA polymerase, which is different from that of other kinds of organisms.

RNA AND RNA VIRUSES

RNA is a polynucleotide that differs from DNA in several fundamental ways. For example, it utilizes the pentose ribose as its sugar, and the pyrimidine uracil is found in place of thymine as the complement of adenine. Additionally, in some RNA molecules a variety of "unusual" bases occurs, which are formed secondarily as chemical derivatives of uracil or guanine (cf. Chapter 8). Perhaps the most important difference between

RNA and DNA is that in most situations, RNA is single-stranded. Where double-strandedness does occur, it is usually only apparent—namely, one part of the strand has folded back and undergone base pairing with another part. The molecule known as *alanine tRNA*, which has been isolated from yeast and thoroughly characterized, illustrates this point clearly (Fig. 7–18). However, true double-stranded RNA is known to occur in certain viruses (Duesberg and Colby, 1969; Newton, 1970), as described below.

RNA, except for that of RNA viruses, is not self-replicating. It is synthesized as a single strand complementary to portions of a DNA strand, which therefore functions as its template. The RNA-synthesizing enzyme *RNA polymerase* be-

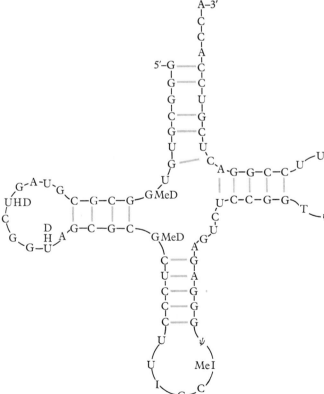

Fig. 7–18. The cloverleaf structure of alanine tRNA. The "unusual" bases are formed after synthesis of the molecule by transformation of usual bases (e.g., by methylation). Those shown here are dihydrouridine (DHU), dimethylguanosine (DMeG), methylinosine (MeI), pseudouridine (ψ), and inosine (I).

Anticodon

comes attached to the DNA molecule; the hydrogen bonds between complementary DNA strands then presumably separate, and ribonucleoside triphosphates are assembled in a sequence that is complementary to one of the DNA strands. As the RNA molecule polymerizes, the hydrogen bonds between the complementary strands of DNA reform, so that the DNA strands are never separated for any great length. When the RNA molecule is completed or "transcribed," it moves away and the RNA polymerase detaches from the DNA (Fig. 7–19).

Evidence that RNA is polymerized in association with DNA is derived from two different experimental approaches—base labeling and determination of base composition. If a cell that is synthesizing RNA is administered radioactively labeled uracil, the label is taken up by the nucleus, just as labeled thymidine is during DNA synthesis. However, uracil is not a normal component of DNA. If the label is followed, it is found to leave

the nucleus and migrate to sites in which RNA is known to concentrate, such as the nucleolus, ribosomes, and endoplasmic reticulum. The second approach involves synthesizing RNA *in vitro*. This can be done by using the enzyme *RNA phosphorylase* and the diphosphate form of the four nucleotides, but the bases will have random distribution along the RNA molecule. In order to synthesize RNA with a consistent and predictable sequence of the four bases, DNA must be used together with RNA polymerase and the triphosphate form of the nucleotides. When this is done, the four bases in the recovered RNA correspond in amount and distribution to the complementary base sequence in one of the strands of the primer DNA.

Perhaps the most convincing evidence of the DNA-RNA relationship is provided by DNA/RNA hybridization studies. If cell-free preparations of DNA are heated to a temperature just below 100°C, the hydrogen bonds between complementary strands are broken and the two strands separate. If the preparation is slowly cooled, complementary strands "find" each other and the double-helical condition is restored. However, if RNA is added to the preparation while the DNA strands are still separated, and the mixture is then slowly cooled, double-stranded DNA/RNA hybrid molecules may form and be recovered (Fig. 7–20). If the RNA used has been enzymatically synthesized with DNA that is homologous to that used for hybridization, a relatively large yield of DNA/RNA hybrid molecules will be recovered

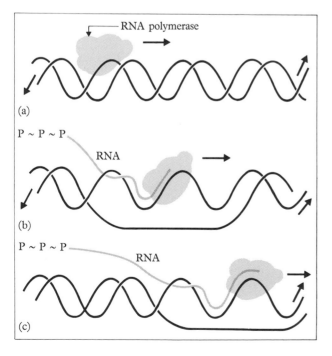

(a)

(b)

(c)

Fig. 7–19. A postulated scheme of RNA transcription. (a) RNA polymerase attaches to a short section of one strand of DNA, separating it from its complementary strand. (b) Ribonucleoside triphosphate precursors undergo base pairing with the free bases on the RNA polymerase-bound strand of DNA, followed by phosphodiester bonding to form a strand of RNA. (c) As the RNA polymerase moves along the DNA, the strands reestablish their hydrogen bonds and the RNA strand "peels off."

(Table 7–4). This procedure has proved to be very useful in assessing the degree of relationship between cells and between species, and it is now being used as a means of mapping chromosomes and of isolating and purifying those genes that code for ribosomal RNA (Brown, et al., 1972; Wimber and Steffensen, 1970). Hall and Spiegelman (1961) observed sequence complementarity between the DNA of T2 phage and T2-specific RNA. They found that no such hybrid formation was obtainable with other DNA even if it had the same overall base composition as T2-DNA.

Table 7–4. Hybridization of RNA with template and nontemplate DNA. This table shows that RNA hybridizes readily with homologous DNA but rarely with nonhomologous DNA.

DNA template for RNA polymerase reaction	Percent of RNA hybridized with:		
	T7 DNA	λ DNA	Blank filter
T7	50	0.2	0.3
λ	0.3	49	0.4

[Courtesy of J. P. Richardson (1969), *Prog. Nuc. Acid Res. Mol. Biol.* **9**:85.]

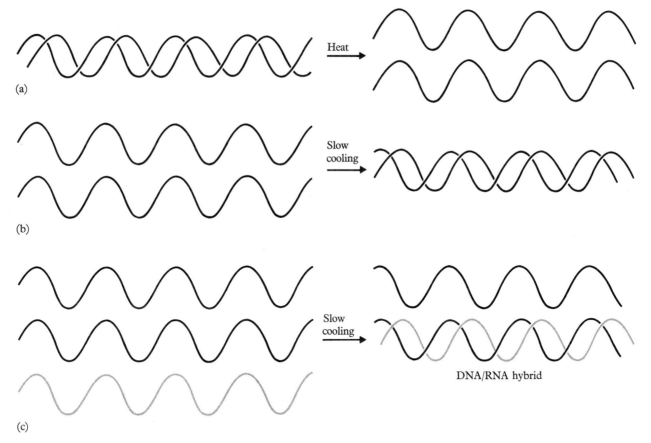

Fig. 7–20. The technique of DNA/RNA hybridization. (a) The complementary strands of DNA separate when heated in a cell-free preparation. (b) They rejoin when slowly cooled. (c) If RNA (shaded lines) is introduced while the DNA strands are separated, DNA/RNA hybrids may appear upon cooling. These hybrids can be separated from nonhybridized DNA and RNA by density-gradient centrifugation or by selective filtration.

It was mentioned above that an apparent exception to the generalization that RNA is not self-replicating occurs in the RNA viruses. A number of RNA viruses have been described, and they include some of the simplest and smallest viruses known. Among the best known are the tobacco mosaic virus (TMV), the viruses of influenza, mumps, yellow fever, poliomyelitis, and several bacterial viruses (F2, R17, Q). Certain other RNA viruses, such as the Gross leukemia virus and the Rous sarcoma virus, have been implicated in the etiology of cancer, and still others are suspect. In most of these viruses the RNA is single-stranded, but in one group—the so-called *Reoviruses*—the RNA is double-stranded (Rhodes and van Rooyen, 1968; Newton, 1970; Watanabe and Graham, 1968).

The RNA viruses consist of a molecule of RNA enclosed in a protein shell. Unlike the shell of many DNA viruses, this shell may consist of only one kind of protein (Fig. 7–21). The virus particle may have one or two additional proteins, however. For example, the RNA bacteriophages have a molecule of A protein, which facilitates attachment of the virus to the bacterial cell wall.

That RNA is the genetic molecule of the RNA viruses was clearly demonstrated by Fraenkel-Conrat and Williams in 1955. They chemically dissociated the RNA and the protein of TMV and attempted to infect tobacco plants with the two fractions separately. The protein proved to be inactive, but the RNA did infect and new virus particles were produced, complete with protein shells. Thus the RNA was shown to be the only chemical link between generations. In another series of experiments, Fraenkel-Conrat (1956) assembled a "hybrid" TMV by combining the RNA of one strain with the protein coat of another. The initial strains differed from each other in their serologically distinguishable proteins. The progeny virus obtained by infecting plants with the "hybrid" had

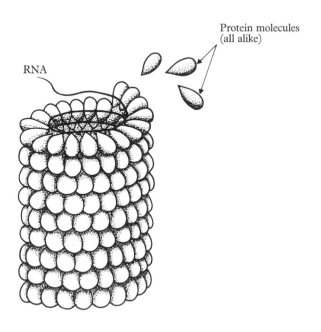

Fig. 7–21. The structure of the tobacco mosaic virus (TMV). [Inset photograph courtesy of Carl Zeiss, Inc.]

the serological characteristics of the strain that supplied the RNA.

The RNA of RNA viruses has a very limited number of "genes." The RNA bacteriophages, for example, have typically only four or five—one is the *A* segment for coding the synthesis of the A protein, another is the *CP* segment for the coat protein, and a third is the *SYN* segment for RNA synthetase (Fig. 7–22). Other genes are presumed to exist on the basis of the number of nucleotides present in the RNA molecule, but their function has not yet been determined. On infection, the viral RNA invades the host cell and, in the case of single-stranded RNA, immediately attaches to the host's ribosomes. Utilizing the host cell's enzymes and amino acids, the viral RNA directs the synthesis of A protein, CP protein, and RNA synthetase. The RNA synthetase then binds to the viral RNA after it is released from the ribosomes, and replication occurs, producing double-stranded RNA. The complementary strand thus produced, the − strand, functions as a template for the replication of the infecting strand, the + strand. These replicas then engage in further protein synthesis and replication until thousands of new particles are formed. Each particle consists of a + strand of RNA and protein and, when released from the host cell, is capable of initiating a new infection. Double-stranded RNA viruses undergo the same sequence of events except that, already having the + strand, they can start replicating sooner.

Still another mechanism has been described in certain tumor viruses—Rauscher mouse leukemia virus and the Rous sarcoma virus (Baltimore, 1970; Temin and Mizutani, 1970). In these cases, the infecting virus introduces single-stranded RNA and DNA polymerase into the host cell, whereupon specific provirus DNA is synthesized with the viral RNA as the template. The DNA polymerase that does this in these cases has been found to be specific for this kind of "reverse transcription" and has been isolated and purified (Ross, et al., 1971). From the DNA thus synthesized, new viral RNA is polymerized in turn for inclusion in new virus particles. Hence in effect, the genome of these viruses appears to be RNA outside the host cell and DNA inside the host cell.

The discovery of RNA-dependent DNA synthesis (reverse transcription) has opened new and exciting avenues of investigation, for the principle involved has provided insight into the actions of oncogenic (cancer-producing) viruses and has suggested a mechanism of specific gene amplification in normal differentiation and development (Temin, 1971) (cf. Fig. 11–2).

Although RNA can constitute the genetic material in the RNA viruses, its role in all other cases known occupies an intermediate position between DNA and gene expression. In this role, it appears in several forms which differ in molecular size, molecular configuration, and function. The principal kinds of RNA are *messenger RNA (mRNA), transfer RNA (tRNA),* and *ribosomal RNA (rRNA).* The structure and function of each of these forms of RNA are described in Chapter 8.

Fig. 7–22.   A genetic map of the RNA virus R17. The number of nucleotides are shown, starting from the 5′ end of the molecule. [From J. D. Watson (1970), *Molecular Biology of the Gene* (2nd edition). Menlo Park, Calif.: Benjamin. By permission.]

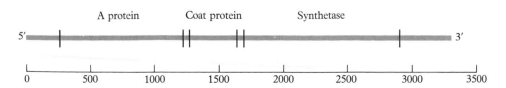

QUESTIONS

1. Define:
   Nucleoprotein   Pyrimidine
   Nucleic acid   Pentose
   Nucleotide   Phosphate group
   Purine

2. Why were proteins originally thought to be the genetic material?

3. During the search for the chemical identity of the gene, it was assumed that whatever the substance might be, certain criteria must be met. These criteria were (1) a relatively high degree of molecular stability, (2) an ability to mutate or change on occasion, (3) a capacity for self-replication, (4) the possibility for wide structural variation, and (5) transmissibility to the next generation. How did the data and conclusions of the experiments performed by each of the following scientists relate to the criteria necessary for identifying the hereditary material?

   a) Griffith
   b) Avery, MacLeod, and McCarty
   c) Hershey and Chase
   d) Meselson and Stahl
   e) Chargaff
   f) Watson and Crick

4. In which ways do temperate phages differ from virulent ones?

5. Why is base-pair complementation basic to the concept of DNA as the hereditary material?

6. Why cannot DNA polymerase be the sole enzymatic agent for DNA replication?

7. Outline the differences in structure between DNA and RNA.

8. What evidence points to the fact that RNA is synthesized on a DNA template? Why is it self-replicating only in certain, exceptional cases?

9. The major chemical variations in RNA structure greatly exceed those observed for DNA. Such variations include (1) the ability to incorporate exceptional bases, (2) the ability to fold back into loops, thus resulting in a partial state of double-strandedness, and (3) the capacity to act as an intermediary molecule in most living things, but still retain the ability to function as hereditary material in certain viruses. Why do you think that evolution has led to such variability in structure and function of RNA, but has not tolerated such differences in DNA?

8

Gene Function: Proteins and Enzymes

Metabolic processes, those chemical processes that characterize living matter, are almost always mediated by catalytic substances called *enzymes.* Enzymes have been shown in all instances to be protein, and their outstanding characteristic—specificity—is directly attributable to their protein structure. The basis of enzyme-substrate specificity has been likened to a lock-and-key relationship (Fig. 8–1), which has proved to be a very apt analogy. The substrate or substrates with which an enzyme will form a complex—a necessary condition of enzyme action—depends upon the shape of the enzyme molecule and upon both the positions and kinds of atomic configurations in the enzyme that facilitate such combining forces as hydrogen bonds, van der Waal bonds, hydrophobic bonds (Atlas, et al., 1970) and, in some cases, even covalent bonds to form between enzyme and substrate (Fig. 8–2). These points of bonding are called the *active sites* of the enzyme. Some enzymes require a non-protein component (prosthetic group) to provide one or more groupings for an active site. This requirement may be served by something as simple as a divalent cation, such as Ca^{++}, Co^{++}, Mg^{++}, Mn^{++}, or Zn^{++}, or by something highly complex such as nicotinamide adenine dinucleotide (NAD) (Fig. 8–3).

Enzymes work by binding substrate in such a way that a chemical reaction is facilitated that otherwise might either be highly improbable, require a high activation energy, or proceed at a slow rate. They do this in a variety of ways depending upon what substances and what kinds of reactions are involved, but these ways all amount to lowering the energy of activation between those substances. Figure 8–4 schematically illustrates the relationship between an enzyme and the reaction it catalyzes.

The complex formed between enzyme and substrate is of amazingly short duration, in some cases lasting less than one ten-thousandth of a second. It is precisely because of its short duration

163

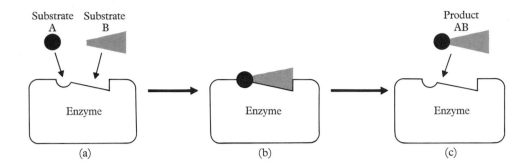

$$\text{Enzyme} \quad + \quad \text{Substrate} \longrightarrow \text{Acyl-enzyme} + \text{Byproduct} \\ \text{intermediate}$$

◀ **Fig. 8–1.**   The lock-and-key model of enzyme action is based on the complementary shapes of substrate and enzyme in the region of the enzyme's active site (a), which juxtaposes the substrate components (b), and thereby facilitates chemical bonding and the formation of the product (c).

◀ **Fig. 8–2.**   During the formation of an enzyme-substrate complex, a strong, though temporary, covalent bond may be formed to hold the union together as shown between the oxygen atom of the enzyme and the carbon atom of the substrate. For other enzyme systems, the bond may be a weaker one such as a hydrogen bond.

◀ **Fig. 8–3.**   Nicotinamide adenine dinucleotide (NAD) functions as a co-enzyme by being able to accept hydrogen atoms, thus converting it to $NADH_2$. $NADH_2$ can then be used by the cell as a hydrogen donor. In this manner, NAD is reconstituted. [After J. D. Watson (1970), *Molecular Biology of the Gene* (2nd edition). Menlo Park, Calif.: Benjamin, p. 44.]

tion. (b) In this series of diagrams, the lip of the shelf is an "enzyme"—a pivot with a magnet embedded in one end and a weak spring mounted on the other. (1) The steel ball, the "substrate," is at rest in starting position A. (2) The ball is attracted by the magnet to position B. (3) The weight of the ball in position B imposes an imbalance of forces in the system, so that the "enzyme" tips down, giving potential energy to the spring by compressing it. (4) The weight of the ball is greater than the force of the magnet, so the ball falls off (the "reaction") and the spring restores the "enzyme" to its initial position. The specificity of this "enzyme" is relatively low, because it will react in this way with any ferrous ball with a mass that is low enough so that its inertia is less than the force of the magnet within the distance between positions A and B, but great enough so that the force of gravity operating on it will exceed the magnetic force, allowing the ball to fall.

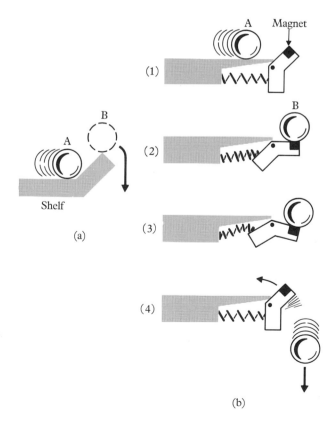

(a)

(b)

Fig. 8–4.   Schematic representation between enzyme, substrate, and activation energy. (a) In order for a steel ball in position A to fall off the shelf, it must be raised to position B. The energy required to do this represents the "activation" energy and can be calculated as the product of the mass of the ball and the vertical distance between positions A and B. Once the ball drops over the lip of the shelf by being raised to position B, it will fall and its potential energy will be transformed to kinetic energy as it falls to the floor. The falling is analogous to the reac-

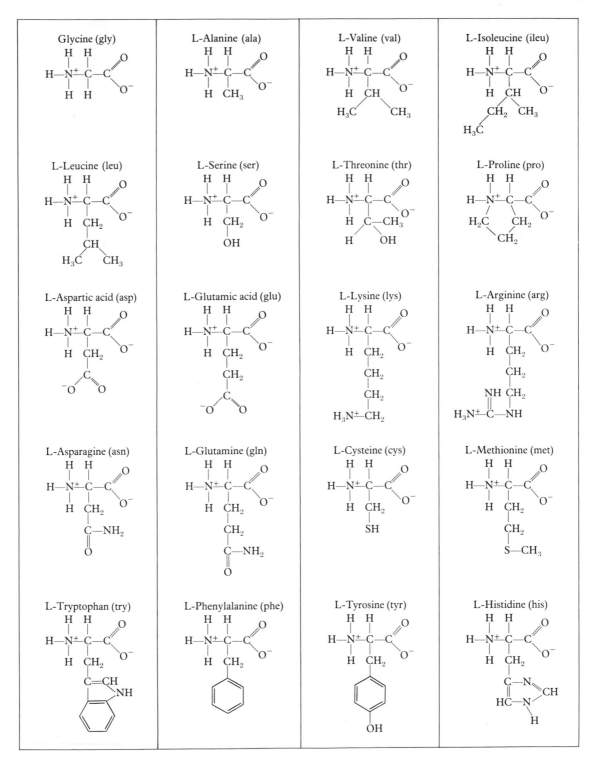

Fig. 8–5. The twenty common α-amino acids found in native proteins. The characteristic amino group (NH_2) and carboxyl group (COOH) are shown in their ionized form.

that the possibility of covalent bonds was rejected until the evidence for their existence in some instances became overwhelming (Doonan, et al., 1970). The rate at which enzyme-substrate complexes are formed and separated can be markedly depressed or even reduced to zero by anything that interferes with the position and/or function of the enzyme's active sites. An active site may be a prosthetic group, as mentioned above; it may be a specific sequence of amino acids in the enzyme molecule, such as the sequence Asp-Ser-Gly (cf. Fig. 8–5) in trypsin, chymotrypsin, and thrombin (Table 8–1); or it may be the sulfydril (SH) group on a cysteine residue. It becomes apparent, therefore, that altering the position of certain critical amino acids, or substituting others for them, can have significant consequences. This being the case, the detailed structure of the entire enzyme molecule becomes important. Therefore, the basic features of protein structure are reviewed below.

PROTEIN STRUCTURE

Proteins found in nature are heteropolymers of the twenty amino acids represented in Fig. 8–5. These are all α-amino acids, which means that the characteristic amino (NH_2) group is on the same carbon atom as the equally characteristic carboxyl (COOH) group. The α-carbon atom also has a hydrogen atom occupying one of its valences—the various amino acids differ from one another in terms of what atom or groups of atoms are bonded to the one remaining valence of the α-carbon. The twenty amino acids, therefore, all have the same atomic configuration at one end:

$$
\begin{array}{c}
\quad H \quad H \quad O \\
\quad | \quad\ | \quad\ \| \\
H-N-C-C \\
\quad\quad | \quad | \\
\quad\quad R \quad O-H
\end{array}
$$

where R represents the variable side chain. Both the NH_2 and the COOH can dissociate to become reactive, and two amino acids can combine with the loss of one molecule of water between them:

$$
\begin{array}{c}
H \quad H \quad O \qquad\qquad H \quad H \quad O \\
| \quad\ | \quad\ \| \qquad\qquad\ | \quad\ | \quad\ \| \\
H-N-C-C \quad + H-N-C-C \quad \rightarrow \\
\quad\quad | \quad | \qquad\qquad\qquad | \quad | \\
\quad\quad R \quad O-H \qquad\qquad R \quad O-H
\end{array}
$$

$$
\begin{array}{c}
H \quad H \quad O \qquad H \quad O \\
| \quad\ | \quad\ \| \qquad | \quad\ \| \\
H-N-C-C-N-C-C \qquad + \quad H_2O \\
\quad\quad | \qquad\ | \quad | \quad | \\
\quad\quad R \qquad H \quad R \quad O-H
\end{array}
$$

Table 8–1. The amino-acid sequence in the active site of selected enzymes that have serine as the central amino acid.

Enzyme	Sequence	Source
Chymotrypsin	-Gly-Asp-SerP-Gly	Turba and Gundlach (1955); Schaffer, et al. (1957)
Trypsin	-Gly-Asp-SerP-Gly	Schaffer, et al. (1958); Oosterbaan, et al. (1956); Dixon, et al. (1958)
Elastase	-Gly-Asp-SerP-Gly-	Naughton, et al. (1960)
Thrombin	-Gly-Asp-SerP-Gly-	Gladner and Laki (1958)
Subtilisin	-Gly-Thr-SerP-Met-	Sanger and Shaw (1960)
Aliesterase	-Gly-Glu-SerP-Ala-	Cohen, et al. (1955)
Pseudocholinesterase	-Gly-Glu-SerP-Ala-	Cohen, et al. (1955)
Phosphoglucomutase	-Thr-Ala-SerP-His-Asp-	Milstein and Sanger (1960)

[Adapted from Oosterbaan and Cohen (1964), *Structure and Activity of Enzymes.* New York: Academic Press.]

and form a *dipeptide*. The link

$$-\overset{\overset{\displaystyle O}{\|}}{C}-\underset{\underset{\displaystyle H}{|}}{N}-$$

is called an *amide linkage*, and the covalent bond between the C and the N is called a *peptide bond*. The dipeptide shown still has an amino group at one end and a carboxyl group at the other, and can form additional peptide bonds with other amino acids. Thus the amino acids lend themselves to infinite polymerization—that is, to forming chains that, no matter how long, still have an amino group at one end and a carboxyl group at the other. A chain of many amino acids linked by peptide bonds constitutes a *polypeptide*, which may attain a molecular weight as high as 100,000. Chains larger than this, or molecules comprised of two or more polypeptide subunits, are known as *proteins.*

The number and sequence of amino acids that constitute a protein represents that protein's *primary structure*, which is essentially a chain. The disposition in space of the atoms that make up the ''backbone'' of the chain,

$$-\overset{\overset{\displaystyle H}{|}}{\underset{\underset{\displaystyle R}{|}}{C}}-\overset{\overset{\displaystyle O}{\|}}{C}-\overset{\overset{\displaystyle H}{|}}{\underset{\underset{\displaystyle H}{|}}{N}}-\overset{\overset{\displaystyle H}{|}}{\underset{\underset{\displaystyle R}{|}}{C}}-\overset{\overset{\displaystyle O}{\|}}{C}-\overset{\overset{\displaystyle H}{|}}{\underset{\underset{\displaystyle H}{|}}{N}}-\overset{\overset{\displaystyle H}{|}}{\underset{\underset{\displaystyle R}{|}}{C}}-\overset{\overset{\displaystyle O}{\|}}{C}-\overset{\overset{\displaystyle H}{|}}{N}-,$$

is such that this backbone acquires a twist or helical configuration that imparts a precise geometry known as the *secondary structure* of the protein. As a consequence of both primary and secondary structure, the side chains of the component amino acids project out from the helix in certain ways, and some of these ways result in interaction between two or more side chains. Specifically, a covalent interaction will occur spontaneously if the SH groups of two cysteine residues are at the right proximity, forming a strong disulfide bond (Fig. 8–6). This puts a permanent kink or fold in the

chain, and leads to other kinds of bonding between other side chains. Hydrophobic bonds, for example, form between the side chains of the so-called *nonpolar amino acids* (Fig. 8–5), and hydrogen bonds and ionic bonds can form between others. As a consequence, the protein molecule can become highly contorted and folded, a state that is known as its *tertiary structure*. The tertiary structure of a given protein is a constant charac-

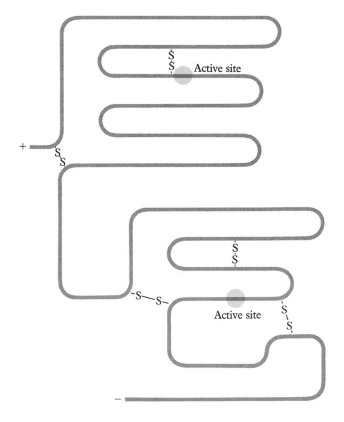

Fig. 8–6.   The tertiary structure of the protein chymotrypsinogen is caused by the formation of disulfide bonds between sulfur-containing amino acids. [After Watson (1970), p. 173.]

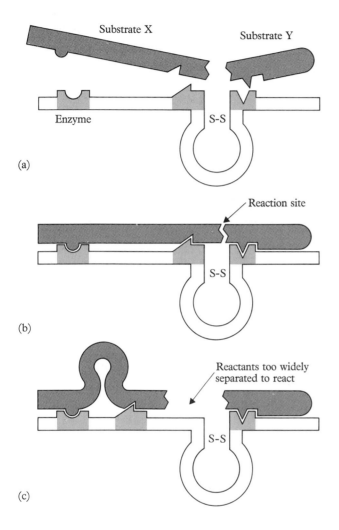

Substrate X

Substrate Y

Enzyme

S-S

(a)

Reaction site

S-S

(b)

Reactants too widely
separated to react

S-S

(c)

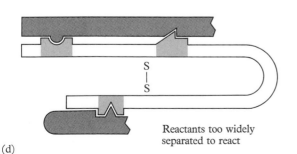

S
|
S

Reactants too widely
separated to react

(d)

teristic, for it is a necessary consequence of that protein's primary structure. But important to the present discussion is the fact that *the positions of the active sites of an enzyme are determined not only by the enzyme's primary structure, but also by its tertiary structure.* Obviously, if an active site on an enzyme centers around the amino acid histidine, and the position of histidine in the enzyme's primary structure is altered, this active site will appear in a different position in the tertiary structure as well. On the other hand, if the histidine position is un-altered in the enzyme's primary structure, but the tertiary structure is changed by a shift in the posi-tion of, say, cysteine, then the position of histidine relative to the enzyme's geometry is again affected (Fig. 8–7).

In substance then, if an enzyme is to function properly, the amino acids that comprise it must be of the right kind assembled in the right order. If we accept the propositions that (1) the differences and similarities between individuals and between species are basically attributable to differences and similarities in the structure of enzymes and other proteins, (2) these differences and similarities are largely inherited, and (3) DNA is the only signifi-cant and consistent link between generations, then it appears reasonable to conclude that DNA is im-

Fig. 8–7. Tertiary structure and its implications on en-zyme function. (a) A hypothetical enzyme and hypothet-ical substrates. Active sites are shaded. (b) The enzyme-substrate complex, showing how the active sites both position the substrate molecules and facilitate their interaction. (c) The consequence of changing the position of the active site relative to the amino-acid se-quence (primary structure). (d) The consequence of changing the tertiary structure of the enzyme. Although the active sites are still in their normal position relative to the amino-acid sequence, their position in space rela-tive to the contours of the enzyme molecule has shifted.

plicated in the regulation of protein (enzyme) structure. Moreover, the effect of regulation is the precise ordering of the amino acids that constitute the primary structure of those enzymes. The observation that the primary structure of DNA also involves a precise ordering of its subunits, the nucleotides (which differ by virtue of their four different bases), suggests a correspondence between base sequence in DNA and amino-acid sequence in protein (Crick, 1958). This correspondence was later confirmed by the colinearity studies of Yanofsky, et al. (1964).

Long before the Watson-Crick theory of DNA structure and the dramatic and immediate insights that it inspired, however, some indications of a very direct relationship between genes and enzymes had been observed. As mentioned in Chapter 1, Garrod (1909) showed that the heritable disease alkaptonuria is attributable to an enzyme deficiency. But it was not until 1941, with the brilliant studies of Beadle and Tatum, that any serious and productive steps were taken to correlate genes and biochemical functions. Although there was no hint of the nature of the correlation, for the structure of DNA had not yet been proposed, they conceived that for every enzyme there is a corresponding gene, expressed tersely by the dictum, "one gene—one enzyme." In 1944, Srb and Horowitz placed this postulate on a firm base by their investigations of the genetic control of arginine synthesis in *Neurospora crassa*.

Neurospora Culture and Life Cycle. The breadmold *Neurospora* is an ascomycete—that is, a fungus that produces its reproductive spores in little containers called asci (singular: ascus). The asci of *Neurospora* are narrow cylinders with the same cross-sectional diameter as the zygotes they contain, and each ascus has only one zygote. The zygote forms by the fusion of two haploid cells produced by opposite mating types. The diploid zygote undergoes two meiotic divisions and one mitotic division, thereby producing eight haploid cells called ascospores. In due course the ascus bursts,

liberating the ascospores, which then give rise to new individual colonies of mycelia. Because of the geometry of the cylindrical ascus, the eight ascospores lie in a single row in an order that directly reflects the sequence of each cell division that produced them. The generation of "ordered" division products such as these has proved to be of enormous value in analyzing segregation, linkage, and crossing-over, as discussed in Chapter 4.

Beadle and Tatum (1941) developed techniques of germinating *Neurospora* spores and growing them on agar medium in culture tubes, a modification of the technique described earlier for bacterial culture. They found that wild-type *Neurospora* will grow on a "minimum" medium of inorganic salts, sucrose, and the plant growth factor biotin. They induced mutations in the sex cells of conidia by means of X-rays, then crossed opposite mating types, and finally cultured the resulting ascospores on "complete" medium, which includes malt and yeast extract in addition to the simple nutrients in minimum medium. This was done in order to build up a growth of mycelia from each ascospore that was sufficiently large to test by subculturing on a variety of culture media. Subculturing was done in two steps (Fig. 8–8). (a) The first step was to transfer a sample of mycelia to minimum medium. If the cells showed essentially normal growth, it was concluded that the parent ascospore was free of any nutritional mutations. If the cells either did not grow or showed highly abnormal growth on minimum medium, it was concluded that the ascospore was a nutritional mutant and was so identified. Then the second step in subculture was taken. (b) In the second step, samples of mycelia from the mutant culture growing on complete medium were transferred to a large variety of culture tubes, each of which contained minimum medium plus one other nutritional substance which differed in each tube (Fig. 8–8). For example, one tube contained minimum medium plus the vitamin thiamine; another tube contained minimum medium plus the amino acid methionine, and so on. One-by-one the various extra nutri-

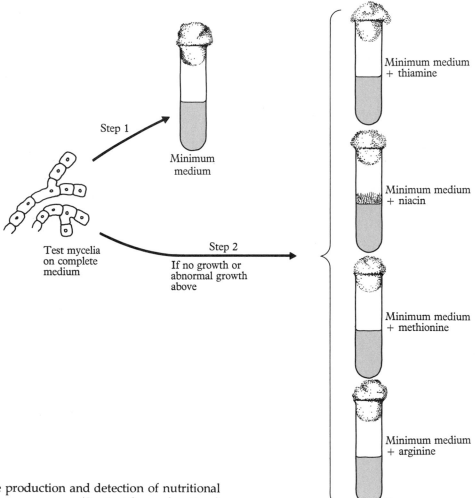

Fig. 8–8. The production and detection of nutritional mutants in *Neurospora crassa*. (See text for explanation.)

tional components contained in the yeast and malt extracts used for complete medium were tested. Figure 8–8 shows that growth has occurred in the culture supplemented with niacin, thus identifying the induced mutation in this hypothetical example.

Making use of these techniques, Srb and Horowitz isolated seven mutant strains that would grow on minimum medium, provided that the medium had been fortified with the amino acid arginine. Of the seven mutants, they found that six could also be grown on minimum medium that was fortified with a related substance, citrulline, and that four of these would accept as a substitute another related substance, ornithine. One of the seven would accept no substitute. The substitutes that Srb and Horowitz used were known to be metabolic precursors of arginine (which is why they used them), and their results enabled them to reconstruct a small segment of the metabolic pathway for the biosynthesis of arginine (Fig. 8–9). Having shown by other means (described in Chapter 9) that the seven mutant strains in question involved seven different genes, they were able to predict on the basis of the one gene—one enzyme theory that there are at least seven different enzymatic conversions involved in the arginine synthesis pathway. This prediction has since been borne out, as shown in Fig. 8–10, which identifies the seven intermediate metabolites in the pathway

between glutamic acid and arginine (Fincham and Day, 1965).

This kind of genetic analysis has made it possible to work out the metabolic pathways in many biosyntheses. But more important than this in the present context, such analyses have consistently confirmed the direct relationship between genes and enzymes. A readily testable prediction of the one gene—one enzyme theory remains, however —if a mutant gene expresses itself as a functional deficiency of an enzyme, then the cell should show some tendency to accumulate the substrate(s) of that enzyme. More specifically, if the enzymatic reaction in question represents a step in a metabolic pathway, and this step for some reason (mutation) is blocked, then there should be an accumulation of the metabolite produced by the preceding step. This prediction has been verified in many different in-

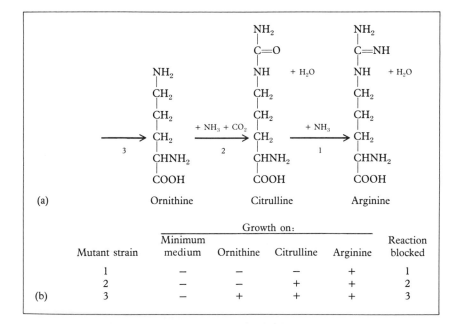

Fig. 8–9. (a) A segment of the pathway of arginine biosynthesis in *Neurospora crassa*. (b) Data that led to the deduction of the pathway.

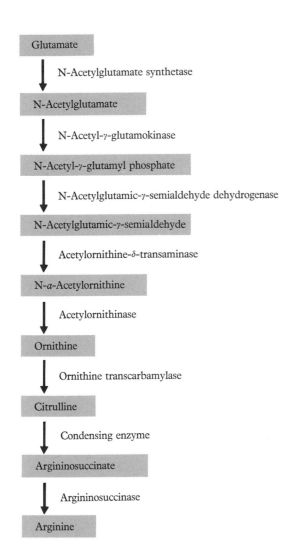

Fig. 8–10. Pathway of arginine biosynthesis in *E. coli.*

involved and other genetic deficiencies associated with it are presented in Fig. 8–11.

The argument developed above clearly involves DNA in the determination of protein structure. In the case of mutations expressed as deficiencies in enzyme function, how is one to know what the nature of the deficiency is? Is it that the enzymes are improperly formed in terms of primary structure; is it that the enzymes are not formed at all; or is it something else? Answers to these questions should provide some insight into how the genes themselves function.

Sickle-Cell Anemia.   Native populations in certain areas of Africa are characterized by a high incidence of the disease called sickle-cell anemia. Red blood cells of people afflicted with this disease spontaneously change their shape from biconcave discs to half-moon or sickle shape, as shown in Fig. 8–12. The disease is inherited as a simple Mendelian recessive. The blood of individuals who are heterozygous for the gene does not undergo this changed morphology spontaneously, but can be induced to do so by reduced oxygen concentration or by certain chemical reducing agents. They are thereby identified as carrying the sickle-cell "trait." Because the heterozygotes have a certain selective advantage, as discussed in Chapter 13, the gene has a fairly high frequency in these African populations.

Pauling, et al. (1949) compared the hemoglobins of people with sickle-cell anemia and normal people using electrophoresis and other methods. Electrophoresis indicates the relative net charge of large molecules by measuring the rate and direction of their migration in an electric gradient. The results obtained by Pauling and his colleagues (Fig. 8–13) showed a clear difference in electrophoretic response; they also revealed that individuals heterozygous for the sickle-cell gene synthesize both types of hemoglobin. Ingram (1956) devised a technique, which he called "fingerprinting," by which he was able to distinguish

stances. One confirmation described by Garrod in 1909 is the case of alkaptonuria, which is a clinical manifestation of accumulation of the metabolite homogentisic acid. This causes the urine to turn dark when voided, which is harmless. The pathway

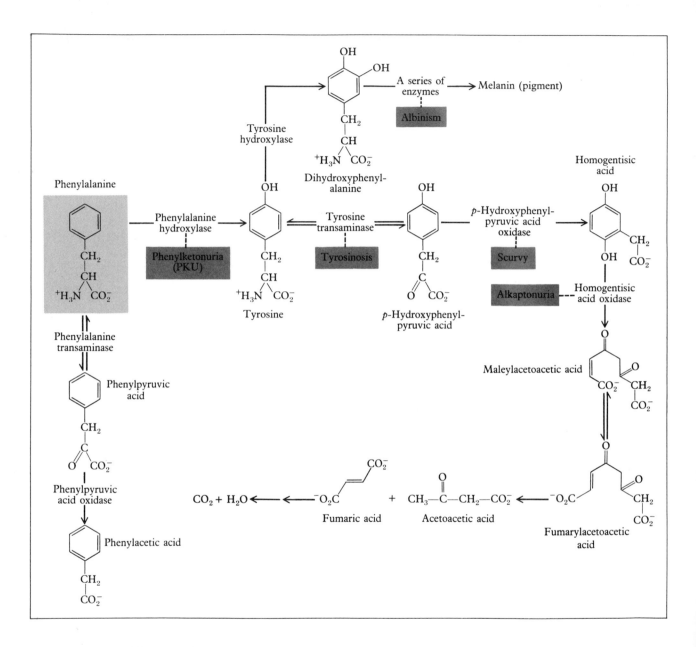

Fig. 8–11.   Phenylalanine and tyrosine metabolism in human beings. Different mutations result in a variety of enzyme blocks that are expressed as diseases such as phenylketonuria (PKU), albinism, and alkaptonuria. The critical enzymes, coded by normal genes, are indicated. The defect that results from a deficiency of each enzyme is labeled within the shaded box.

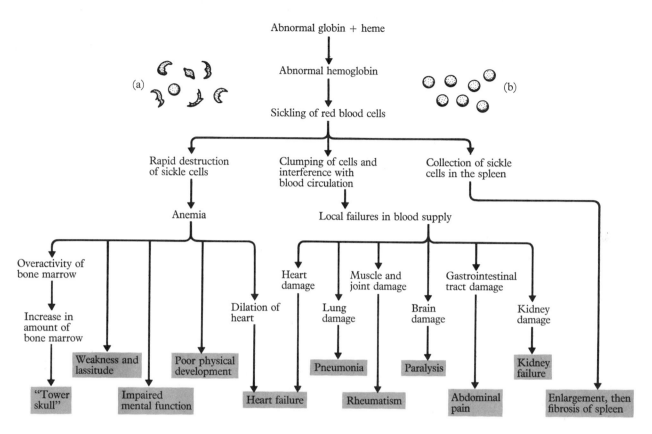

Fig. 8–12. The manifold (pleiotropic) effects of sickle-cell anemia. This disease is caused by a single amino-acid substitution in hemoglobin. Inset: The microscopic appearance of sickled erythrocytes (a) as compared to that of normal erythrocytes (b). [After J. V. Neel and W. J. Schull (1954), *Human Heredity.* University of Chicago Press. By permission.]

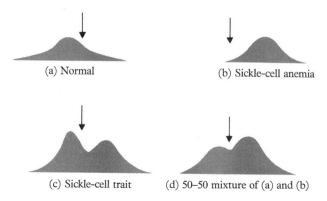

(a) Normal

(b) Sickle-cell anemia

(c) Sickle-cell trait

(d) 50–50 mixture of (a) and (b)

Fig. 8–13. Comparison between the electrophoretic patterns of hemoglobin from (a) normal individuals, (b) those with sickle-cell anemia, and (c) those with sickle-cell trait. The arrows indicate the initial position of hemoglobin before the application of electric current. Normal hemoglobin can be seen to migrate in a direction opposite to that of sickle-cell hemoglobin. [From L. Pauling, et al. (1949), *Science* **110:**545. By permission.]

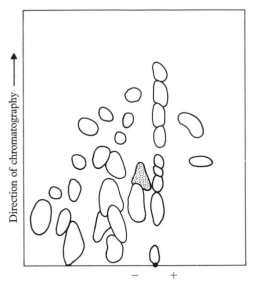

Sickle-cell hemoglobin

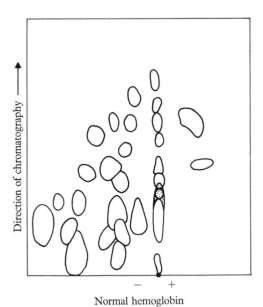

Normal hemoglobin

Fig. 8–14. The "fingerprints" of normal and sickle-cell hemoglobin. Each spot represents the accumulation of one kind of peptide. Note the single difference (stippled area) between the hemoglobins. [From V. M. Ingram (1956), *Nature* **178**:792–794. By permission.]

Fig. 8–15.   Various hemoglobin variants differ from the ▶ normal molecule in a single amino-acid location. Arrows identify the amino-acid substitution. The changes in charge indicated in each case are significant. [After Watson (1970), p. 245.]

between the various peptides that can be obtained by digesting hemoglobin with trypsin (Fig. 8–14). The technique is a combination of electrophoresis and paper chromatography, and it showed that sickle-cell hemoglobin differs from normal hemoglobin in only one peptide. If this peptide is removed from the preparation (by simply cutting out the spot and redissolving it) and analyzed for its amino-acid composition, it is found that it differs from its normal counterpart only in one amino-acid residue. In sickle-cell hemoglobin (Hb[s]), the amino acid valine is substituted for one of the glutamic acid residues in one of the polypeptide chains. Other heritable hemoglobin abnormalities have been subsequently investigated by similar techniques and have also been found to represent alterations in a single amino-acid residue in one of the two kinds of polypeptide chains that comprise adult hemoglobin (Fig. 8–15).

The significance of these findings is twofold. First, a change in only one amino-acid residue in a protein—and hemoglobin is a large and fairly complex protein (Fig. 8–16).—can have profound and often diverse effects (Fig. 8–12). Second, there is a direct correlation between protein structure and genotype—that is, genes determine the precise amino-acid composition of proteins, amino acid by amino acid. These findings also answer in part one of the questions posed above—a mutation's abnormality of function can be attributed to a fault in the primary structure of a protein.

Position	Alpha chain							
	1	2	16	30	57	58	68	141
	Val	Leu---Lys+---Glu- ---Gly				His+---AspN --- Arg		
Hb variant								
Hb I			Asp-					
Hb G Honolulu				GluN				
Hb Norfolk					Asp-			
Hb M Boston						Tyr		
Hb G Philadelphia							Lys+	

Position	Beta chain									
	1	2	3	6	7	26	63	67	125	150
	Val	His+	Leu--Glu-	Glu--Glu--His+--Val--Glu--His+						
Hb variant										
Hb S				Val						
Hb C				Lys+						
Hb G San José					Gly					
Hb E						Lys+				
Hb M Saskatoon							Tyr			
Hb Zürich							Arg+			
Hb M Milwaukee-1								Glu-		
Hb D β Punjab									GluN	

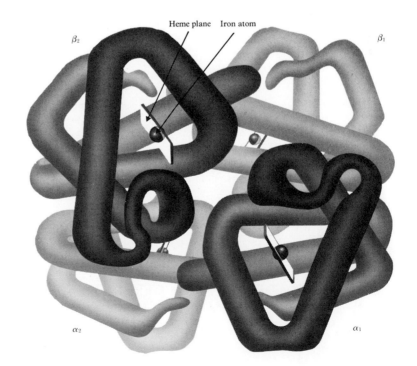

Fig. 8–16. The structure of the hemoglobin molecule, consisting of two kinds of polypeptide chains, α and β. Each chain is represented twice. [Courtesy of Irving Geis.]

COLINEARITY OF GENE STRUCTURE AND PROTEIN STRUCTURE

Watson and Crick (1953b) immediately appreciated the genetic implications of base-pair sequence in DNA. In 1958, Crick proposed that the sequence of nucleotides in DNA and of amino acids in protein are *colinear*. Since this means that there is a direct correspondence between the base-pair sequence in DNA and the amino-acid sequence in the corresponding protein, then one must conclude that one base pair or one group of contiguous base pairs "codes" a specific amino acid. Because there are at best only four kinds of base pairs (A-T, T-A, G-C, and C-G) but twenty kinds of amino acids, obviously a minimum of three base pairs is required to constitute the "code" for any one amino acid. (Two base pairs can code only sixteen (4^2), and three can code sixty-four (4^3).) In 1961, Crick and his collaborators at the University of Cambridge demonstrated that in T4 phage, the genetic code is indeed a triplet code. They did this by making various crosses between mutant strains that differed in terms of either having an extra base pair or set of base pairs inserted in a particular sequence, or lacking a base pair or set. How crosses of this kind are made and analyzed is described in Chapter 10. Interpretation of results is shown in Fig. 8–17.

Proof of colinearity between genes and the proteins or polypeptides coded by them depends, however, on knowledge of both the amino-acid sequence in the protein structure and the linear composition of the corresponding gene. Yanofsky, et al. (1964) isolated a large number of mutants of *E. coli* that were deficient in their ability to synthesize an active form of the enzyme tryptophan synthetase. This enzyme is a protein consisting of two polypeptide moieties, A and B. By a series of genetic crosses, Yanofsky and his collaborators located the position of each of the mutations in the gene that codes the synthesis of polypeptide A. They also analyzed the amino-acid sequence of each of the mutant forms of polypeptide A. When the two "maps" were compared, they were found to be colinear (Fig. 8–18). In the same year, Sarabhai, et al. demonstrated colinearity between the head protein of T4 virus and the gene that codes that protein (see Chapter 7).

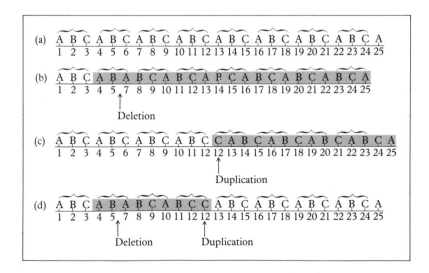

Fig. 8–17. The effect of deletions and insertions in a nucleotide chain. If the code is read from left to right, the shaded area represents the material that is out of phase, resulting in frameshift mutations. (a) A normal chain. (b) Deletion of nucleotide pair in position 6. (c) Duplication of nucleotide pair of position 12. (d) Deletion of 6 and duplication of 12. [Adapted from Crick, et al. (1961), *Nature* **192**.]

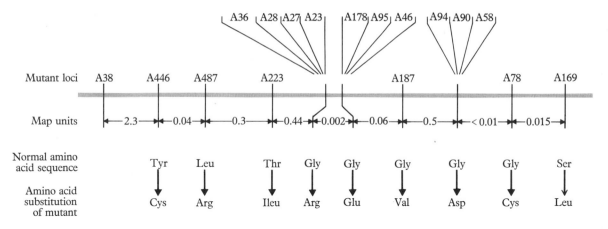

Fig. 8–18. Amino-acid substitutions in the A proteins of various mutants. Colinearity is apparent between the order of the mutant genes and the corresponding sequence of amino-acid substitutions. [Adapted from Yanofsky, et al. (1964), *Proc. Nat. Acad. Sci.* **51**.]

PROTEIN SYNTHESIS AND THE GENETIC CODE

Although the structure of protein is dictated by the base-pair sequence of DNA, DNA itself evidently does not enter directly into the process of protein synthesis. For example, if the nucleus is removed from a cell, protein synthesis does not cease immediately, but may continue for some time. Hemoglobin synthesis in mammalian red blood cells continues after extrusion of the nucleus in the normal course of events. What is immediately affected by loss or inactivation of DNA is the synthesis of RNA. Moreover, cells that are characterized by high levels of protein synthesis, such as pancreatic cells and antibody-producing cells, have a correspondingly high content of RNA, whereas the amount of DNA in these cells does not measurably differ from that of cells characterized by low rates of protein synthesis. The major difference in amount of RNA of high protein-producing cells is found in the cytoplasm, which is also the major site of protein synthesis. The use of labeled amino acids has revealed that the site of protein synthesis in cells is the ribosomes, which are the major repositories of cytoplasmic RNA. Finally, if RNA is selectively hydrolyzed by the enzyme RNase, protein synthesis immediately stops. Collectively, such evidence clearly suggests that DNA controls the synthesis of protein through the intermediary nucleic acid RNA. This relationship constitutes the so-called "central dogma" of molecular biology, namely, "DNA makes DNA and RNA; RNA makes protein," expressed as DNA→RNA→protein. As with all dogmas, this one has become weakened by certain exceptions that have been recently discovered, and as a result has changed its status from dogma to rule. Although DNA generally comes only from preexisting DNA, and RNA comes only from DNA as a template, it appears that in the reproduction of certain RNA viruses at least, DNA may be synthesized on a template of RNA, and in still others RNA may be synthesized from other RNA (cf. Chapter 7).

RNA Synthesis

How does RNA work as an intermediary between DNA base-pair sequence and amino-acid sequence? As described briefly in the preceding chapter, RNA is polymerized from ribonucleotides (actually ribonucleoside triphosphates) by the enzyme RNA polymerase on one of a pair of complementary strands of DNA as a template. RNA polymerase "recognizes" and becomes attached in some way to certain points along the DNA molecule. These points, distinguished presumably by unique base-pair sequences, are the starting points of RNA polymerization. In addition to the enzyme, other factors may be involved in initiating RNA polymerization. For example, Eron, et al. (1971) have found that a specific protein and the nucleotide cyclic adenosine monophosphate (cAMP) are essential for expression of the genes involved in lactose metabolism in *E. coli* (cf. p. 205), both *in vivo* and *in vitro*. Once polymerization is initiated, it always proceeds from 5' to 3' (Ochoa, 1968) just as with DNA polymerization. This was derived from the fact that RNA molecules analyzed before completion of polymerization always have the triple phosphate at the 5' end and the newly inserted nucleotides at the 3' end.

The sequence of bases on DNA recognized by RNA polymerase is not known, but in prokaryotes it is suspected to be a repetition of pyrimidines, either poly-T or poly-C, because of the correspondence between the number of such clusters in the genome and the number of genes (Richardson, 1969). In any event, the first triplet of bases in DNA that is ultimately translated in *E. coli* is T-A-C, the corresponding mRNA base sequence being A-U-G. How universal this sequence is as the leading triplet in polypeptide synthesis is not known as yet. The A-U-G codon is probably preceded by a series of nucleotides that serve solely for recognition and are not translated (Steitz, 1969). Before transcription of a gene is completed, another protein factor is believed to bind to RNA polymerase (Darlix, et al., 1971) and bring about termination of transcription at the appropriate site on the DNA template. In some cases of mRNA transcription, the molecule produced is secondarily cleaved into two or more molecules of mRNA.

As mentioned in the preceding chapter, several kinds or "species" of RNA have been identified. They differ from one another in terms of molecular size, tertiary structure, and their role in protein synthesis. A useful unit for distinguishing between different sizes of macromolecules is the *Svedberg Unit* (S). This is determined by measuring the rate at which a given kind of molecule sediments in a density gradient under a standard centrifugal force. Because rate of sedimentation is a function of both the size and shape of the molecule, it is not a direct measure of molecular weight and is not usually additive. For example, bacterial ribosomes consist of two subunits differing in size and slightly in shape. One of these has an S-value of about 50 and the other of about 30, but when they are joined to form a single particle, as they are during protein synthesis, this particle has an S-value of 70. Similarly, a typical eukaryotic ribosome, such as that found in animals and plants, has a value of about 80S, with subunit values approximating 60S and 40S.

1. Messenger RNA (mRNA). Messenger RNA is a single-stranded molecule that is comprised of a broad range of nucleotide numbers (300 to 12,000, but mostly 900 to 1500 in *E. coli*) with correspondingly high molecular weights (up to about 4×10^6). The bases are all A, G, U, or C. Functionally, mRNA molecules code the sequence of amino acids in protein synthesis, and so they represent transcriptions of the so-called "genes" *per se*. Although the molecule is single-stranded, base pairing is believed to occur between segments, producing secondary structure in the form of hairpin loops (cf. Fig. 8–23).

2. Transfer RNA (tRNA). Transfer RNA molecules consist of about 75 to 80 nucleotides and have molecular weights on the order of 2.5×10^4. They are characterized by a complexly folded tertiary structure (Kim, et al., 1971) (Fig. 7–18), which is a consequence of hydrogen bonding between a number of the constituent bases, and by possession of a number of "unusual" bases. These bases, which presumably are initially the A-U-G-C variety when tRNA is first synthesized, are mostly methylated derivatives of guanine or uracil (Fig. 7–18). The unusual bases are situated in the curves of the molecule where no base pairing occurs. It is believed that they are incapable of forming energetically stable hydrogen bonds with other bases (Watson, 1970), and therefore it is because of their position that the characteristic tertiary structure is formed. The molecular structure of one of them, dihydrouridine (DHU), is believed to actually promote loop formation (Sundaralingam, et al., 1971). That all tRNA molecules so far described have a similar distribution of unusual bases and accordingly a similar tertiary structure implies that the shape has some fundamental importance in tRNA function. An additional characteristic of all tRNA molecules is the presence of the unpaired bases A-C-C at the 3′ end (Fig. 7–17). Three of the unpaired bases in the loop that projects diametrically opposite the A-C-C end constitute the so-called "anticodon," a triplet of bases that undergoes base pairing with a complementary triplet of bases, the "codon," in mRNA during protein synthesis, as described below. The roles of the other loops and the A-C-C end will also be considered below.

3. Ribosomal RNA (rRNA) and Ribosomes. Most of the RNA of cell cytoplasm is in the form of rRNA. There are three kinds of rRNA molecules that can be distinguished—light, medium, and heavy. In bacterial cells (the source of most of our knowledge of ribosomes and rRNA), these molecules are characterized as 5S, 16S, and 23S, respectively, with corresponding molecular weights of about 4×10^4, 6×10^5, and 1.2×10^6. Each kind is single-stranded, but some folding has taken place due to hydrogen bonding between complementary bases, giving the molecule a very irregular structure (Fig. 8–19). The rRNA becomes conjugated with protein in the ribosomes, which are bipartite particles that consist of two subunits (Fig. 8–20). In bacteria, the larger subunit has a sedimentation constant around 50S and is comprised of one molecule of 23S rRNA, one molecule of 5S rRNA, and about thirty different molecules of protein (Kurland, 1970). The smaller ribosomal subunit in bacteria is about 30S and is comprised of one molecule of 16S rRNA and about twenty different molecules of protein (Kurland, 1970). When Traub and Nomura (1968) separated 30S subunits into their component RNA and proteins, purified these components, and then mixed them together again, they recovered reconstituted 30S subunits that were fully functional in protein synthesis. This example of self-assembly (Kushner, 1969) is remarkable because of the structural and functional complexity of ribosomes.

The rRNA and ribosomes of other prokaryotic microorganisms (rikketsia; bluegreen algae) and of mitochondria and plant plastids are like those of bacteria in terms of size and functional detail (Margulis, 1970; Rabbitts and Work, 1971). But the ribosomes of eukaryotic cells, including fungi as well as higher animals and plants, show important differences, although much less is known about them. Eukaryotic ribosomes have a sedimentation constant of about 80S, with subunit values approximating 40S and 60S, and corresponding rRNA molecules of 18S and 28S.

4. Other RNA. The nuclei of eukaryotic cells contain a large amount of RNA of kinds different from the foregoing. Some of it is precursor tRNA;

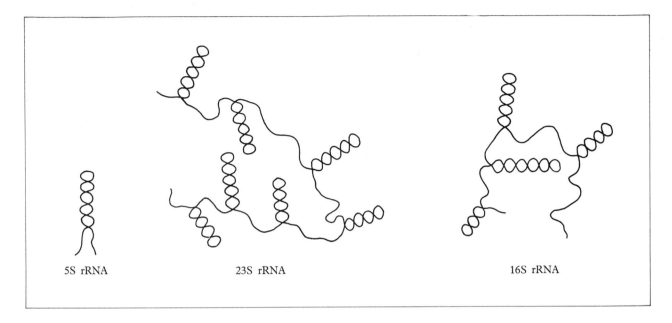

Fig. 8–19. The structure of ribosomal RNA (rRNA) of *Escherichia coli.* The light (5S) and heavy (23S) molecules are combined with 30 to 35 different molecules of protein to constitute the 50S ribosomal subunit. The medium (16S) molecule is associated with about 20 different molecules of protein in a 30S ribosomal subunit.

some of it is precursor rRNA. The bulk of nuclear RNA, however, is a "heavy" variety with sedimentation values as high as 90S. The function of this RNA is not precisely known, but it has been proposed that it partially functions as a precursor of mRNA and may be concerned with the regulation of gene function (Davidson, 1968; Britten and Davidson, 1969; Darnell, et al., 1973), as will be discussed later and in Chapter 11).

Protein Synthesis

1. Amino-Acid Activation. The amino acids present in the cytoplasm are activated by ATP to form amino-acyl adenosine monophosphate

(AA–AMP) plus inorganic phosphate (P_i). This activation is mediated by an enzyme, *amino-acyl synthetase*, which exists in specific form for each kind of amino acid and that amino acid's corresponding tRNA. The sugar group of the terminal adenylic acid at the 3' end of the tRNA molecule then bonds covalently with the carboxyl group of the enzyme-bound amino acid, displacing AMP:

$$AA + ATP \xrightarrow[\text{AA-synthetase}]{} AA\text{–}AMP + (2P_i)$$

$$AA\text{–}AMP + tRNA \xrightarrow[\text{AA-synthetase}]{} AA\text{–}tRNA + (AMP)$$

$\longrightarrow$ ATP

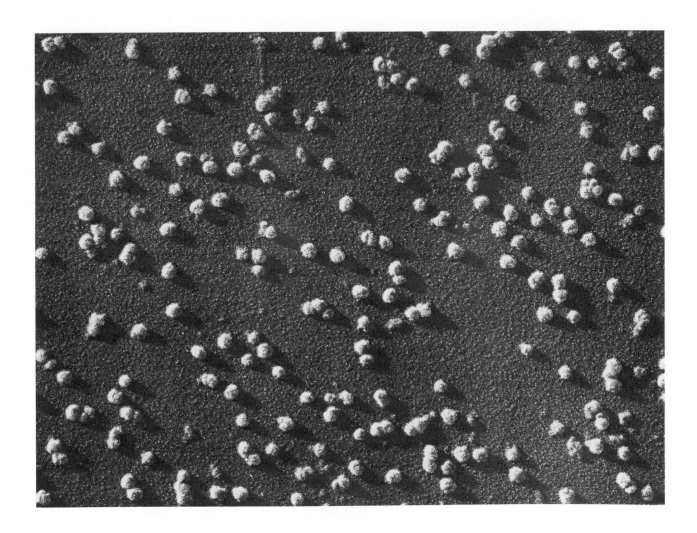

Fig. 8–20.   Electron micrographs of freeze-dried 70S ribosomes of *E. coli.* The ratio of shadow length to object height is $3^1/_2$:1. Magnification X135,000. Inset: Enlargement of one 70S ribosome showing detailed structure. [Courtesy Dr. V. D. Vasiliev (1971), *FEBS Letters,* **14**:203.]

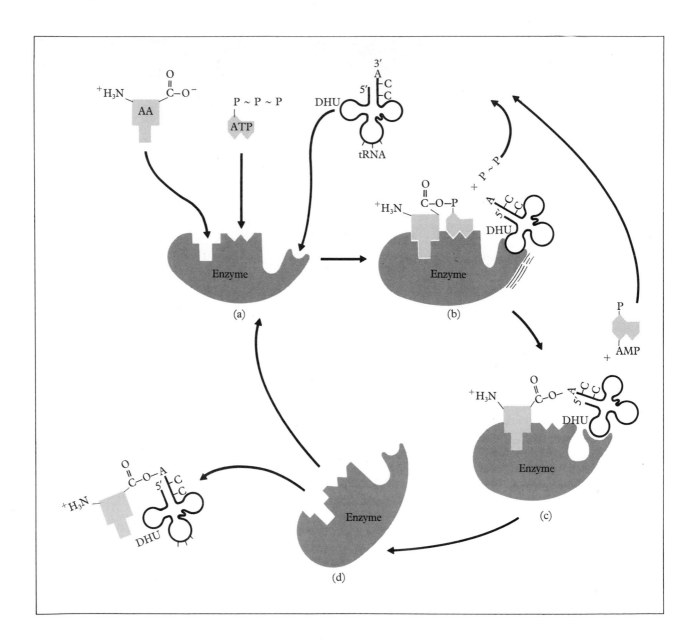

Thus it appears that there are at least as many kinds of AA-synthetase as there are amino acids—that is, at least twenty. Each molecule of AA-synthetase has at least three active sites—one that binds its specific amino acid, another that holds ATP in position, and a third that binds its specific tRNA—holding each of these kinds of molecules in the proper stoichometric position for reaction to occur between them. The DHU loop of tRNA is believed to be the part that is recognized and bound by the enzyme. It positions the tRNA so that its 3' adenylic acid residue juxtaposes the

◀ **Fig. 8–21.** Hypothetical mechanism of amino-acyl-tRNA activation. (a) The substrates—a specific amino acid (AA), a molecule of ATP, and a specific molecule of tRNA—recognize their corresponding amino-acyl-synthetase enzyme. (b) The substrates occupy their appropriate positions on the enzyme, and the energy level is raised by the conversion of ATP to AMP. (c) The 3' end (A) of the tRNA displaces AMP and bonds with the amino acid. The liberated AMP combines with the inorganic phosphate (P ~ P) to restore ATP. (d) The activated amino-acyl-tRNA is carried to a ribosome and released, and the enzyme is free to enter another activation cycle.

Fig. 8–22.   Steps in the formylation of methionine.

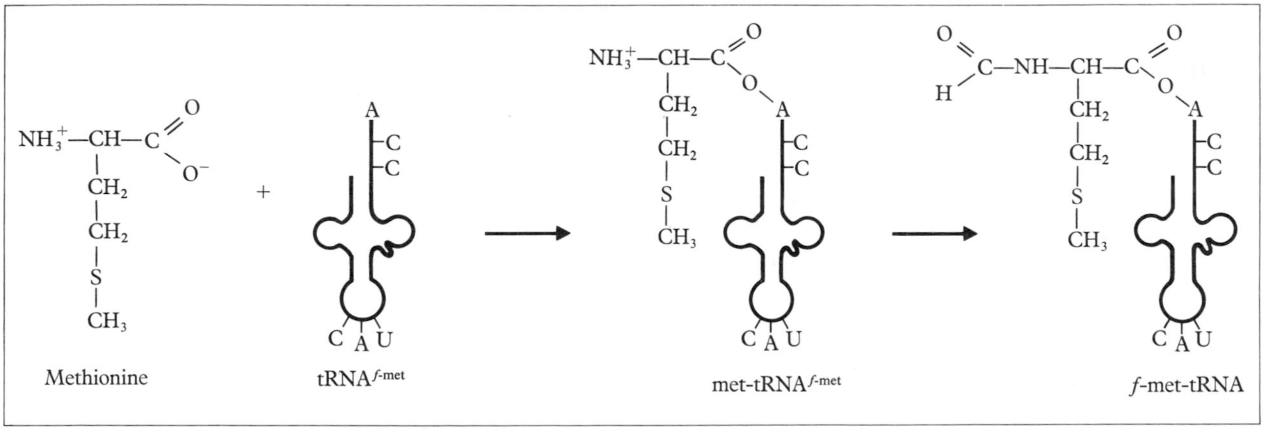

| Methionine | tRNA$^{f\text{-met}}$ | met-tRNA$^{f\text{-met}}$ | f-met-tRNA |

ATP site and displaces AMP. As soon as this occurs, the AA–tRNA is released and the enzyme is ready to catalyze another similar reaction. The amino-acyl bond between tRNA and the amino acid residue is energetically favorable for subsequent polypeptide condensation (Fig. 8–21).

2. The Polypeptide Chain Initiator. In *E. coli* and eukaryotes, the first AA–tRNA to enter into polypeptide synthesis is formyl-methionine-transfer RNA (*f*-met-tRNA) (Adams and Capecchi, 1966; Brown and Smith, 1970). Methionine tRNA is enzymatically formylated (Fig. 8–22) before it becomes associated with the ribosome. In its formylated condition, the amino acid cannot form a peptide bond at its amino end; it is chemically committed to being the "lead off" amino acid or *initiator* in polypeptide polymerization. A polypeptide initia-

Fig. 8–23. *Overleaf.* Proposed mechanism of polypeptide chain initiation. The initiation factor F1 promotes the binding of *f*-met-tRNA to the 30S subunit. At the same time or slightly earlier, the dissociation factor F3, which is complexed with the 30S subunit, promotes the binding of mRNA to the same subunit. The combination formed is known as the "initiation complex." Energy for these reactions is provided by the hydrolysis of guanosine triphosphate (GTP). The F3 then separates from the initiation complex while the second initiation factor (F2) effects the binding of the remaining members of this complex to the 50S subunit. Energy for this reaction is also derived from GTP hydrolysis, and the 70S complex is formed. Following this, translation of mRNA proceeds with the ultimate release of a polypeptide, the tRNA's, and mRNA, leaving a free 70S ribosome. Dissociation of 30S and 50S components is effected by F3, and the cycle is completed.

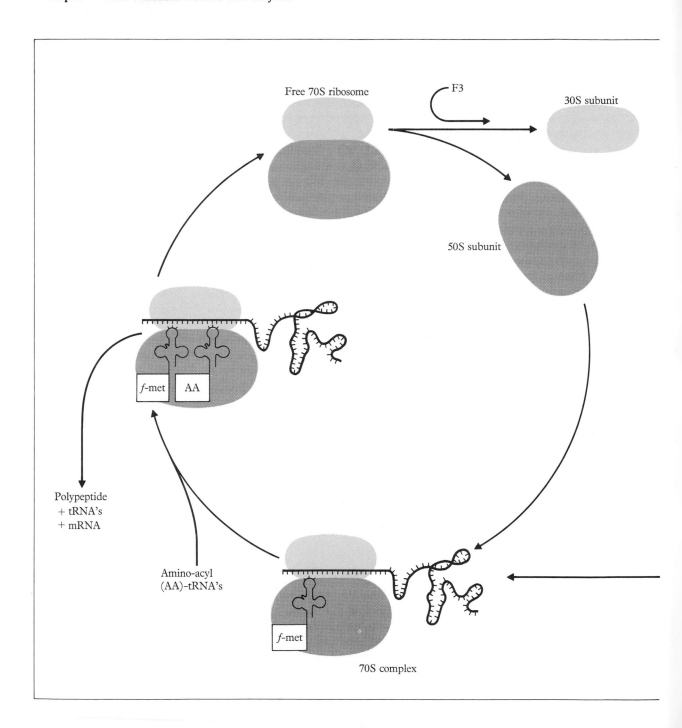

Free 70S ribosome

F3

30S subunit

50S subunit

70S complex

f-met

f-met   AA

Polypeptide
+ tRNA's
+ mRNA

Amino-acyl
(AA)-tRNA's

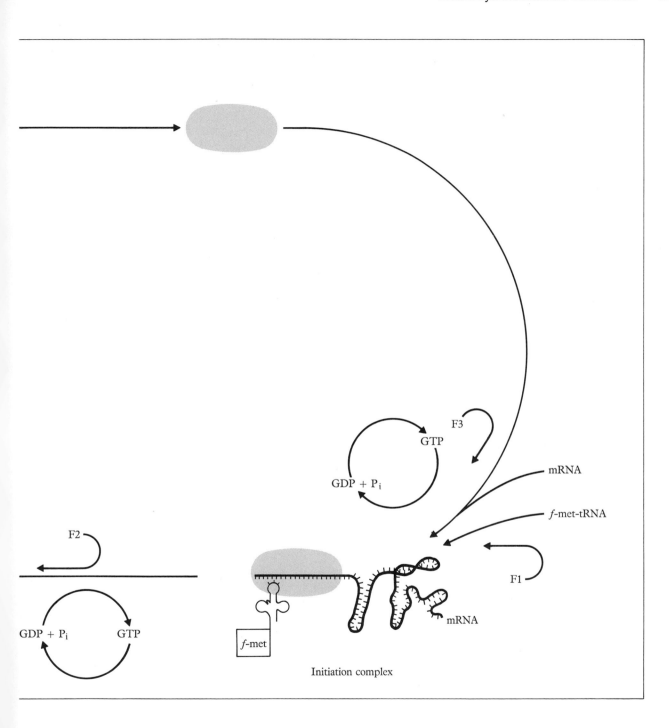

GTP

GDP + P$_i$

F3

mRNA

f-met-tRNA

F2

F1

GDP + P$_i$ GTP

f-met

mRNA

Initiation complex

tion factor, designated F1, forms a complex with *f*-met-tRNA and a 30S ribosomal subunit. Either at the same time or slightly in advance of this, another initiation factor F3 (also known as the dissociation factor DF) binds a molecule of mRNA to the same 30S subunit. Then a third initiation factor F2 forms a complex with the 30S subunit and a 50S subunit to complete the formation of a 70S complex (Davis, 1971; Dubnoff and Maitra, 1971). Phosphorylation of the triphosphonucleoside *guanosine triphosphate* (GTP) to guanosine diphosphate (GDP) provides the energy for these associations. The 70S complex consists then of *f*-met-tRNA, mRNA, and the two ribosomal subunits 30S and 50S (Fig. 8–23).

After the polypeptide chain has started polymerizing, as will be described in the next section, the formyl group is enzymatically removed from the *f*-met-tRNA (Takeda and Webster, 1968). This is followed in many cases by the removal of the entire methionine residue (Lengyel and Söll, 1969). Inasmuch as methionine residues occur in subterminal positions in polypeptides, which they could not do if they were formylated, it follows that not all met-tRNA's are formylated. This suggests that there may be two different tRNA's for

Fig. 8–24.   Comparison between the transfer RNA's for formylated methionine and for methionine. Note that the anticodons (C-A-U) are the same. [Adapted from S. K. Dube, et al. (1969), *Cold Spring Harbor Symp. Quant. Biol.* **34.**]

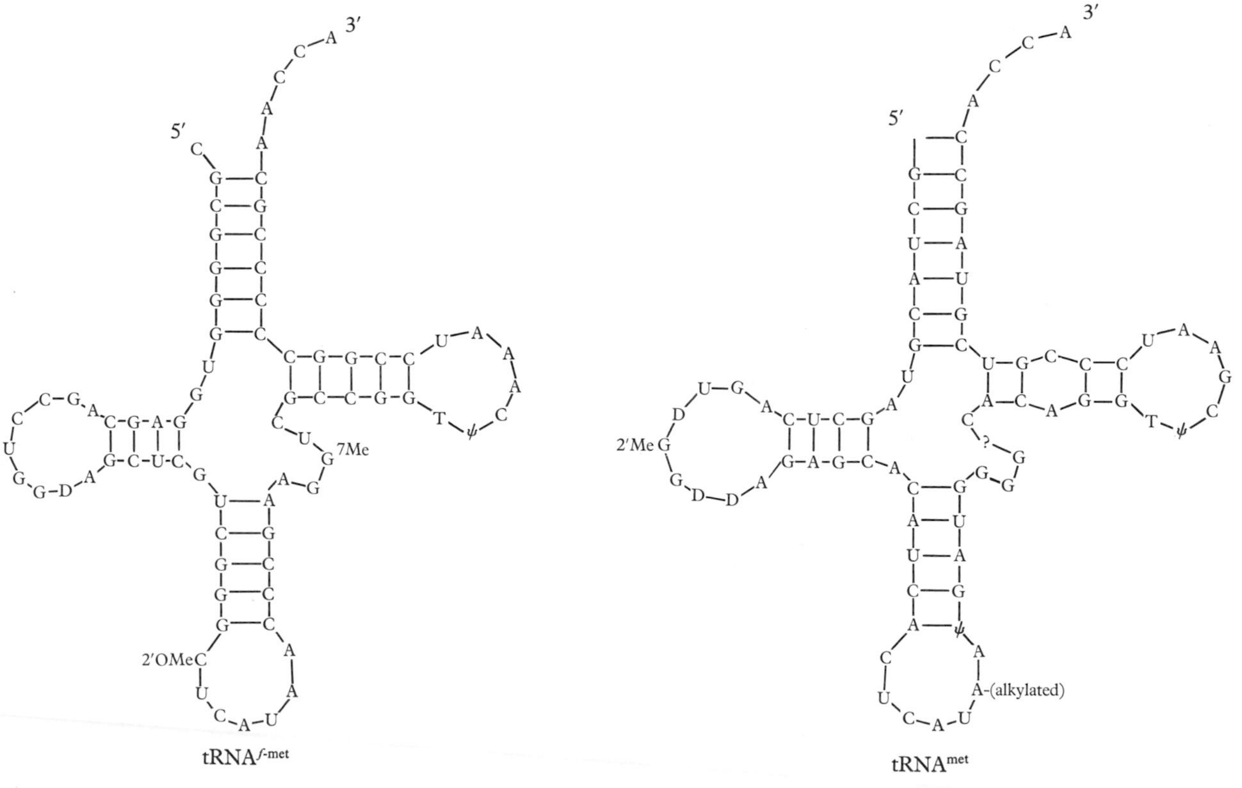

methionine, one of which facilitates formylation and the other of which does not. As we shall see shortly, methionine is one of the only two amino acids for which there is a unique code triplet; therefore the two met-tRNA's would have to differ in some way other than in their anticodon. This has been confirmed, the differences having been found to reside in other base components of the tRNA (Dube, et al., 1969; Smith and Marcker, 1970) (Fig. 8–24).

3. Positioning of mRNA. Prior to becoming bound to the ribosome, mRNA is believed to be in something of a tangle, due to internal hydrogen bonding, except along one short stretch located near the 5′ end (Fig. 8–25). In phage R17 RNA, the first triplet of bases in the 5′ end of this stretch is A-U-G (Steitz, 1969), which is the mRNA code for methionine. The other bases in the stretch appear to be variable. It is this stretch that is recognized by F3 and becomes bound to the 30S subunit (Rudland, et al., 1971). Inasmuch as the first amino acid residue is coded by the triplet A-U-G in bacteria and eukaryotes as well, this mechanism is probably universal.

4. Binding of AA–tRNA. With the positioning of mRNA and formation of the 70S complex in *E. coli*, at least two AA–tRNA binding sites are established. They lie side-by-side across the line of conjunction of the two subunits and are bounded on the 30S side by the molecule of messenger RNA (Figs. 8–23, 8–26). One of these binding sites, which is usually called the *amino-acyl site* (A-site), receives AA–tRNA from the cytoplasm and binds it in position for complementary base pairing with the corresponding codon of the bound mRNA. Base pairing between anticodon and codon is, of course, antiparallel, and because it occurs only between complementary bases, only the proper AA–tRNA will be accepted by the A-site. Binding of the appropriate AA–tRNA to the A-site is accomplished by a binding enzyme T1 (transferase 1), which derives its energy for this from the con-

version of GTP to GDP (Lipmann, 1969; Schreier and Noll, 1971).

5. Peptide Bond Formation. The second AA–tRNA binding site that has been demonstrated in *E. coli* is the *peptidyl site* (P-site). Like the A-site, the P-site is shared between the two ribosomal subunits and is bounded on the 30S side by mRNA (Fig. 8–26). Which site is occupied first by the polypeptide chain initiator *f*-met-tRNA is not known, but by the time the next AA–tRNA comes into position in the A-site, *f*-met-tRNA is in the P-site. When both sites are occupied, the enzyme *peptidyl transferase* catalyzes the formation of a peptide bond between the carboxyl group of the amino-acid residue on the tRNA in the P-site and the amino group of the AA–tRNA in the A-site (Fig. 8–26), utilizing energy released by phosphorylation of GTP → GDP. This enzyme is one of the component proteins of the 50S ribosomal subunit (Lengyel and Söll, 1969). When the peptide bond forms, the tRNA in the P-site is released and either moves out directly or moves to a third site, the *donor* site, from which it is then released into the cytoplasm. The P-site is thereby opened, and the tRNA is free to circulate in the cytoplasm to come into contact with another AA-synthetase–amino-acyl-AMP complex.

6. Translocation. Coincident with peptide bond formation, and possibly causally related to it, is the action of another transfer factor T2, which moves or "translocates" the AA–tRNA from the A-site to the vacated P-site, using the phosphorylation of another molecule of GTP as the source of energy. When translocation has been accomplished, the A-site is again opened and is immediately filled by the next AA–tRNA, one with an anticodon that is complementary to the next mRNA codon in line.

7. Chain Elongation. Codon-by-codon repetition of the foregoing sequence of events is made possible by the fact that the molecule of mRNA

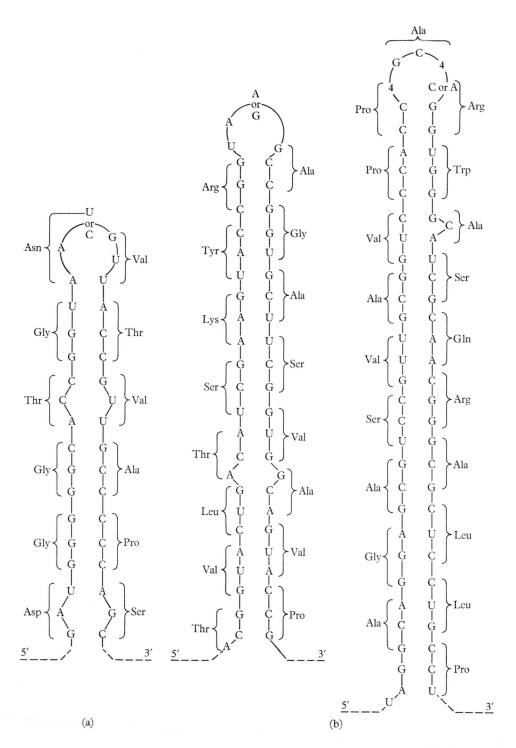

Fig. 8–25. Possible base-paired loops in mRNA from two disparate sources, (a) bacteriophage R17 coat protein, and (b) α-chain human hemoglobin. The number 4 indicates that any one of the four bases may be present. [Adapted from H. B. White, et al. (1972), *Science* **175:** 1264–1266. Copyright © 1972 by the American Association for the Advancement of Science. By permission.]

(a)

(b)

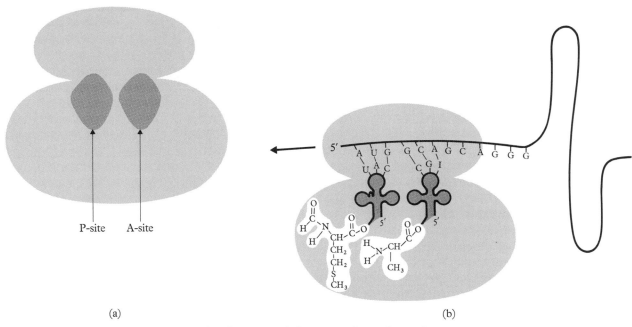

(a)   (b)

Fig. 8–26. The postulated ribosome binding sites. At least two sites exist on the ribosome for binding amino-acyl-tRNA during mRNA translation. (a) These are the peptidyl (P) site and the amino-acyl (A) site. (b) The positions on the ribosome are shown for mRNA, tRNA, and two amino acids just prior to peptide-bond formation. The arrow indicates the direction of movement of the mRNA strand during translation. The lead-off amino acid is formylated methionine; the second is alanine.

also moves across the ribosome codon-by-codon. What enzymes or factors are required for this movement is not clear as yet. Hardesty, et al. (1969), in their analysis of chain elongation in mammalian reticulocytes (immature red blood cells), postulated that the mRNA becomes buckled by the action of T2, and the force involved in this action moves the codon opposite the A-site to the position opposite the P-site as the previous codon in the latter position moves out. This action is reminiscent of "inchworm locomotion" (Fig. 8–27).

8. Chain Termination. The foregoing sequence of events continues for as long as there are AA–tRNA's that correspond to successive codons of mRNA as the latter move across the ribosome. The rate of movement, and accordingly of

polypeptide synthesis, has been estimated at about 20 to 30 codons per second in *E. coli* at 37°C. Polymerization ceases when any one of three "terminator" codons enters the ribosome. These are codons for which there is no corresponding tRNA (Table 8–3). The terminator codons have been shown in mammals to activate one or more protein "release" factors, which release the terminal amino-acid residue of the polypeptide chain from the tRNA by which it is attached to the ribosome (Scolnick and Caskey, 1969; Beaudet and Caskey, 1971). Some molecules of mRNA include only one initiator and one terminator codon, but others may include two or more of each. In the latter case, the mRNA molecule passes continuously across the ribosome, the translation of one polypeptide following that of another until the en-

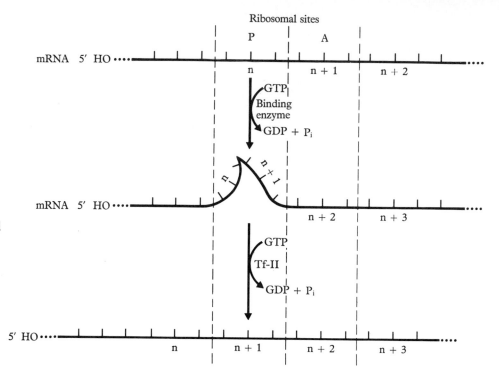

Fig. 8–27. Conformational changes in mRNA during translocation according to the inchworm hypothesis. [From Hardesty, et al. (1969), *Cold Spring Harbor Symp. Quant. Biol.* **34**:342. Copyright © 1969, Cold Spring Harbor Laboratory. By permission.]

tire molecule of messenger RNA has been "read." The DNA segments that are templates for long mRNA molecules such as these are called *polycistronic operons*. After a molecule of messenger RNA has completed translation, it separates from the ribosome, and then the ribosome subunits dissociate from one another (Kaempfer, et al., 1968). This separation has also been shown to be dependent upon the protein factor F3, which was involved in formation of the initiation complex (Dubnoff and Maitra, 1971; Davis, 1971). In mammals and many other higher organisms, a single strand of mRNA often becomes associated with a number of ribosomes in a cluster called *polyribosomes* (Rich, et al., 1963) (Fig. 8–28). By this device, one molecule of messenger RNA can translate as many polypeptide copies as there are ribosomes in the cluster almost simultaneously, a mechanism that would gladden the heart of any assembly-line foreman. The various protein factors involved in polypeptide-chain initiation, elongation, and termination are summarized in Table 8–2 (Lipmann, 1969).

Table 8–2. Protein factors for initiation, elongation, and termination of polypeptide chains in *E. coli*.

Protein factor	Probable location	Function
F1	Ribosome surface	Binds *f*-met-tRNA to 30S subunit.
F2	Ribosome surface	Promotes association of initiation complex with 50S subunit.
F3	Ribosome surface	Binds mRNA to 30S subunit.
T1	Cytoplasm	Binds subsequent amino-acyl-tRNA's to A-site.
T2	Cytoplasm	Translocates AA–tRNA from A-site to P-site.
Peptidyl transferase	50S subunit	Catalyzes formation of peptide bonds in polypeptide synthesis.
R1, R2	Cytoplasm	Release factors that terminate translation.

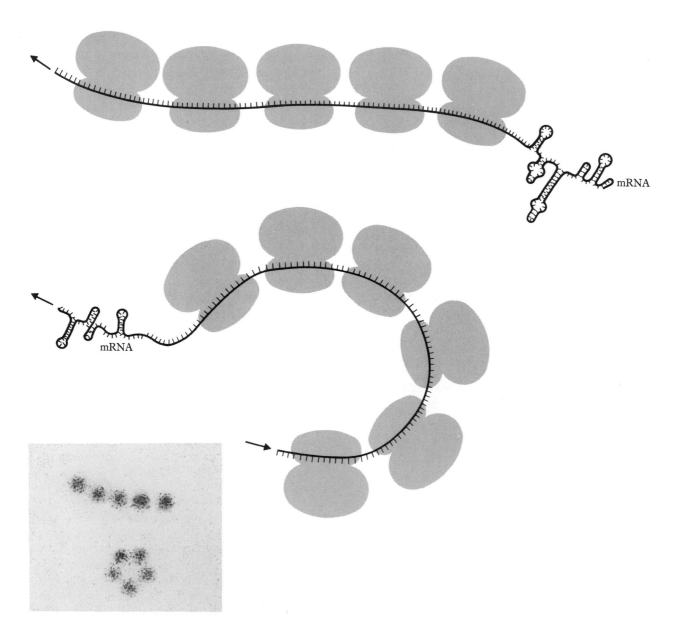

mRNA

mRNA

Fig. 8–28.   Polyribosomes. A single strand of mRNA may pass across several ribosomes in close succession, thereby linking them together in a loose cluster. This figure shows two polyribosome clusters as they appear by electron micros-copy (inset) and the physical relationship between them and a strand of mRNA.

It is quite evident from the foregoing description that the translation of genetic information is both precise and complex. Superficially, this degree of complexity argues against attempts to explain the origin of life as the fortuitous coincidence of molecular "accidents." But it must be borne in mind that the processes that we have been considering have had 3000 million years or more in which to become refined to the extent observed today. Obviously the cells that make fewer mistakes in transcription and translation are the ones that are more likely to have successful progeny. Therefore, just as the so-called "forces" of natural selection have operated at the organismic level, they have operated at the molecular level. The most fundamental aspects of genetic processes must have achieved their present evolutionary state at least 1000 million years ago, when eukaryotic cells first appeared, for they are the same in all terrestrial forms of life as far as is known. Hence the evolution of DNA, the three major species of RNA, ribosomes, nucleic acid polymerases, adaptor enzymes, transfer factors, initiation factors, and the genetic code was essentially completed within the first 2000 million years of terrestrial life. Few and relatively minor modifications have occurred since that time. Two-thousand million years is really a very long time in terms of the rates of molecular reactions—it would be even more surprising if the mechanisms of genetic transcription and translation were less complicated then they are.

The Genetic Code

One would think that when the genetic code was first postulated, biochemists would have set out immediately to try to "break" the code. Crick, et al. (1961) had already demonstrated that it is most likely a triplet code—that is, that a sequence of three contiguous bases in DNA corresponds to one amino acid, that the code does not overlap, and that it does not include any "commas." But it was believed that before the code could be deciphered,

it would be necessary to know precisely the sequence of bases in a gene and the sequence of amino acids in the protein that was coded by that gene. This would represent years of painstaking work, and was not the sort of thing that could be done overnight. Actually, this has been accomplished during the intervening years, but the first breaks in the code came long before that. In fact it was in 1961 that Nirenberg and Mathaei, utilizing the enzyme RNA phosphorylase, demonstrated that the mRNA sequence U-U-U positioned the amino acid phenylalanine. In normal cell metabolism, RNA phosphorylase hydrolyzes the phosphodiester bonds of RNA to yield diphosphoribonucleosides, the unit building blocks of RNA. However, this is an equilibrium reaction, so that if the reacting system is charged with an excess of "product" (the diphosphoribonucleosides), the reaction will reverse itself and effect the *synthesis* of RNA. Unlike the action of RNA polymerase, RNA phosphorylase functions without a template, and so the ordering of the nucleotides is random. Nirenberg and Mathaei charged their *in vitro* system with molecules of the single diphosphoribonucleoside diphosphouridine, and derived polyuridilic acid (poly-U)—that is, RNA that consisted solely of uridine nucleotides (U-U-U-U-U-U-U-U- . . . -U). They then added this synthetic RNA and a mixture of all the amino acids to a cell-free protein-synthesizing system and analyzed the protein that was produced. It was a polypeptide that consisted of only one kind of amino acid—phenylalanine.

In a similar way, the synthetic RNA's poly-A and poly-C were prepared, and from them it was concluded that A-A-A codes lysine and C-C-C codes proline. Poly-G presented problems of internal hydrogen bonding that interfered with translation, and so the "meaning" of G-G-G could not be determined by this method. By synthesizing RNA's from two and later from three and four different diphosphoribonucleosides, other elements of the code were suggested, although their specific order could not be determined. In other words, it was not possible to distinguish between A-A-G, A-G-A, or G-A-A.

In 1964, Nirenberg and Leder devised an ingenious technique whereby they could correlate the precise 5′-to-3′ orientation of trinucleotides with a specific amino acid by synthesizing any nucleotide combination and determine its "meaning" in terms of specific amino acids. It was done in the following way. A cell-free, protein-synthesizing system from *E. coli* was allowed to proceed until it stopped of its own accord. Previously it had been

shown that such systems will stop protein synthesis in a very short time (4 to 5 minutes), because mRNA is very short-lived and is quickly degraded, although both tRNA and ribosomes are long-lived. Then the system was charged with synthetic trinucleotides of known composition (e.g., 5′–G-U-A–3′) and with a mixture of amino acids that included one that had been labeled with radioactive carbon (^{14}C) (Fig. 8–29). The trinucleotides become bound by ribosomes, the amino acids become activated by tRNA as AA–tRNA, and those AA–tRNA molecules with anticodons

Table 8–3. The genetic code of mRNA.

First base	Second base →								Third base
	U		C		A		G		
U	U-U-U	Phe	U-C-U	Ser	U-A-U	Tyr	U-G-U	Cys	U
	U-U-C	Phe	U-C-C	Ser	U-A-C	Tyr	U-G-C	Cys	C
	U-U-A	Leu	U-C-A	Ser	U-A-A	term.	U-G-A	term.	A
	U-U-G	Leu	U-C-G	Ser	U-A-G	term.	U-G-G	Trp	G
C	C-U-U	Leu	C-C-U	Pro	C-A-U	His	C-G-U	Arg	U
	C-U-C	Leu	C-C-C	Pro	C-A-C	His	C-G-C	Arg	C
	C-U-A	Leu	C-C-A	Pro	C-A-A	GluN	C-G-A	Arg	A
	C-U-G	Leu	C-C-G	Pro	C-A-G	GluN	C-G-G	Arg	G
A	A-U-U	Ileu	A-C-U	Thr	A-A-U	AspN	A-G-U	Ser	U
	A-U-C	Ileu	A-C-C	Thr	A-A-C	AspN	A-G-C	Ser	C
	A-U-A	Ileu	A-C-A	Thr	A-A-A	Lys	A-G-A	Arg	A
	A-U-G	Met	A-C-G	Thr	A-A-G	Lys	A-G-G	Arg	G
G	G-U-U	Val	G-C-U	Ala	G-A-U	Asp	G-G-U	Gly	U
	G-U-C	Val	G-C-C	Ala	G-A-C	Asp	G-G-C	Gly	C
	G-U-A	Val	G-C-A	Ala	G-A-A	Glu	G-G-A	Gly	A
	G-U-G	Val	G-C-G	Ala	G-A-G	Glu	G-G-G	Gly	G

Legend

Ala = Alanine
Arg = Arginine
Asp = Aspartic acid
AspN = Asparagine
Cys = Cysteine
Glu = Glutamic acid
GluN = Glutamine

Gly = Glycine
His = Histidine
Ileu = Isoleucine
Leu = Leucine
Lys = Lysine
Met = Methionine (initiator)
Phe = Phenylalanine

Pro = Proline
Ser = Serine
term. = Chain terminator
Thr = Threonine
Trp = Tryptophan
Tyr = Tyrosine
Val = Valine

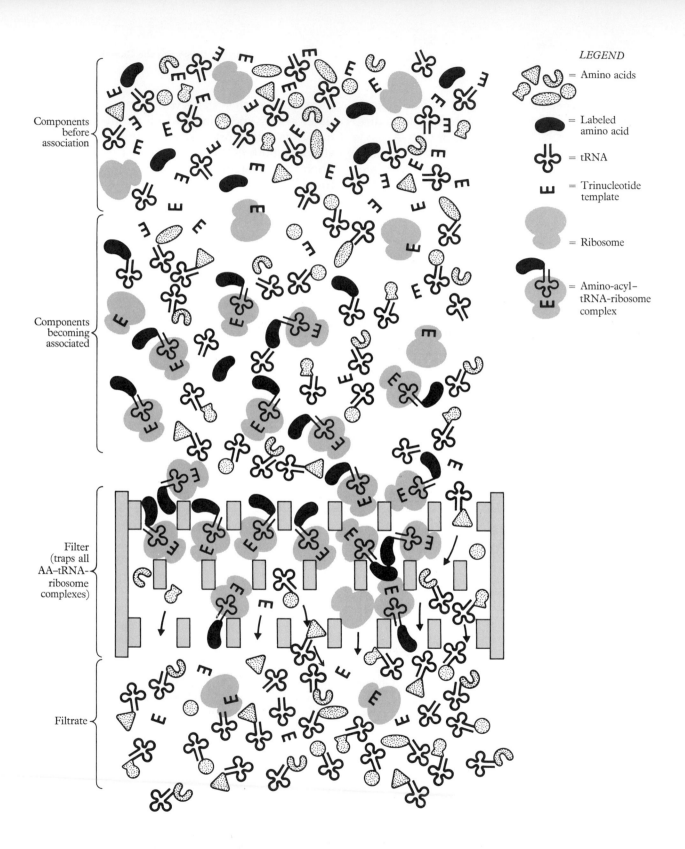

Components before association

Components becoming associated

Filter (traps all AA–tRNA-ribosome complexes)

Filtrate

LEGEND

= Amino acids

= Labeled amino acid

= tRNA

E = Trinucleotide template

= Ribosome

= Amino-acyl–tRNA-ribosome complex

◀ **Fig. 8–29.**   Scheme to illustrate the method by which Nirenberg and Leder deciphered the genetic code in cell-free, protein-synthesizing systems derived from *E. coli*. (See text for explanation.)

complementary to the trinucleotide will then also be bound by the mRNA-ribosome complex. Other AA–tRNA's will not. After binding was given time to occur, the system was passed through millipore filters with 0.45μ pore size. Ribosomes without bound AA–tRNA will pass through filters of this pore size, but ribosomes with bound AA–tRNA will be trapped. By comparing the radioactivity of the filtrate with that of the trapped ribosome component, it was possible to match specific amino acids with specific trinucleotides—that is, codons. By means such as this, the complete genetic code for mRNA has been ascertained and is represented in Table 8–3.

It will be noted that all amino acids except methionine and tryptophan are coded by more than one codon, and most of them by more than two. This compensates for the differential that exists between the number of different kinds of amino acids in nature (20) and the number of possible codons in a triplet code of four characters (64). There are certain advantages that result from this arrangement. For example, a gene that codes for the amino acid proline in a specific position in a polypeptide can mutate in any way with respect to the base that corresponds to the third position in the proline codon, and it will not make any difference. The same would be true for the codons for valine, threonine, alanine, and glycine, all of which are abundant in most proteins. The codons for the amino acids leucine, serine, and arginine show even greater flexibility. A code system that provides more than one code for a given word or meaning (in this case, amino acid) is called "degenerate." The genetic code is degenerate in this sense. The English alphabet is degenerate to some extent in that the "code" that means "radiant energy in the visible range of the spectrum" can

be either the sequence of letters L-I-G-H-T or I-L-L-U-M-I-N-A-T-I-O-N. Language is not a triplet code, and one would be unwise to carry too far any analogies between it and a severely limited code such as the genetic code. Purists might well argue, however, that if our cells were as careless and sloppy with the genetic code as most people are with their language, we would not only be a race of intellectually confused people, we would be congenital monstrosities as well.

One of the very interesting findings relative to the genetic code was a group of triplets that correspond to none of the amino acids. These were called "nonsense" codons, and a number of hypotheses were proposed to account for their function. It is now known that they represent the termination codons described earlier in this chapter (p. 191). This function was inferred when it was learned that mutations resulting in a foreshortening of polypeptide chains represent changes in a codon from any of those that do have corresponding amino acids to either U-A-A, U-A-G, or U-G-A. There are no tRNA's with anticodons for these triplets—instead they are "recognized" by the specific protein release factors already described.

Because there are sixty-one different codons that code amino acids, and all of them are used in natural mRNA, one would infer that there must be sixty-one different varieties of tRNA to correspond with each of them. This is not necessarily so. In most cases in which an amino acid has a plurality of codons, the first two bases of the codon are the same; it is the third that may vary. Crick (1966) proposed that the third base in the anticodon (the base at the 5' end) is slightly askew in relation to the other two so that it can "wobble" about a bit when effecting hydrogen bonding with its corresponding codon. Accordingly certain mismatches can readily occur. Specifically, G in the wobble position of the anticodon can pair with either C or U in the codon; U can pair with either A or G; and the unusual base I, which is in the wobble position in the anticodons of some tRNA's, can pair with either A, U, or C. Hence the amino acid glycine, for example, could get by with only two species of tRNA, provided one had the anticodon C-C-I and the other one had the anticodon C-C-C. A cell that has tRNA with I in the wobble position is not necessarily required to have a correspondingly reduced number of species of tRNA, however. For example, one or more of the anticodons for which C-C-I can function may also be present. These possibilities permit a range of variation in the numbers of different species of tRNA from one cell to another, or from one species of organism to another. This variation may be significant in regulating rates of polypeptide synthesis, if one assumes that the anticodon C-C-A will effect stable hydrogen bonding with the codon G-G-U more rapidly than the anticodon C-C-I will and, contrariwise, C-C-I in the absence of C-C-A will slow down the rate of translation because it will take longer to pair.

Universality of the Genetic Code

Perhaps the most unifying phenomenon in all of nature is the genetic code, for it has been found to be the same in all forms of life in which it has been investigated, including the viruses. Not only are the codes for the amino acids the same, but also the codes for chain initiation (Brown and Smith, 1970) and chain termination (Beaudet and Caskey, 1971) are the same. Evidence for the universality takes several forms, but perhaps the most persuasive is the demonstration that protein synthesizing systems (tRNA, ribosomes, transfer factors, etc.) derived from bacteria will translate mRNA from a variety of sources (phage, TMV, birds, various species of mammals) into the polypeptide or protein that is characteristic of the species from which the mRNA was derived (Ycas, 1969). It is conceivable that variations of a very minor sort may one day be found, but it is highly doubtful if they will be found in eukaryotic organisms. The reason for this is that any mutation altering the code for a

particular amino acid would alter every protein that includes that amino acid. In viruses that do not contain very many proteins, this poses a minor problem. But in eukaryotes, which are comprised of thousands of different proteins, this would constitute a change of such astronomical proportions that it is inconceivable that it would be anything but lethal. The differences between different species are surely not differences in genetic code, but in the kinds and numbers of proteins and polypeptides provided for by that code. This is reflected in the total amounts of DNA characteristic of a cell, and often in the relative frequencies with which certain triplets are used, as approximated by A+T:G+C ratios.

REGULATION OF GENE ACTION

Virtually every metabolic process of living cells is catalyzed and regulated by enzymes. The kinds of metabolic processes taking place within a cell vary according to that cell's nutrition and to the particular functions that the cell is called upon to perform by influences external to it. A bacterial cell such as E. coli, for example, can grow on a wide variety of nutrients, and the enzymes that it produces vary according to the nutritional composition of the medium in which it finds itself. If E. coli is fed lactose, it will synthesize the enzymes appropriate to lactose absorption and metabolism. If lactose is removed from the medium, the cell will immediately stop producing these enzymes. The general principle that enzymes are synthesized on demand is broadly applicable to all forms of life, and it implies the existence of some mechanism or mechanisms for regulating the transcription and/or translation of the genes that correspond to the enzymes. Nowhere is the regulation of gene activity more pronounced or dramatic than during differentiation and development, as is discussed at length in Chapter 11. But the best insights into some of the basic mechanisms have again been gained from microorganisms, particularly bacteria and some of the fungi.

Gene regulation can be either positive or negative. Positive gene regulation implies the action of some substance on either the transcriptional or translational process, or both, whereby the synthesis of the gene's "product" is stimulated. This kind of mechanism is analogous to that of an automobile—to make it go you step on the accelerator; the harder you push on it, the faster it will go; and when you take your foot off, it will slow down and soon stop. Negative gene regulation implies the removal or inhibition of some substance or substances that prevent either transcription or translation. This is analogous to releasing the brake of an automobile that is parked on a hill—you may release the brake part of the way, in which case the car will roll slowly, or all the way; but you cannot release it more than all the way.

1. Positive Gene Regulation.   There have not been many cases of positive gene regulation described as yet. One of the best documented examples is that of the arabinose metabolic pathway in *E. coli.* Three consecutively functioning enzymes in this pathway (Fig. 8–30) are coded by the genes *a, b,* and *d.* According to Englesberg, et al. (1969), the transcription of these genes is regulated by the product of a fourth gene *c.* All four genes are contiguous and function as a unit. Mutant strains that are deficient in gene *c* product have very low activity of the products of the other three genes, and they do not respond to increasing levels of arabinose, as wild-type strains do. According to Englesberg's group, the product of gene *c* (presumably a protein) exists in two functional states, P1 and P2. These bind to the end of the gene se-

Fig. 8–30.   The relationship of the arabinose (*ara*) operon of *E. coli* to the arabinose metabolic pathway. [After Watson (1970), p. 457.]

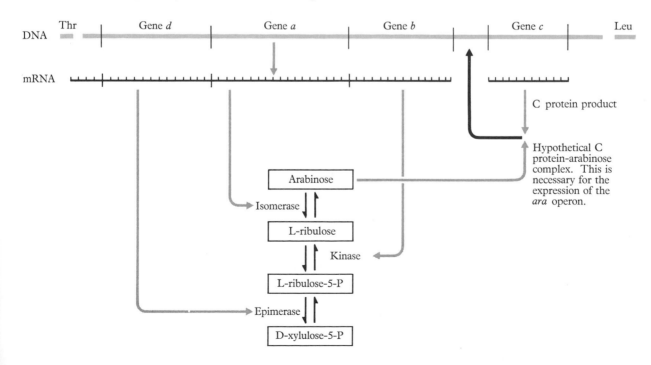

quence where transcription is initiated. In the presence of arabinose, the equilibrium that exists between P1 and P2 is shifted to the right and transcription is stimulated. The higher the level of arabinose, the greater the amount of P2, and the more are genes *a, b,* and *d* stimulated. However, until the protein product of gene *c* is isolated, many aspects of this system will remain obscure and controversial.

Another example of positive gene regulation is reputed to be in nitrate metabolism in the mold *Aspergillus nidulans.* Nitrate is reduced to ammonia in *Aspergillus* in two enzymatically regulated steps:

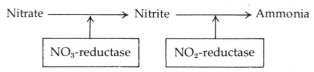

If the level of nitrate is zero, only small amounts of these enzymes are detectable. If nitrate is added, these enzymes are synthesized proportionately up to a maximum level, to which can be assigned the arbitrary value of 100 units. If ammonia is added, synthesis of both enzymes is depressed to zero. The genes that code these enzymes (genes $n3$ and $n2$, respectively) have been identified and are located about ten map units apart on the same chromosome. Another gene, nr, has been found that has a regulatory action on both $n3$ and $n2$, and it is located about 30-40 map units from them on the same chromosome. One mutant strain of *Aspergillus, nr^c,* produces NO_3-reductase at the level of 100 units and NO_2-reductase at the level of 50 units without nitrate stimulation. If nitrate is added, the level of NO_3-reductase remains at 100, but the level of NO_2-reductase increases to 200 units. Both genes are still repressible by ammonia, but not to the same degree as wild type. Diploids of *Aspergillus* that are heterozygous for nr^c (nr^c/nr^+) show an effect intermediate between homozygous mutants and wild type.

On the basis of these observations and those obtained with other mutant alleles of the nr locus,

Pateman and Cove (1967) concluded that the nr^+ gene produces a repressor for genes $n3$ and $n2$ which is transformed in the presence of nitrate or nitrite to an inducer for these genes (Fig. 8–31).

The mechanism of gene regulation in higher animals and plants is undoubtedly more complicated than in lower forms. More complicated mechanisms are necessary because of the variety of cells that are produced by cellular differentiation in multicellular forms. In all probability, hormones and cyclic AMP are intimately involved in the process of regulating animal genes. Furthermore, all the genes of a bacterium are readily available for transcription at a moment's notice; they are all in a perpetual state of responsiveness to the environment. In higher forms, especially in higher animals, this is not the case at all—many kinds of differentiated cells have very limited genetic responsiveness. Most of the genes of such highly differentiated cells as muscle cells and neurons are permanently refractory to any known stimulus and will not transcribe. It is believed that genes such as these are covered with a "shell" of protein which more or less permanently shields them from RNA polymerase. This shell is presumed to be largely histone, a group of basic proteins peculiar to the nuclei of eukaryotic cells (Georgiev, 1969). Each kind of tissue has groups of genes that do respond to environmental change, however, but these genes appear to be scattered throughout the genome. By what mechanism they are switched on and off in a coordinated fashion is not known. The operon scheme, which has been so thoroughly characterized in bacteria and is described in the next section, does not seem to be generally applicable to higher forms. According to one theory (Britten and Davidson, 1969), higher cells have "sensor" genes that react with specific hormones or with proteins released by the cell as a consequence of hormone action. This reaction releases adjacent "integrator" genes which are transcribed into "activator RNA." This is a kind of RNA that differs from every other type because it functions

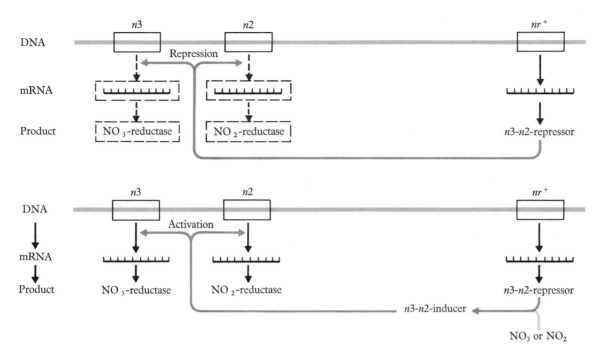

Fig. 8–31. Hypothetical mechanism of the nitrate operon in *Aspergillus nidulans*. (See text for explanation.)

solely within the nucleus. Its function is to recognize specific genes, regardless of how widely dispersed in the genome they may be, and to activate them. In addition to regulation at the level of mRNA synthesis, it is believed likely that additional control points exist in eukaryotic cells. For example, more mRNA may be made than is used. Intrinsic cellular factors, as yet unknown, would determine which mRNA and how much of it would be translated into protein at any given time. Control over the output of protein translated from a given amount of mRNA may occur as well (Darnell, et al., 1973). These hypotheses represent attempts to account for the enormous quantities of nuclear RNA having no known function and characteristic of eukaryotic cells. They are also at-

tempts to provide a mechanism for manifold, yet integrated responses to a specific stimulus. The bearing of these concepts on cellular differentiation is discussed in Chapter 11.

2. Negative Gene Regulation. Negative gene regulation has been superbly demonstrated in intestinal bacteria, stemming from the brilliant researches of Jacob and Monod (1961). The first explanation of the mechanism was associated with three contiguous genes in *E. coli* that are concerned with lactose metabolism. One of these (z) codes the synthesis of the enzyme, β-galactosidase, which hydrolyzes lactose to its simple sugar components, glucose and galactose. Next to this gene is a gene (y) that codes galactoside permease, an enzyme that facilitates the transport of lactose across the cell wall. The third in order is the gene ac, the product of which is thiogalactoside transacetylase, an enzyme whose function *in vivo* is uncertain. In the absence of lactose, only trace amounts of these enzymes are produced, but when lactose is introduced, the organism responds within 3 to 5 minutes by synthesizing large amounts of each of them.

Enzymes that can be induced to appear in a manner such as the foregoing are called, quite reasonably, *inducible enzymes*, and the substance that induces them is called an *inducer*. In general, it is the enzymes concerned with catabolic processes that are inducible, and their substrates are the inducers. On the contrary, enzymes of anabolic pathways, such as those concerned with amino-acid synthesis, are generally *repressible*—that is, their production or activity can be repressed and the substance that brings about their repression is usually the end product of their activity. A well-established repressible system of enzymes in *E. coli* is that of the biosynthetic pathway for the amino acid histidine. The enzymes of this pathway are synthesized and activated in a coordinated fashion during normal metabolism and growth, but if histidine accumulates in the cell or if it is added exogenously, these enzymes are repressed and further synthesis of histidine soon stops.

Jacob and Monod studied a mutant strain of *E. coli* that produces the enzymes of lactose metabolism in the absence of inducer—that is, the genes z, y, and ac are active irrespective of any need for their products by the cell. Utilizing techniques that effect gene recombination, as described in Chapter 10, they determined that the site or locus of mutation in this strain is quite separate from genes z, y, and ac. Since its action appeared to be concerned with the induction of activity on the part of the other genes, it was called the i gene (Fig. 8–32). It was evident, then, that in wild-type bacteria the i gene, or the product of its transcription or translation, in some way prevents

genes z, y, and ac from being expressed except in the presence of inducer. Hence, using the superscript + to designate normal or wild-type alleles, and − to designate mutant alleles, cells having the genotype $i^+z^+y^+ac^+$ produce the β-galactosidase series of enzymes only in the presence of inducer, and cells with the genotype $i^-z^+y^+ac^+$ produce the enzymes "constitutively," that is, irrespective of presence or absence of inducer. Jacob and Monod called the i gene a *regulator gene*, and they called the other genes *structural genes*, because they code the structure of specific enzymes.

In an attempt to learn the mechanism by which the regulator gene controls the activity of its corresponding structural genes, Jacob and Monod utilized a mutant (z^-) which produces an inactive form of β-galactosidase that can be detected by immunological means, and they used genetic "tricks" whereby cells can be made diploid for certain sections of DNA. These tricks, which are described in greater detail in Chapter 10, involve the use of Hfr and F⁻ strains (cf. p. 264) so that exconjugants that are partially diploid are produced. In this way, Jacob and Monod derived cells that had the genotypes shown in Table 8–4. Genotype 3 in this table shows that the wild-type allele i^+ is dominant. Genotype 5 shows that the regulator acts not directly on the structural genes, but through a diffusible substance which could move from the DNA strand $i^+ z^+ y^+ ac^+$ to the strand $i^-z^-y^+ac^+$. This diffusible substance, called a *repressor*, was believed to be a protein and was believed to act by binding to the structural genes in such a way as to prevent their transcription. Subsequent work has established that the repressor is indeed a protein (Gilbert and Müller-Hill, 1967) and that it does bind to the DNA in the structural gene region (Riggs, et al., 1968). A mechanism by which the inducer might act then suggested itself to Jacob and Monod—that is, that the inducer competes with the structural genes for the repressor, thereby removing it and enabling transcription to proceed.

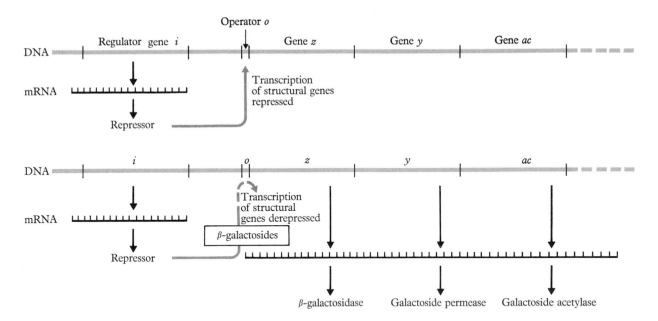

Fig. 8–32. The *lac* operon of *E. coli*. (See text for explanation.)

Table 8–4. An analysis of the genetic regulation of the *lac* operon in *E. coli*.

	Genotype	Phenotype
1.	$i^+ z^- y^+ ac^+$	Produces inactive form of β-galactosidase and normal products of y and ac, but only in the presence of inducer.
2.	$i^- z^- y^+ ac^+$	Produces inactive form of β-galactosidase, normal products of y and ac, all constitutively.
3.	$\dfrac{i^+ z^+ y^+ ac^+}{i^- z^+ y^+ ac^+}$	Produces all normal enzymes, but only in the presence of inducer.
4.	$\dfrac{i^- z^+ y^+ ac^+}{i^- z^- y^+ ac^+}$	Produces both forms of β-galactosidase constitutively, as well as normal products of y and ac
5.	$\dfrac{i^+ z^+ y^+ ac^+}{i^- z^- y^+ ac^+}$	Produces both forms of β-galactosidase, as well as normal products of y and ac, but only in the presence of inducer.

[After Jacob and Monod (1961).]

The next question to be resolved was that of the precise part of the DNA molecule to which the repressor binds. Jacob and Monod discovered a constitutive mutant that, on genetic analysis, gave results that differed from those obtained with i^- mutants. They found that when F$^-$ cells with the new constitutive mutation were made partially diploid by conjugating with Hfr cells with the genotype $i^- z^- y^+ ac^+$ to produce the partial diploid heterozygote (more properly termed a *heterogenote*)

$$\frac{i^? z^+ y^+ ac^+}{i^- z^- y^+ ac^+},$$

only the active form of β-galactosidase was produced constitutively. (Normal y and ac products were produced both constitutively and by induction.) This result implies that the z^- gene was repressed, which further implies that the $i^?$ allele is in fact i^+. The location of the new mutation was then mapped, and it was found to be immediately adjacent to the z locus. Jacob and Monod called this new locus the *operator* (*o*) and postulated that it is the site at which repressor is normally bound. The new mutant allele o^c ("c" for constitutive) is incapable of binding repressor, and so the associated structural genes are transcribed constitutively, irrespective of the presence of i^+ and repressor. The genotype of the foregoing heterogenote should therefore be written:

$$\frac{i^+ \ o^c \ z^+ \ y^+ \ ac^+}{i^- \ o^+ \ z^- \ y^+ \ ac^+}.$$

Jacob and Monod coined the term *operon* to apply to a group of genes whose transcription is controlled by an associated operator. This particular operon, being concerned with the enzymes of lactose metabolism, is known as the *lac* operon (Fig. 8–32). The original work of Jacob and Monod, which represents a milestone in genetic research, has been extended, and several other operons have been described. These include operons for the metabolism of galactose and glycerol phosphate; for the synthesis of the amino acids arginine, histidine, isoleucine, leucine, and tryptophan; and for protein synthesis in phage lambda (Martin, 1969). Some of the more interesting and significant findings that have emerged from investigations of these operons are discussed below.

a) The Promoter. According to the original Jacob-Monod model, an operon has a site for binding repressor and a site for mRNA initiation—that is, for binding RNA polymerase. There is no logical reason why the two sites cannot either be the same, overlap, or be entirely separate. The site for binding repressor has been identified as the operator. There is genetic evidence that the site for binding RNA polymerase is a separate promoter site and, for the *lac* operon at least, lies between the operator and the regulator (Beckwith and Zipser, 1970). Evidence for a promoter site takes two forms.

First, deletion of either the operator or the regulator results in constitutive transcription of the *lac* operon. Second, "leaky" mutations have been described which are characterized by slowed rates of *lac* operon transcription and which map between the i and o loci.

The promoter need not be located at the very end of the operon, as it appears to be in the *lac* operon. It can be inserted between the operator and the first structural gene, and there is some evidence that the tryptophan operon in *Salmonella typhimurium* has two promoters, one of which is located near the middle of the operon (Bauerle and Margolin, 1966).

As mentioned earlier, cyclic AMP plays a role in regulation of the *lac* operon (Fig. 8–33). This role has been shown to involve a specific protein, CAP, that binds to the operon and is a requisite for transcription (Eron, et al., 1971). De Crombrugghe, et al. (1971), who developed a cell-free system that includes only the components of the *lac* operon, found that CAP binds at the promoter site. They suggest that the *lac* promoter consists of two parts or subsites—one that binds RNA polymerase, as previously shown, and one that binds CAP. They also found that binding CAP is a necessary condition for binding RNA polymerase.

b) Number of Genes in an Operon. There is little doubt that a bacterial operon transcribes a single molecule of mRNA, and that this molecule has as many "stops" coded by terminator codons as there are genes within the operon and, correspondingly, polypeptides to be translated (Martin, 1963; Imamoto, et al., 1965). The *lac* operon includes the three structural genes described, as well as sequences of nucleotides that represent promoter and operator regions. The leucine operon of *S. typhimurium*, on the other hand, includes four structural genes (Burns, et al., 1966), and the histidine operon in *E. coli* includes at least ten. Operons that include more than one structural gene are known as *polycistronic operons*.

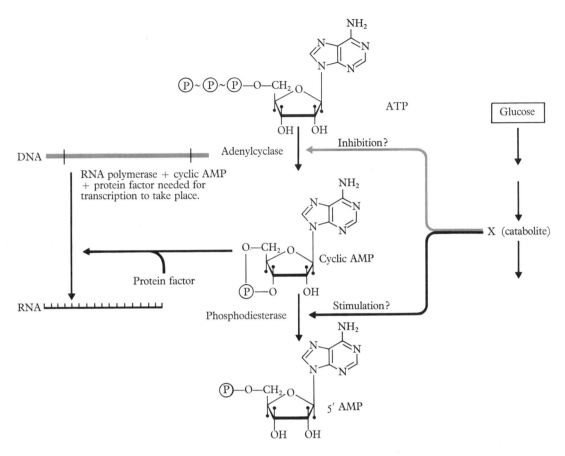

Fig. 8–33. The role of cyclic AMP in regulation of the *lac* operon in bacteria. [From Watson (1970), p. 460. By permission.]

c) Number of Operons per Regulator. Since regulator genes code the synthesis of protein repressors, which are diffusible in the cytoplasm, it is conceivable that a single regulator can affect a number of different operons, provided that there is an advantage to the coordinated transcription of the elements of these operons. The genes directly involved in the synthesis of the related amino acids, isoleucine and valine, appear to be distributed among three operons in *E. coli*, and they are all under the influence of the one regulator (Ramakrishnan and Adelberg, 1965).

d) Regulation of Repressible Enzymes. As mentioned earlier, enzymes that are concerned with anabolic processes are often repressible—that is, they are "turned off" in the presence of the substance that is normally the end product of their activity. A classic example of this is the group of enzymes coded by the histidine operon in *S. typhimurium* (Ames and Hartman, 1963). It was postulated that the regulators for repressible operons such as this code inactive forms of repressors which are incapable of binding to their respective operators. The inactive repressors are made active, however, when they combine with substances called *corepressors*. The corepressors are the end products of the biosynthetic pathways that they are instrumental in repressing, or they are substances derived from these end products. The corepressor of the histidine operon, then, is histidine itself; the corepressor of the arginine operon is arginine (Fig. 8–34).

e) Glucose Inhibition. A number of operons are inhibited by the presence of free glucose. These include the operons that are concerned with the catabolism of other carbohydrates such as lactose,

maltose, arabinose, and galactose. Hence, *E. coli* that are grown in a mixture of glucose and another sugar will utilize only the glucose. The mechanism of glucose inhibition appears to involve a reduction in available cyclic AMP, which is essential for transcription of the glucose-sensitive operons (Magasanik, 1970).

f) Translational Gradients. The order or sequence of structural genes in polycistronic operons is probably not unimportant, for it appears that they are translated decrementally. Translation of the *lac* operon, for example, results in the synthesis of

Fig. 8–34. A comparison between mechanisms of regulation of (a) inducible and (b) repressible enzyme systems.

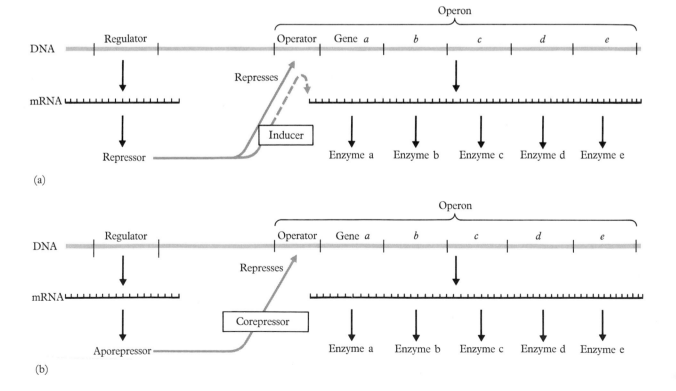

more molecules of β-galactosidase than of galactoside permease, and more of the latter than of the transacetylase (Beckwith and Zipser, 1970). The mechanism by which such differential production is controlled is not known. It may involve a certain chance probability of a ribosome dropping off at each chain termination signal, thus accounting for the regularity of decrement, or it may be something more precise. In any event, it appears to adjust for the relative numbers of enzymes needed in a given pathway, perhaps compensating for differences in stability (longevity) and/or turnover rates among them.

g) Nonregulated Genes and Feedback Control. Enzyme function can be blocked at any one of three different levels—transcription, translation, or specific activity—with equal effectiveness. Enzymes that are coded by regulated genes are, of course, blocked at the transcription level when their activity is not needed. Others either are not blocked at all, as is evidently the case for the enzymes involved in glucose degradation, or are blocked by feedback inhibition. This mechanism appears to play a very prominent role in many biosynthetic pathways and may apply even in the presence of an operon system (Ames and Hartman, 1963). In the biosynthesis of arginine, for example, as this amino acid begins to accumulate in the cell, it combines with the first enzyme in its own pathway of biosynthesis, thereby blocking that pathway at its beginning. It is believed to combine with the enzyme at a site other than the one or ones used in the pathway. In so doing it distorts the enzyme (allosteric effect) so that it cannot function. Superimposed upon this mechanism may be operon repression, as is the case in arginine biosynthesis, but this usually requires a higher concentration of end product than the feedback mechanism does. Between the two mechanisms, therefore, the rate of biosynthesis can be very finely controlled.

QUESTIONS

1. What are the determinants of the primary, secondary, and tertiary aspects of protein structure? How does each of these structural levels act as a determinant of the active site?

2. If a nutritionally deficient strain of *Neurospora* strictly requires arginine for growth (and will accept no other substitutes), in which of the seven genes of the pathway does the mutation exist?

3. What is the minimum number of DNA nucleotide pairs that must be altered to account for sickle-cell anemia?

4. Do both strands of a DNA double helix carry the same genetic information? Explain.

5. Suppose that an extra base pair could be inserted anywhere at will in the region of a DNA double helix that codes for a particular enzyme. Further, suppose that the active site of the enzyme lies in the middle of its single polypeptide chain. What would be the effect if the extra base pair were inserted at the beginning of the gene? middle of the gene? end of the gene? Compare.

6. Would it make any difference if DNA could be transcribed in both directions (i.e., from the 5' and the 3' end)? Explain.

7. Outline the differences in structure and function of the major varieties of RNA. How do the unique features of each contribute to the function of each?

8. Considering the entire process of protein synthesis (DNA → RNA → protein), how many points of biochemical or enzymatic specificity exist? Begin with a length of double-stranded DNA that is 12 base pairs long.

9. How is the amino acid methionine used as opposed to all other amino acids in protein synthesis?

10. In what ways do the nucleotides ATP, GTP, and cyclic AMP function during protein synthesis?

11. Speculate on the evolutionary origin of a degenerate genetic code.

12. What is the logic behind the theory that enzymes that work primarily in catabolic processes are inducible, whereas those that function anabolically are repressible?

13. What would be the behavior of the following *lac* operon genotypes?

a) $i^- z^+ y^+ ac^+$

b) $i^+ z^- y^- ac^-$

c) $\dfrac{i^- z^+ y^+ ac^+}{i^- z^+ y^+ ac^+}$

d) $\dfrac{i^+ z^- y^+ ac^+}{i^- z^- y^- ac^-}$

14. List all of the haploid and diploid genotypes that would result in the *lac* operon enzymes being produced constitutively.

15. What advantages accrue from the grouping of genes into operons?

16. In the *lac* operon, lactose functions both as a substrate for the enzyme and an inducer for the operon. In the histidine operon of S. *typhimurium*, what is (are) the role(s) of histidine?

9

Changes in Genetic Material

In view of the mechanism and fidelity of DNA replication, it is apparent that if one, two, or any number of base pairs are altered, the alteration will be as faithfully reproduced as the original sequence. We have already seen that the substitution of a single base pair in the DNA that codes the β-polypeptide of human hemoglobin can result in sickle-cell anemia, and that this altered base composition is reproduced in subsequent generations. But because of the degree of degeneracy of the genetic code (Table 8–3), certain base-pair substitutions can be tolerated, particularly in the third position of a triplet, with no alteration of the amino-acid sequence in the product. Moreover, amino-acid substitutions can be introduced at certain positions in a polypeptide molecule without noticeable modification of that polypeptide's function. For example, the amino-acid sequences of the insulins of various mammals (cattle, sheep, horses, swine, human beings) differ sufficiently to permit immunological differentiation between them, but there is no clinically detectable difference in function among them, and they are freely interchangeable (Sanger, 1955). Thus it is evident that there is a sufficient margin for error in the base composition of DNA, so that some alterations net no phenotypic change and others result in changes that have little or no functional importance. However, in this chapter the concern is with changes that are not only heritable, but also result in clearly distinguishable phenotypic modification.

Any change that is heritable and essentially permanent is called a *mutation*. Because the only consistent link between generations of cells or organisms is DNA (except, with qualification, in the RNA viruses), it follows that mutations represent changes in DNA. But what are the components of DNA that are amenable to change without making the molecule something other than DNA? The sugar cannot be changed, as far as we know, without drastically altering the molecule. Similarly, the phosphate cannot be changed. And none of the four bases can be permanently changed, as we shall see below. In effect, the four kinds of nucleotides that comprise DNA cannot be changed without resulting in a molecule that is no longer DNA. The only thing that can be changed and still preserve the vital, replicable characteristics of DNA is the number, sequence, and linear arrangement of its component nucleotides. Therefore, the kinds of mutations that are possible and that have been encountered are classified according to the kind and/or magnitude of nucleotide rearrangements that characterize these mutations. Changes that involve one or very few base pairs constitute what are known as *gene mutations,* whereas those that represent rearrangement of large blocks of DNA constitute *macrolesions* and *chromosomal anomalies.*

GENE MUTATIONS

Base-Pair Substitutions

Perhaps the most frequent kind of mutation at the molecular level is the substitution of one base pair for another due to an error in replication. Molecules are rarely static entities; rather they are more or less constantly undergoing internal rearrangement of some of their protons, electrons, and (consequently) bonds. For example, most of the time a molecule of thymine has the molecular configuration represented in Fig. 9–1(a), but it may

rarely have the so-called "enol" form shown in Fig. 9–1(b), owing to a shift of a proton (hydrogen) from the N in position 1 to the O of the keto group on the carbon in position 6. Alternatively, thymine may lose a proton and become ionized, as shown in Fig. 9–1(c). In either case, the hydrogen-bonding proclivities of thymine are altered so that it will be more likely to pair with guanine than with adenine, as shown in Fig. 9–2.

Tautomeric forms of the other bases of DNA and their pairing tendencies are represented in Fig. 9–3. The bearing that these tautomers have on mutation, although still largely hypothetical, is shown schematically in Fig. 9–4. Here we see that during replication of a DNA strand, one of the thymines happens to have switched to its rare enol form at the moment of exposure for hydrogen bonding. In this circumstance, guanine will likely be inserted in place of adenine, and the base sequence of the replicated strand will differ from the strand that it replaced. The most important result of the defect is that the modified strand will be perpetuated in subsequent generations, giving rise to a mutant cell lineage.

Transitions.   The kind of base substitution that is described above, that is, the substitution of a base pair by a totally different base pair (such as A-T→G-C, T-A→C-G, G-C→A-T, or C-G→T-A) is called a *transition* (Crick, et al., 1961). Transitions

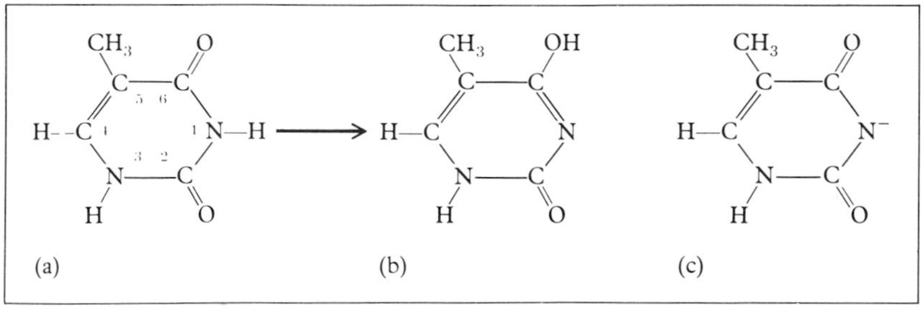

Fig. 9–1.   The various forms of the pyrimidine base thymine. (a) The common keto form. (b) The tautomeric enol form. (c) The ionized form.

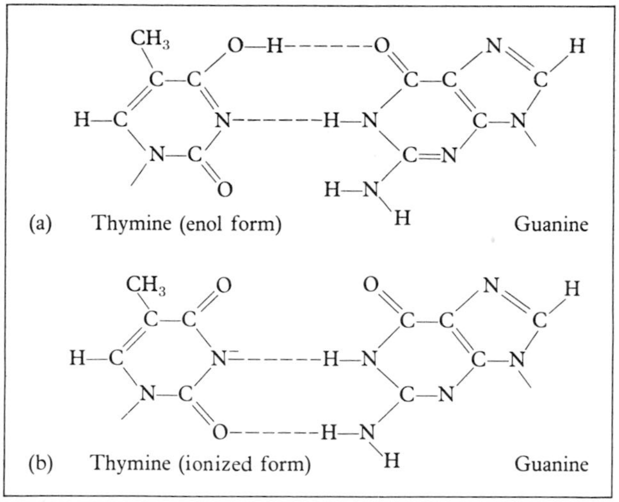

Fig. 9–2. The pairing tendencies of thymine in its (a) enol and (b) ionized forms.

Fig. 9–3. Tautomeric forms of the bases adenine, guanine, and cytosine, and the bases with which these forms tend to establish stable hydrogen bonds. The unusual base pairs that result are symbolized in the right-hand column.▼

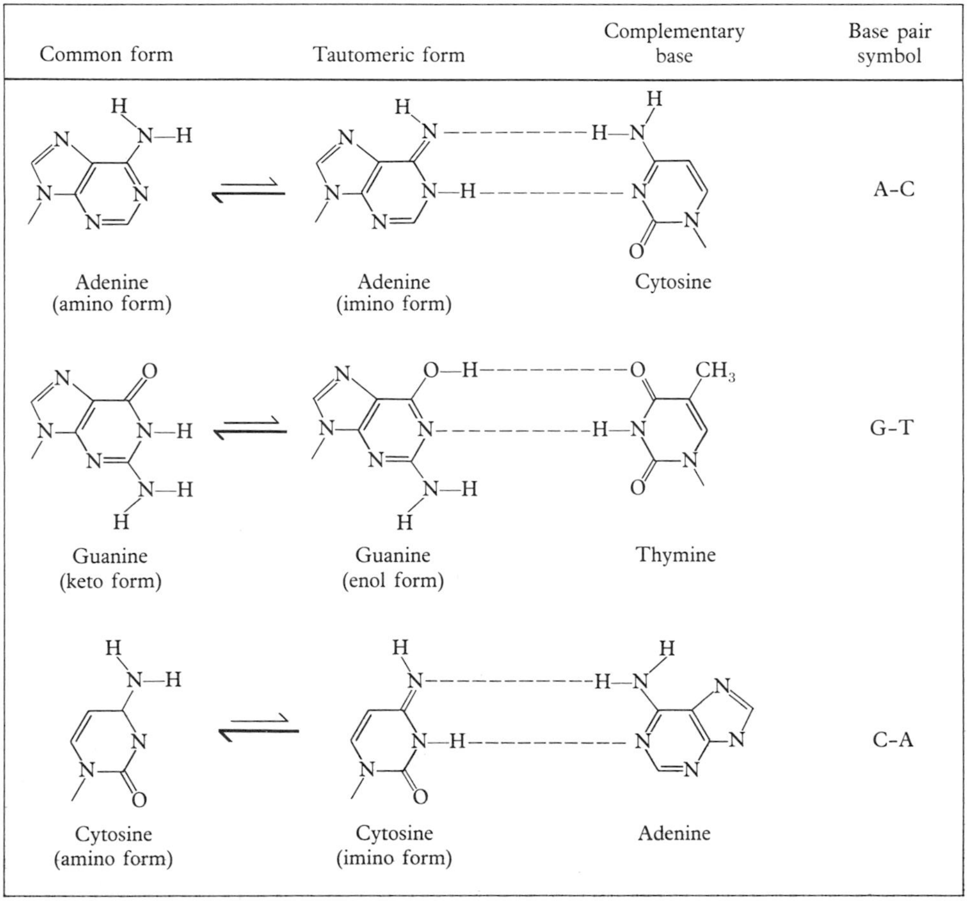

Common form	Tautomeric form	Complementary base	Base pair symbol
Adenine (amino form)	Adenine (imino form)	Cytosine	A–C
Guanine (keto form)	Guanine (enol form)	Thymine	G–T
Cytosine (amino form)	Cytosine (imino form)	Adenine	C–A

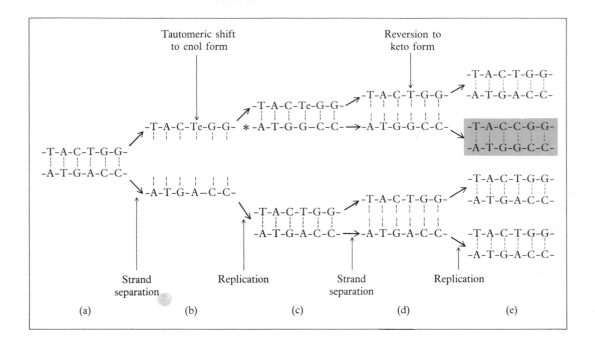

Fig. 9–4. The establishment of a mutant molecule of DNA as a consequence of tautomeric shift in a thymine component during replication. (a) A two-codon segment of wild-type DNA. (b) Strands have separated; the thymine indicated by the arrow has undergone a tautomeric shift from the keto to the enol form (Te). (c) Each strand has replicated, but the strand designated by the asterisk shows that the A in the fourth position has been substituted by a G. (d) Strands have separated preparatory to a second replication. The thymine in the fourth position in the top strand has reverted to its keto form. (e) Second replication has been completed. Three of the DNA generations have remained wild type, but the fourth (shaded) incorporates a base-pair substitution that can be perpetuated and would constitute a mutation. The amino-acid sequence that would be translated from the upper strand of the mutant DNA is *met-ala* (by way of A-U-G-G-C-C mRNA), whereas the wild-type sequence is *met-thr* (by way of A-U-G-A-C-C mRNA). If one visualizes the kind of replicative pattern shown here occurring during the last spermatogonial division and meiosis of sperm formation, one of the sperm cells would incorporate the mutation.

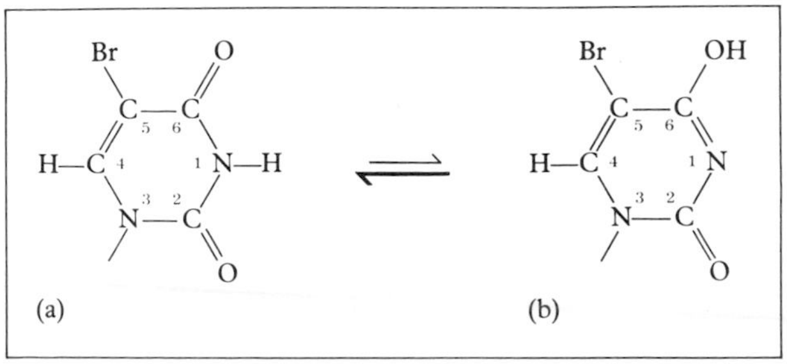

Fig. 9–5. Structure of the base analog 5-bromouracil (5-BU). This molecule is derived from uracil by substitution of bromine for the hydrogen on the carbon in the number 5 position. (a) In its normal keto form, it forms two hydrogen bonds analogous to those of thymine, and thus it will bond with adenine. (b) In its less common enol form, it forms three hydrogen bonds analogous to cytosine, and thus it will bond with guanine.

can come about spontaneously by either a tautomeric shift, as described here and postulated by Watson and Crick (1953c), or by base ionization, as proposed by Lawley and Brookes (1962). They can be induced experimentally by several chemicals known as *chemical mutagens* (Auerbach and Robson, 1946), of which 5-bromouracil (5-BU), 2-aminopurine, nitrous acid (HNO₂), hydroxylamine, and nitrosoguanidine (NG) have proved to be especially effective. The first two of these are base analogs, that is, they are molecules that, although not usually encountered in nature, are accepted by cells as substitutes for naturally occurring bases and do not prevent replication. The base analog 5-BU, for example, combines readily with

deoxyribose and phosphoric acid in cellular metabolism to form the nucleotide 5-bromo-deoxyuridilic acid. This nucleotide has hydrogen-bonding characteristics that are sometimes like those of thymine and at other times like those of cytosine, depending upon whether the 5-BU component is in the keto or the enol state (Fig. 9–5). It undergoes this tautomeric shift more readily than either thymine or cytosine, and thus is more likely to lead to transition substitutions, as shown in Fig. 9–6.

Nitrous acid brings about oxidative deamination of purines and pyrimidines, with consequent alteration of their hydrogen-bonding proclivities. Specifically, adenine is transformed to hypoxan-

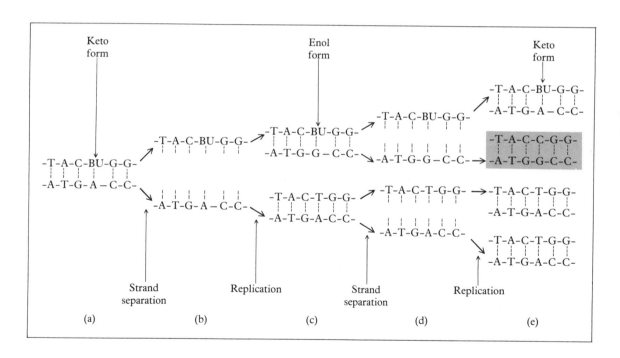

Fig. 9–6. Mutagenic action of 5-bromouracil (5-BU). (a) In its keto form, 5-BU has been incorporated into DNA synthesis to form a 5-BU–A base pair. (b) Strand separation occurs preparatory to replication. (c) 5-BU changes to its enol form and accepts G as its complement during replication. (d) Strand separation preparatory to the next replication occurs. (e) A mutant molecule like that described in Fig. 9–4 is established (second from top, shaded).

thene which tends to pair with cytosine; guanine becomes xanthene which still pairs with cytosine, but through two H-bonds rather than three; and cytosine is deaminated to uracil which pairs with adenine (Fig. 9–7). The consequences of such alteration of these bases is the induction of transition mutations (Freese, 1959).

Hydroxylamine is believed to alter cytosine so that it pairs with adenine (Freese, et al., 1961), whereas NG is thought to work by forming an unstable but highly mutagenic base analog or by altering DNA polymerase (Drake, 1969, 1970).

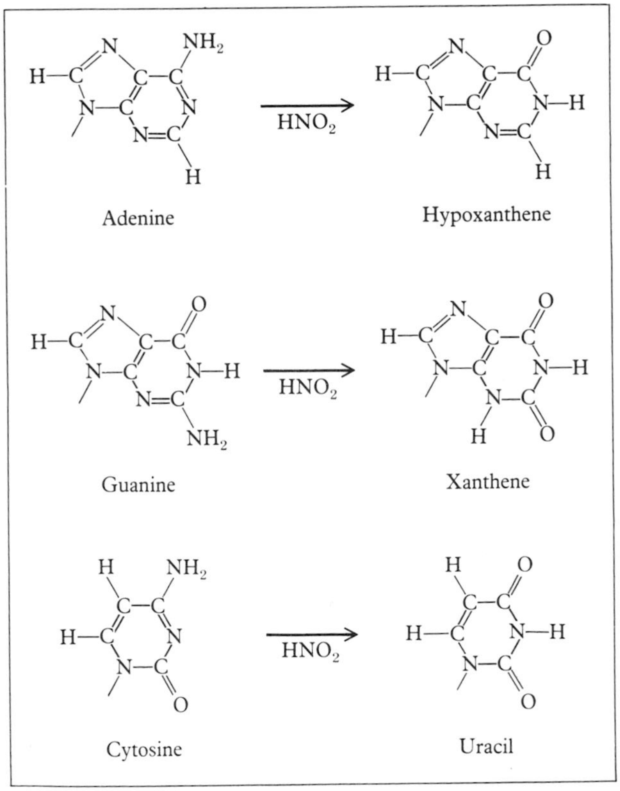

Transversions. Whereas in transitions the purine-pyrimidine axis is preserved, in transversions it is reversed. Hence transversions would include any one of the following changes: A-T↔T-A, G-C↔C-G, A-T↔C-G, and T-A↔G-C. Originally postulated by Freese (1959), molecular mutations of this class are believed to arise spontaneously but are difficult to induce chemically, although there is some evidence that NG may be effective (Cerda-Olmedo, et al., 1968; Drake, 1969). At best, they would be difficult to detect and could easily be confused with insertion or deletion of one or more base pairs.

Microlesions and DNA Repair

A small segment of DNA amounting to perhaps one or two nucleotide pairs may be lost or duplicated, constituting what is known as a microlesion. A frequent cause of this anomaly is believed to be the induction of pyrimidine *dimers* by such mutagens as ultraviolet light (Fig. 9–8). According to Witkin (1969), such dimers cause most of the lethal and mutagenic effects that have been observed in some strains of bacteria. They do this by interfering with both transcription and replication. The evolution of prokaryotes took place before a protecting layer of ozone appeared in the upper atmosphere, consequently during a time in the earth's history when ultraviolet levels were exceptionally high. The molecular structure of DNA is such that its internal bonds are highly susceptible to damage by irradiation in the ultraviolet frequency range. Microorganisms that were exposed to high intensities of ultraviolet irradiation were obliged to acquire some sort of protective mechanism if they were to survive. Those that survived did so because they developed enzymatic mechanisms to *repair* DNA, not to prevent DNA damage.

Two mechanisms for repairing damage caused by the formation of pyrimidine dimers are known in bacteria (Witkin, 1969)—*photoreactivation* and *excision-repair*. The first of these involves production of an enzyme that specifically splits pyrimidine dimers. Activation of this enzyme has

Fig. 9–7. Oxidative deamination by nitrous acid of the amino purines adenine and guanine and the pyrimidine cytosine.

been shown to occur only in the presence of light in the visible range of the spectrum. The second mechanism, which is not light-dependent, utilizes three enzymes that act in succession, as follows. (1) An exonuclease is used to excise the section of one of the DNA strands that incorporates an error, such as a pyrimidine dimer. The excised segment may include only the offending nucleotide or, more likely, a short series of nucleotides. (2) DNA polymerase then directs the resynthesis of the segment that has been excised. (3) DNA ligase joins the repaired segment to the rest of the DNA strand by phosphodiester bonds (Fig. 9–9). Excision-repair mechanisms, unlike the photoreactivation enzyme, are not specific for eliminating pryimidine dimers, but are effective for other abnormalities as well.

Processes of DNA repair occasionally lead to either insertion or deletion of one or more nucleotides. Inserted nucleotides almost invariably represent one or more duplications of the intact nucleotide on either side of the point of insertion. This is to say that if in the nucleotide sequence A-T-G-C a nucleotide is inserted between T and G, it will be either a T or a G nucleotide, depending upon the 3'-to-5' orientation of the strand. If it is a double insertion, both nucleotides will be identical, in this case being either T-T or G-G.

Frameshift Mutations. Crick, et al. (1961) reasoned that if the reading of DNA takes place by triplets from a fixed point, either the insertion or deletion of one or two nucleotide pairs would produce a shift in the "reading frame," as diagrammed in Figs. 8–17 and 9–10, and would represent a mutation. This hypothesis has been confirmed several times (Terzaghi, et al., 1966; Okada, et al., 1966; Brammar, et al., 1967) by inducing frameshifts. A family of compounds called the *acridines* have been found to be especially effective frameshift muta-

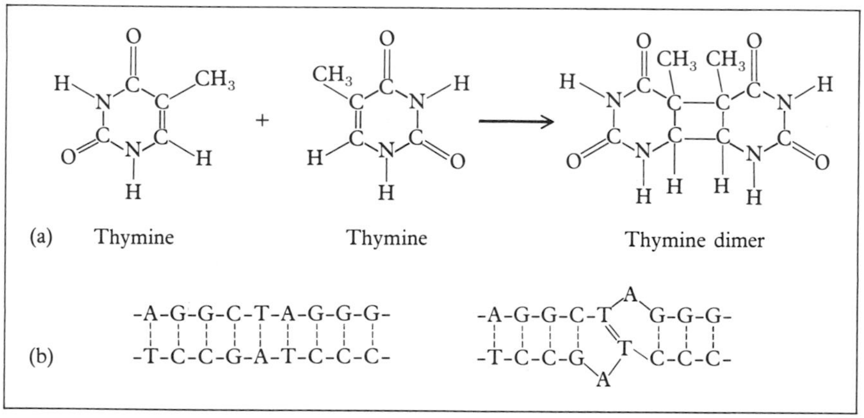

Fig. 9–8. The dimerization of thymine. (a) Two molecules of thymine form a dimer by covalent bonding. (b) The distorting action of a thymine dimer on a molecule of DNA. This distortion may interfere with mRNA transcription or replication, or may lead to errors in both of these processes. The hydrogen bonds are represented by broken lines. The covalent bond of the dimer is represented by a double solid line.

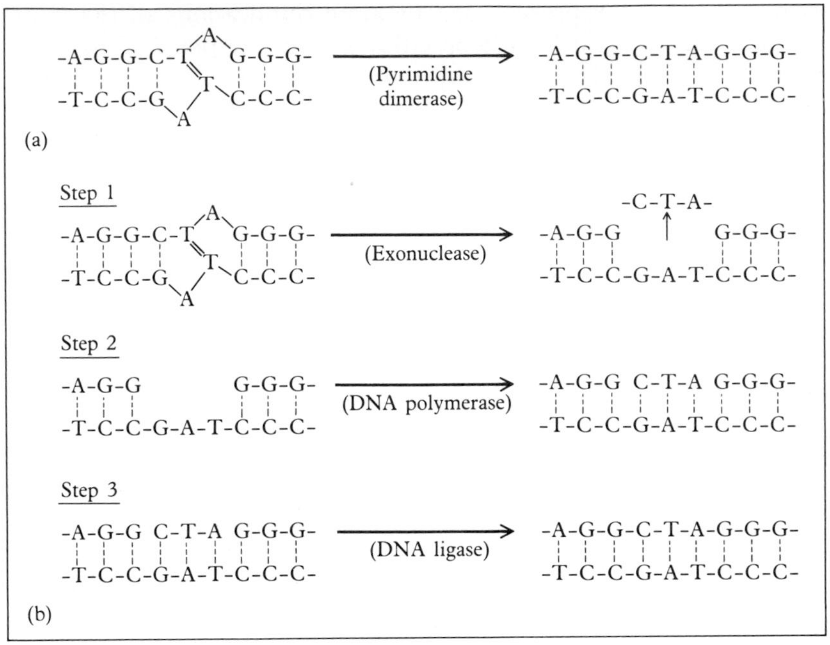

-A-G-G-C-T︵A︵G-G-G-   ──────────→   -A-G-G-C-T-A-G-G-G-
  ┊ ┊ ┊ ┊ ┊   ╲   ┊ ┊ ┊   (Pyrimidine   ┊ ┊ ┊ ┊ ┊ ┊ ┊ ┊ ┊
-T-C-C-G╱ ╲T╲C-C-C-   dimerase)   -T-C-C-G-A-T-C-C-C-
                ︶A︶

(a)

Step 1

                                                           -C-T-A-
-A-G-G-C-T︵A︵G-G-G-   ──────────→   -A-G-G   ↑   G-G-G-
  ┊ ┊ ┊ ┊ ┊   ╲   ┊ ┊ ┊   (Exonuclease)   ┊ ┊ ┊   ┊ ┊ ┊
-T-C-C-G╱ ╲T╲C-C-C-   -T-C-C-G-A-T-C-C-C-
                ︶A︶

Step 2

-A-G-G   G-G-G-   ──────────→   -A-G-G C-T-A G-G-G-
  ┊ ┊ ┊   ┊ ┊ ┊   (DNA polymerase)   ┊ ┊ ┊ ┊ ┊ ┊ ┊ ┊ ┊
-T-C-C-G-A-T-C-C-C-   -T-C-C-G-A-T-C-C-C-

Step 3

-A-G-G C-T-A G-G-G-   ──────────→   -A-G-G-C-T-A-G-G-G-
  ┊ ┊ ┊ ┊ ┊ ┊ ┊ ┊ ┊   (DNA ligase)   ┊ ┊ ┊ ┊ ┊ ┊ ┊ ┊ ┊
-T-C-C-G-A-T-C-C-C-   -T-C-C-G-A-T-C-C-C-

(b)

Fig. 9–9.   Two mechanisms of DNA repair. (a) Direct splitting of pyrimidine dimers. (b) A three-step process of DNA repair by excision and resynthesis. Three enzymes are utilized.

	Base sequence of transcriptional strand of DNA	Amino acid sequence coded	Comment
(a)	T-A-C-T-G-G-T-G-G-T-G-G-T-	Met-Thr-Thr-Thr-etc.	Wild type
(b)	(insert) T-A-C-(T)-T-G-G-T-G-G-T-G-G-	Met-AspN-His-His-etc.	Mutant
(c)	(insert) T-A-C-(T-T)-T-G-G-T-G-G-T-G-G-	Met-Lys-Pro-Pro-Pro-etc.	Mutant
(d)	(delete (T)) T-A-C-G-G-T-G-G-T-G-G-T-	Met-Pro-Pro-Pro-etc.	Mutant
(e)	(delete (T-G)) T-A-C-G-T-G-G-T-G-G-T-	Met-His-His-His-etc.	Mutant
(f)	(delete (T)) (insert) T-A-C-G-G-(G)-T-G-G-T-G-G-T-	Met-Pro-Thr-Thr-Thr-etc.	Pseudo wild type

gens, as has ultraviolet light. The chemical frameshift mutagens are believed to act not so much by causing breaks in the DNA strands, although doubtless they do this, as by adversely affecting the process of DNA repair in such a way that either deletions or redundancies of nucleotides occur (Streisinger, et al., 1966).

Macrolesions and Chromosomal Anomalies

Disruption or rearrangement of nucleotide sequences on a more extensive scale than those discussed above constitute *macrolesions* (Drake, 1969). Ordinarily, macrolesions involve more than the equivalent of one or two genes and frequently embrace more or less considerable sections of the chromosomes. The magnitude of the latter is sufficiently great to be observable by means of an ordinary compound microscope (Fig. 9–11). Macrolesions that involve only a few score or even a few hundred nucleotides are less easily observable. Little is known about the mechanisms by which macrolesions are caused beyond what has been learned by the use of ionizing radiation in eukaryotes and ultraviolet light or HNO_2 in prokaryotes. The categories of macrolesions are *deletions, duplications,* and *segmental rearrangements.*

Deletions. Deletions (also known as deficiencies) represent the most frequent and undoubtedly the most severe form of macrolesion. Deletion of sev-

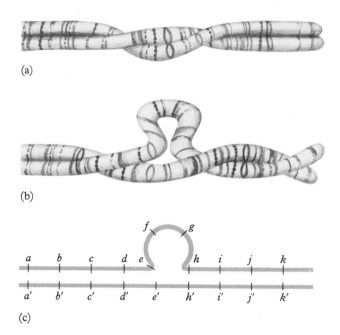

Fig. 9–11. Localization of a deletion in a giant chromosome of a dipteran larva. (a) Normal pair of chromosomes—homologous components are in continuous synapsis. (b) Chromosomes in which one homologue is characterized by a deletion. A loop in the upper chromosome appears opposite the location that would normally be occupied by the deleted segment. (c) A schematic explanation of loop formation in deletion heterozygotes. Loci *f* and *g* are excluded from synapsis for lack of homologous loci.

◀ **Fig. 9–10.** Base-pair deletion and/or insertion in DNA. For simplicity, only the strand that transcribes mRNA is shown. (a) Hypothetical wild-type sequence that codes an isopolymer of threonine, starting with the initiator methionine. (b) If thymine is inserted after the initiator codon, the frames are shifted one position to the left, coding a polymer of histidine after methionine and asparagine. (c) If two bases are inserted, all the frames to the right are shifted two positions to the left, resulting in a polymer of proline. (d) Deletion of a base results in a frameshift one position to the right and codes a polymer of proline. (e) Deletion of two bases results in a frameshift two positions to the right and codes a polymer of histidine. (f) Deletion of one base and insertion of another produces an amino-acid substitution between the respective points, but nets no change from the normal beyond them. Hence in this example, the amino acid proline is substituted for a residue of threonine in one position only; otherwise the polypeptide is wild type. It is unlikely that a single substitution such as this will be phenotypically evident (although it may be), and so the change is termed "pseudo wild type."

eral nucleotides from a gene will essentially render that gene dysfunctional, whereas larger deletions result in the loss of anywhere from one gene to several or even blocks of genes. Deletions of any magnitude large enough to constitute a macrolesion are almost invariably lethal in haploid organisms. Their effect in diploid cells or organisms is variable depending upon the number of genes involved, the quantitative requirements of the products of the genes affected, the position of the genes in an operon or other functionally coordinated gene group, and so on. A diploid cell or organism that is homozygous for a given deletion is in essentially the same situation as a haploid cell—that is, the deletion is likely to be lethal.

The giant or polytenic chromosomes of certain dipteran larvae, which are described in greater detail in Chapter 11, consist of multiples of chromosome strands in synapsis. That is, homologous chromosomes are paired throughout their length. Deletions on the order of 5μ or longer are readily located in these chromosomes when they are heterozygous because the "nondeleted" segment, having no homologous segment with which to pair, appears as a hump or loop on the chromosome. Figure 9–11 illustrates a deletion heterozygote (a cell that has a deletion on only one of a pair of chromosomes) and gives the rationale for the effect it has on chromosome structure. Smaller deletions can be observed by electron microscopy. In order to do this, a mixture of DNA from a deleted and a nondeleted chromosome is "melted" (cf. p. 157 and Fig. 7–20) and then annealed by cooling slowly. Deleted single strands have a 50 percent probability of undergoing base pairing with intact strands, yielding configurations similar to those that characterize deletion heterozygotes in polytenic chromosomes (Davis and Davidson, 1968).

The first genetic proof of chromosomal deletion is attributed to Bridges (1917). In his investigations of sex-linked inheritance in *Drosophila*, Bridges used females having two X chromosomes and a Y (cf. Chapter 5 and Fig. 9–12). These

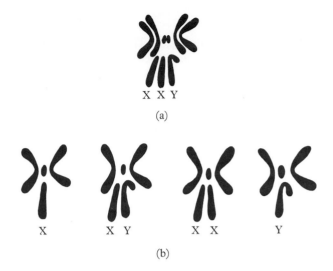

Fig. 9–12.   Karyotype of (a) somatic cells and (b) ova of Bridges's exceptional female fruit fly, *Drosophila melanogaster*.

females produce four types of eggs (X, XY, XX, and Y), and when mated with normal males produce variable offspring in the ratio of two XXY females to one XX female to two XY males to one XYY male. The genotypes XXX and YY doubtless occur, but because of gross deficiencies or lack of genic balance, they are usually inviable and consequently not recoverable. By crossing XXY females with males having specific "markers" (phenotypically recognizable mutations) on the X chromosome, Bridges was able to incorporate these markers in the progeny, which in effect enabled him to trace each X chromosome in subsequent crosses and generations. In this way he developed females having the marker w^e (eosin eye) on the X chromosome. He crossed these females with males having the markers w (white eye) and B (bar eye) on their X chromosome (Fig. 9–13). The wild-type allele w^+ is dominant to either w or w^e, and so these females had red or wild-type eyes. The males, having only one X, expressed both w and B; accordingly they had eyes that were both white and bar-shaped. Knowing that w^e is dominant to w and that B expresses itself when heterozygous, one would expect among the female

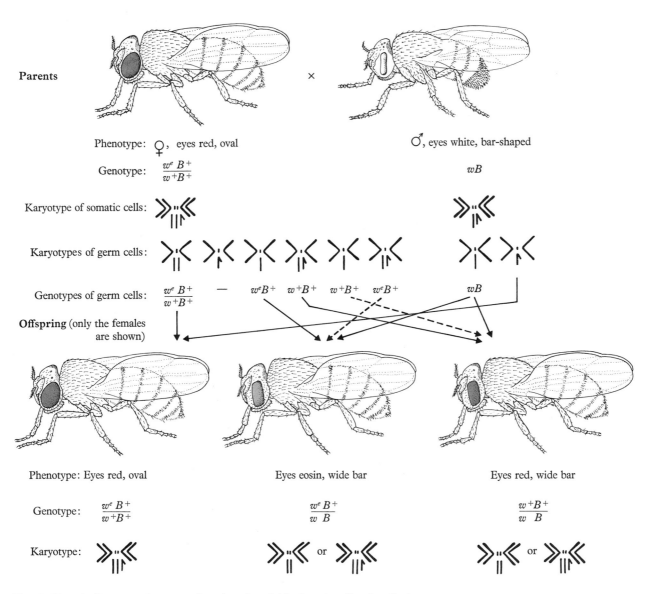

Fig. 9–13. A diagram of a cross showing the viable female offspring that would be predicted from a mating of XXY female fruit flies with the genotype $w^e B^+/w^+ B^+$ and male fruit flies with the genotype wB.

progeny resulting from this cross only the phenotypes shown in Fig. 9–13. These phenotypes were encountered, in fact, *but one female had eyes that were eosin in color but normal in shape.* It had obviously inherited the w^e and B^+ alleles from its

mother, but evidently the w allele and not the B allele from its father. Bridges crossed the female to a normal male and found that while none of the offspring had bar-shaped eyes, they had the sex ratio of two females to one male.

Suspecting that a segment of the chromosome that included the *B* gene had been lost or deleted, resulting in a genetic deficiency lethal in males, Bridges crossed females having the suspected deletion with males that had one of several recessive alleles located adjacent to the bar locus. The cross in which the males carried the recessive allele forked (*f*) resulted in 50 percent forked females (Fig. 9–14). This observation supports the hypothesis that the nature of the abnormality is a deletion, and that it includes the *f* locus as well as

the *B* locus. Subsequent work has provided additional support to Bridges's interpretation, and the extent of the deletion has been mapped. Numerous other deletions have been described and verified in *Drosophila* and in other species from viruses to mammals. Moreover, it has been determined that about 20 percent of all lethal mutations that occur on the X chromosome of *Drosophila*, irrespective of whether they are spontaneous or induced, are attributable to small deletions (Slizynska and Slizynska, 1947). The *cri du chat* syndrome that occurs on occasion among human newborns is believed to be due to a deletion in the fifth chromosome (Philips, et al., 1970; Lejeune, 1964). This syndrome is characterized by abnormal development of the brain and cranium, among other things, and by the fact that the infant's crying sounds are like the mewing of a cat, from which the anomaly received its name. Cri du chat infants rarely live beyond a few hours after birth.

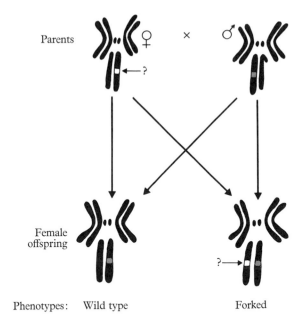

Fig. 9–14. When female fruit flies having a suspected deficiency in one X chromosome were crossed with males having the recessive mutant allele forked (*f*) on its X chromosome, half of the female progeny showed the recessive trait. This is interpreted as evidence that there is a deficiency and that it includes the forked locus. The position of the suspected deficiency is open and the forked locus is indicated by the shaded area.

Duplications.   Segments of chromosomes may be repeated in the genome, either on the same chromosome or on another chromosome. The segment may be as small as a single gene or it may comprise large numbers of genes. A clearly demonstrable example of duplication is the bar mutation in *Drosophila,* mentioned in the preceding section. Bridges (1936) discovered a repeat of six bands in the salivary-gland chromosomes of homozygous bar females (Fig. 9–15); the repeat or "duplication" was in the region of the X chromosome on which the mutation had been previously mapped by genetic means. Sturtevant (1925) discovered that bar can revert to wild type with a frequency that can be explained by oblique or "unequal" crossing-over, as diagrammed in Fig. 9–15. Implicit in this discovery is the fact that the physical basis of the bar mutation is simply a duplication and not one of the gene mutations described earlier in this chapter.

Figure 9–15 also shows two other mechanisms by which duplications may arise. Note that the

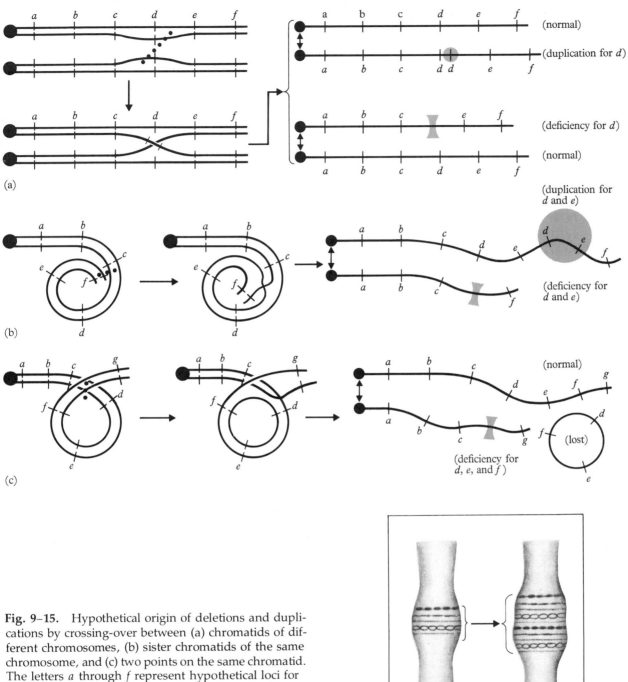

(a)

(b)

(c)

Fig. 9–15. Hypothetical origin of deletions and duplications by crossing-over between (a) chromatids of different chromosomes, (b) sister chromatids of the same chromosome, and (c) two points on the same chromatid. The letters *a* through *f* represent hypothetical loci for reference. The splitting of centromeres is indicated by double-headed arrows. The point of contact and crossing-over is indicated by the dotted mark. Inset compares the "bar" region in salivary-gland chromosomes of wild-type and mutant *Drosophila melanogaster.*

Wild type   Bar

mechanisms that produce a duplication in one chromatid simultaneously effect a deletion in another. Whereas deletions are generally harmful and necessarily lead to evolutionary regression, duplications are believed to be often harmless and to provide genetic redundancy that may actually facilitate evolutionary progression (cf. Ohno, et al., 1968; Britten and Davidson, 1971). They may represent the physical basis for multiple factors (polygenes), for example. Some duplications are deleterious, however. For example, there is a causal relationship between duplication of part of one of the chromosomes (usually chromosome 21)

and Down's syndrome (mongolism) in human beings. The repeat involved here may be occasioned by *trisomy*, in which case the entire chromosome is represented three times instead of the normal twice, or by *translocation* of the significant segment. Both of these situations are described later on in this chapter. The various chromosomal associations of duplications that might occur are diagrammed in Fig. 9–16.

It is difficult to explain the circumstances in which duplications have a phenotypic effect, as they sometimes do. The first thought would be that if genes code enzymes, then at most a duplication

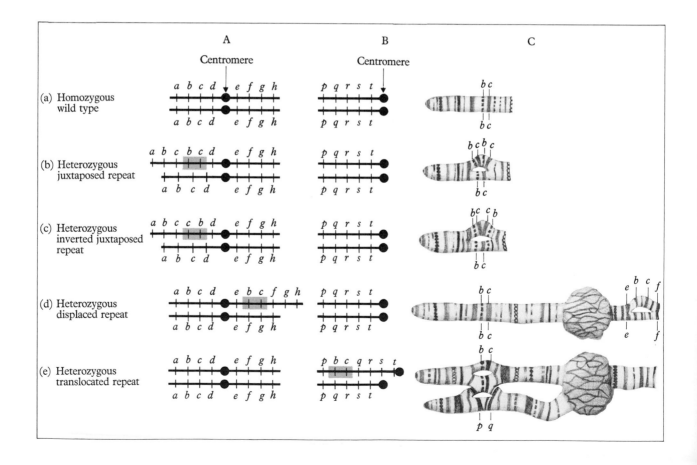

would result in the production of excessive amounts of the enzymes coded by the duplicated genes. As a matter of fact, this might be all that is needed to have an effect for at least two reasons. (1) Many enzymes utilize or "compete for" the same substrate, and any change in the amount of one of the competing enzymes would result in a shift in the relative amounts of the end products of those enzymes, resulting in a phenotypic effect. This certainly would apply to situations in which substrate availability is limited in either time or amount. (2) Enzymatically regulated processes are profoundly influenced by a number of different factors, one of which is enzyme concentration. (Others which perhaps are not significant in this situation are pH, temperature, concentration of substrate and/or product, etc.) Hence a duplication might result in quantities of some metabolic product that are excessive for normal function, and this would be expressed as a phenotypic effect.

If a duplicated sequence is inserted into a "foreign" operon, presumably it would come under the influence of that operon's regulator gene. In this situation, the duplicated gene or genes might be "turned on" by metabolic conditions that normally would not turn them on. In a delicately regulated chemical system, which is what a normal cell's metabolism is, a miscue of this kind could lead to very pronounced consequences. This kind of situation may be the basis of the so-called "position effect," whereby the genetic

neighborhood of one locus can alter the effect of that locus. This sort of situation has been observed and described, and is an effect applicable to the bar mutation itself (Sturtevant, 1925).

Perhaps the most important feature of duplications is that they represent a means of introducing new genes into the genome (Ohno, et al., 1968). These genes would be free to mutate with little chance of deleterious effect, because the enzymes that they initially coded would be taken care of by their prototypes already represented in their original position in the genome. Thus, new enzymes could evolve as additions to a cell's metabolic machinery, and new functions and/or structures could be introduced into the cell's lineage. The fact that many banding patterns in the chromosomes of contemporary species such as *Drosophila* are represented many times throughout the genome supports this idea.

Segmental Rearrangements

Inversions.   If crossing-over occurs within a chromatid at the point where the strand crosses itself in a simple loop formation, the sequence of genes in the internal segment may be reversed (Fig. 9–17). If the inverted segment does not include the centromere, it is called a *paracentric inversion;* otherwise it is known as *pericentric inversion.* A remarkable analysis of a pericentric inversion in the amphibian *Pleurodeles waltlii* is shown in Fig. 9–18. Inversions may or may not be accompanied by phenotypic

◀ **Fig. 9–16.**   Arrangements by which a duplication may be incorporated within a genome. Two pairs of chromosomes are indicated in columns A and B, respectively, as lines with hypothetical gene loci represented by letters *a* through *h* in one pair of chromosomes, and *p* through *t* in the other. The region of the repeat is shaded. In column C, the appearance of polytenic chromosomes in the regions in which the repeated loci occur is represented. Note the cross-tie between nonhomologous chromosomes in the case of a translocated repeat (lower right). In preparations of giant chromosomes from the salivary glands of dipteran larvae, the centromeres of the various chromosomes are usually clustered together and appear as a fuzzy and amorphous region, as shown.

expression, depending perhaps upon whether or not there is a position effect, as was postulated for duplications. Position effect would certainly be expected if either or both ends of the inversion were included in an operon sequence.

The effects of inversions on meiosis in cells heterozygous for the anomaly are both interesting and significant in terms of evolutionary implications. Studies on the giant salivary-gland chromo-somes of a number of races and closely related species of *Drosophila* have revealed that, among other things, inversions represent a consistent and probably important mechanism for isolating one incipient species from another, a necessary condition for speciation. The mechanism in this instance would probably be reduced fertility of the hybrids owing to abnormal products of meiosis, as shown in Fig. 9–19.

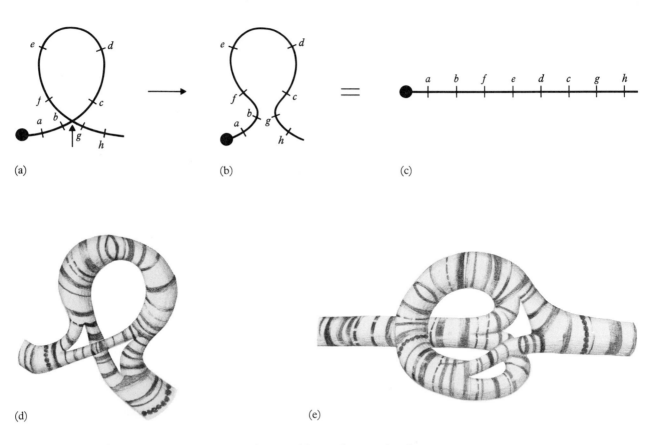

Fig. 9–17. Diagrammatic representation of a possible mechanism for the production of paracentric inversions. Crossing-over is postulated to occur at the point of chromatid crossing (arrow) in a loop (a) resulting in inverted gene order in the intermediate segment (b and c). The appearance of inversion heterozygotes in *Drosophila melanogaster* (d and e).

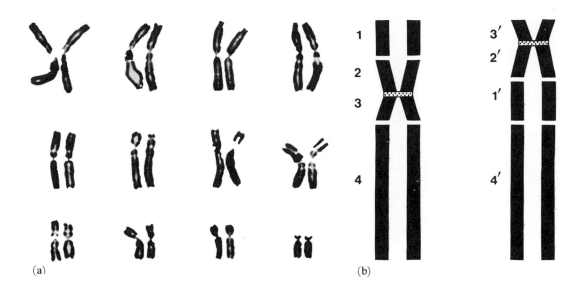

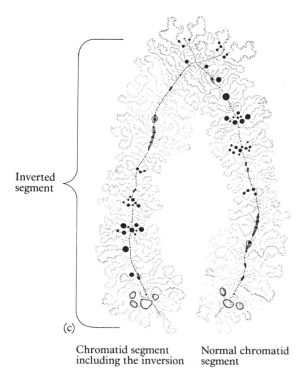

Fig. 9–18. (a) A karyotype of the urodele *Pleurodeles waltlii*. Note the pair of acrocentric chromosomes in the second row, second from the left. One of the members of this pair is normal; the other member of the pair has a pericentric inversion. This is diagrammed in (b) and an analysis of this portion of the lampbrush chromosomes is shown in (c). [(a) and (b) courtesy of A. Jaylet (1971), *Chromosoma* (Berlin) **35**:288–299; (c) from J. C. Lacroix and A. Jaylet (unpublished). By permission.]

Inverted segment

Chromatid segment including the inversion

Normal chromatid segment

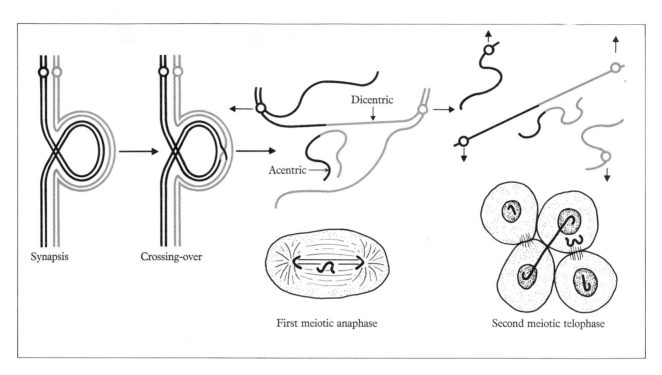

Synapsis   Crossing-over   Acentric   Dicentric   First meiotic anaphase   Second meiotic telophase

Fig. 9–19.   Dicentric and acentric chromatid products resulting from crossing-over within the limits of a paracentric inversion.

Translocations.   The removal of a chromosome fragment to a nonhomologous chromosome represents a *translocation*. The translocated segment may be small or it may be a major part of a chromosome. The observation that translocations are most frequently reciprocal—that is, a segment of one chromosome is exchanged for a segment of another—suggests that they may arise by crossing-over between nonhomologous chromosomes, as shown in Fig. 9–20. Rarely a fragment of a chromosome may be inserted into another without any loss of genetic material by the recipient chromosome, as diagrammed in Fig. 9–16. If the chromosome from which the fragment came and which, accordingly, is deficient in that fragment is also present, there is no net loss or gain of genetic material and any phenotypic effect is likely to be of minor importance (except when a position effect occurs). On the other hand, if the deficient chromosome is replaced by a normal, intact one, then the translocated segment represents a duplication, and the probability of a significant phenotypic effect is increased. Still more rarely, a terminal piece of one chromosome may be translocated to the end of another. It is believed that certain cases of Down's syndrome (mongolism) are a manifestation of this kind of anomaly, in which a segment of chromosome 21 appears translocated onto another chromosome. In an extensive study of Down's syndrome in Japan, 14 out of 321 cases showed translocation of part of chromosome 21 onto one of several other chromosomes (Higurashi, et al., 1969).

The most remarkable cases of translocation are

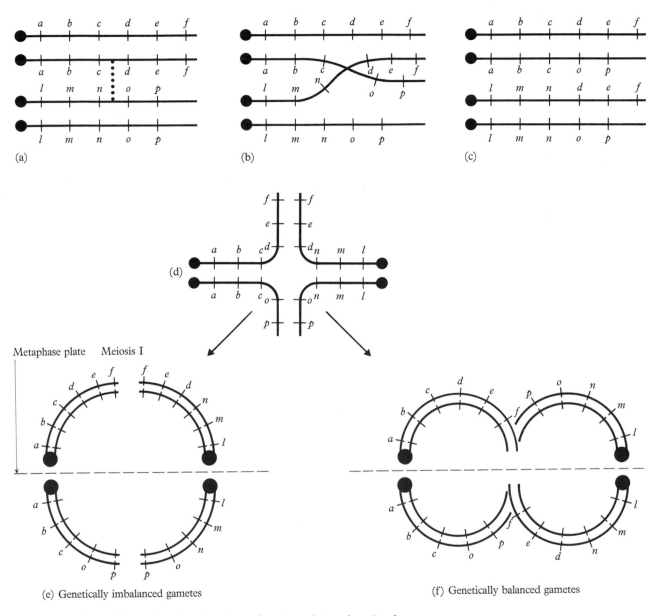

(a)

(b)

(c)

(d)

Metaphase plate   Meiosis I

(e) Genetically imbalanced gametes

(f) Genetically balanced gametes

Fig. 9–20. (a to c) Postulated mechanism of reciprocal translocation by crossing-over between nonhomologous chromosomes. (d) Synaptic figure produced during meiotic prophase I by a translocation heterozygote. (e and f) Two possible ways of alignment of the tetrads of a translocation heterozygote during metaphase I, one resulting in nonviable gametes due to genetic imbalance (simultaneous deletion and duplication) and the other in perpetuation of the translocation.

found in some species of flowering plants. The genus *Oenothera* (evening primrose) includes several "species" that differ from one another primarily in the number and kinds of translocations they have. Moreover, the translocations are maintained in heterozygous condition through incorporation of a series of balanced lethal genes (cf. pp. 121 and 246). In the most extensively studied species, *O. lamarckiana*, translocations involve all but one pair of the fourteen chromosomes, with the result that during the diakinesis stage of meiotic prophase, the tetrads appear in two rings, one of which is formed by twelve chromosomes. It should be pointed out here that events leading to diakinesis frequently (particularly in plants) include separation of pairs of chromatids starting at the centromeres and proceeding to the ends, whereupon separation stops. The pairs of chromatids remain in synapsis only at their ends, a condition known as *telosynapsis*. There are varieties of *Oenothera* that show the ultimate in multiple translocations, for in them all fourteen chromosomes have been involved so that a large, single circle of chromosomes is seen in telosynapsis at diakinesis (Fig. 9–21). According to Cleland (1962), most races of evening primroses found in nature across the United States have such a ring of fourteen chromosomes at diakinesis.

Crossing-over between chromatids of a translocation heterozygote may or may not increase the frequency of genetically unbalanced gametes, depending upon the position of the centromere in relation to the translocated segment and upon the location of the chiasma. If the centromere is situated at or very close to the point of junction of the translocated segment, as is evidently the case in *Oenothera,* crossing-over would neither increase nor decrease the frequency of unbalanced gametes. On the other hand, if there is an appreciable stretch of chromosome between the centromere and the translocated segment and crossing-over occurs within this stretch, the frequency of unbalanced gametes would be affected.

Fig. 9–21.   An analysis of ring formation in a translocation heterozygote of the evening primrose *Oenothera.* If, as suggested by Cleland (1962), the seven pairs of metacentric chromosomes are distinguished by assigning consecutive numbers to their left and right arms, the patterns of synaptic and subsequent diakinesis can be interpreted. (a) A typical chromosome pattern observed in the zygotene stage of meiotic prophase, and (b) its interpretation. (c) The pattern observed at diakinesis, produced by centromeric repulsion and telosynapsis, and (d) its interpretation. (e) Side view of the metaphase plate (metaphase I), and (f) its interpretation.

Changes in Chromosome Number

Since the beginning of man's interest in agriculture and horticulture, unusual varieties of plants have been discovered and exploited to man's benefit. These varieties constitute a vital portion of our agricultural economy, and it is probably no exaggeration to state that were it not for their existence, the shortage of food today would probably be critical in all parts of the world. A large number of these varieties are the consequence of changes in numbers of chromosomes and include such staples as wheat and oats, fruits such as apples, pears, citrus fruits, and tomatoes, many vegetables including cabbages, potatoes, and spinach, and even many varieties of ornamental plants such as roses, petunias, marigolds, and so on.

The kinds of changes in chromosome number that have been encountered and described constitute a continuous series ranging from parts of chromosomes, such as we have already described (duplications and deletions), to whole chromosomes of variable numbers, and to entire sets of chromosomes. In most cases in which extra chromosomes are present, these chromosomes are duplicates of all or certain members of the preexisting basic set, but in some cases they may have been introduced from a different variety or even different species or genus. The various kinds and

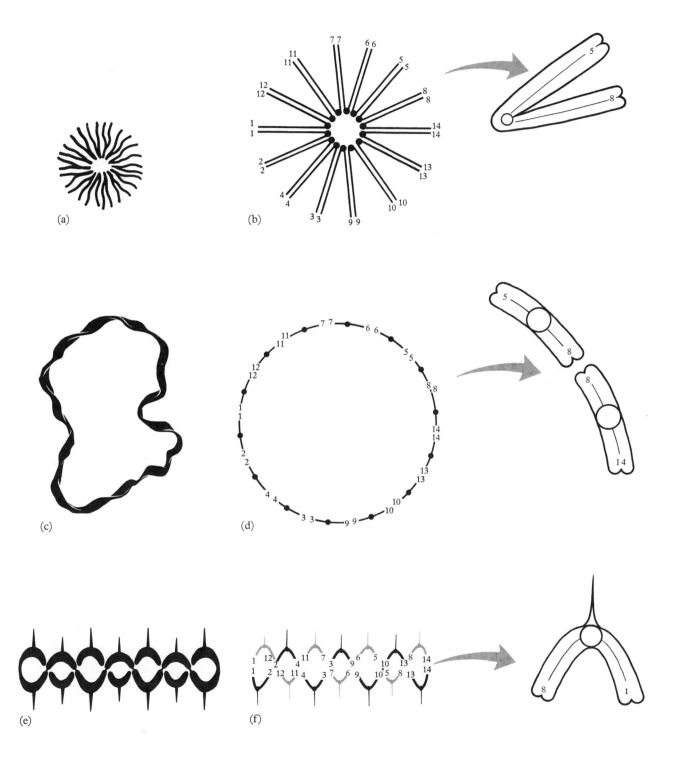

combinations of changes in chromosome number can generally be included in the categories of *aneuploidy* and *euploidy*, the latter including *autoploidy* and *alloploidy*.

Aneuploidy.   Aneuploidy applies to situations in which one or more chromosomes short of an entire set or genome are either lacking or redundant. The kinds of aneuploidy most frequently encountered (Fig. 9–22) are *monosomy*, in which one chromosome of a diploid genome is missing $(2n - 1)$, and *trisomy*, in which three copies of one of the chromosomes exist $(2n + 1)$. The first extensive studies of the phenotypic effects and cytogenetic characteristics of aneuploidy were done by Blakeslee (1934) and others with the flowering plant *Datura stramonium*. This plant grows along the roadside, in fields and grazing lands, and on abandoned farms over most of the United States, but in New England it is cultivated as an ornamental plant. Where it grows wild and uncherished it is known as Jimson weed, whereas in the florist and garden shops of Massachusetts it has the more gracious name of trumpet flower.

Trisomy.   Trisomics are perhaps the most common and, in many respects, the most interesting of the aneuploids. Blakeslee and his collaborators discovered trisomics for each of the kinds of chromosomes of the *Datura* genome $(2n = 24)$ and correlated them with particular phenotypic effects (Fig. 9–23). Subsequently, trisomics have been described in a number of species of animals and plants including *Drosophila*, mice, human beings, dogs, apes, corn, wheat, tomatoes, and many others. The condition is always accompanied by a phenotypic expression of one kind or another, more often detrimental than not, and by reduced fertility.

Trisomy in human beings has received considerable attention. In fact, some maternity and other hospitals have introduced programs of routine cytogenetic screening of newborns in order to facilitate early detection of trisomy and other

Monosomy
$(2n - 1)$

Double monosomy
$(2n - 1 - 1)$

Trisomy $(2n + 1)$

Double trisomy
$(2n + 1 + 1)$

Tetrasomy $(2n + 2)$

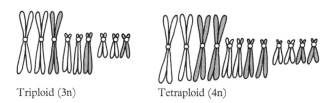

Triploid (3n)

Tetraploid (4n)

Fig. 9–22.   Examples of certain variations in chromosome number in a diploid. Missing chromosomes are indicated by a broken-line silhouette. Redundant chromosomes are indicated by shading.

chromosomal abnormalities (cf. Handler, 1971). Down's syndrome, known commonly as Mongolian idiocy or mongolism, is attributable to trisomy for chromosome 21 (Lejeune, 1964) (cf. pp. 222 and 226). This congenital disease received its unfortunate common name from the slanted eyes and broad face characteristic of the syndrome, traits that Western people associate with the so-called Mongolian race. Down's syndrome is characterized by a variety of physical abnormalities that include mental deficiency, disproportionate

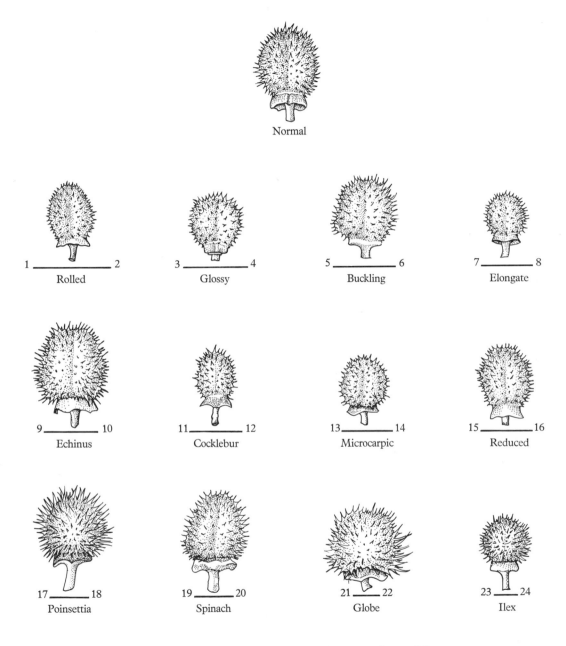

Fig. 9–23. Capsules of normal (2n) and 12 varieties of trisomic (2n + 1) Jimson weed. Each of the trisomics differs from the others in the particular chromosome that is duplicated. [Adapted from A. F. Blakeslee (1934), *J. Hered.* **25**:87. By permission.]

parts of the body, presence of the diagnostic "simian crease" in the palm of the hand, and reduced longevity. The abnormality is most frequently associated with advanced age of the mother at the time of conception (Lejeune, 1964). Presumably the longer an ovum remains in arrested meiotic metaphase, the higher the probability that nondisjunction of certain chromosomes will occur. For reasons not understood, chromosome 21 and the X chromosome are the most prone to nondisjunction (Fig. 9–24), although occasionally trisomy for other smaller autosomes does occur. Examples are the rare Patau's syndrome (trisomy for chromosome 13), Edward's syndrome (trisomy-18), and the extremely rare trisomy-22. Interestingly, Down's syndrome in chimpanzees is associated with trisomy for chromosome 22, which is probably homologous to human chromosome 21 (McClure, et al., 1969). Other well-established examples of trisomy and double trisomy in human beings involve the sex chromosomes and are expressed as Turner's syndrome, Klinefelter's syndrome, and the much-disputed XYY-syndrome, all of which are discussed in Chapter 5.

Other Aneuploids and Mosaicism.   Monosomy (2n – 1), *tetrasomy (2n + 2),* and *nullisomy (2n – 2,* in which both members of a chromosome pair are missing) are less frequently encountered in nature, but they have been studied in experimental

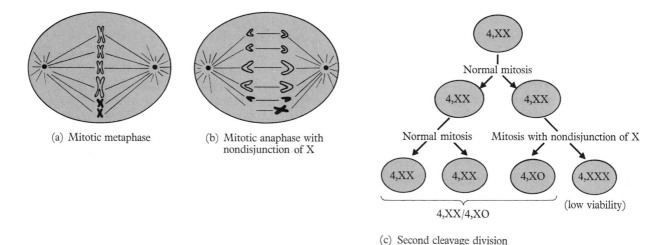

(a) Mitotic metaphase

(b) Mitotic anaphase with nondisjunction of X

(c) Second cleavage division

Fig. 9–24.   Nondisjunction of the X chromosome. In many species, the chromatids of the sex chromosomes are slower to separate (disjoin) from each other at the time of mitotic anaphase. Rarely, the delay results in both chromatids of a pair going to the same pole and subsequently being incorporated in the same daughter cell (a and b). If nondisjunction occurs during the second cleavage division (c), either of two kinds of sex chromosome mosaicism may result. A mosaic of three kinds of cells 4,XX/4,XO/4,XXX) would result if all the products of the abnormal mitosis survive and reproduce. Cells with the karyotype 2A,XXX frequently have reduced viability and would either die and be resorbed, or would contribute so few daughter cells to the final individual as to escape notice, in which case the mosaic found would be 4,XX/4,XO. The X chromosomes involved are shown in solid black.

animals and plants (Sears, 1954). Some forms of Klinefelter's syndrome are associated with tetra- and even pentasomy for the sex chromosomes. Very often these abnormal chromosome numbers occur in some cells of the body and not in others, which may be normal. Such individuals as these are called *chromosome mosaics* and show cells of two, three, or even more types (Steele, et al., 1966) (cf. Chapter 5). Many cases of chromosome mosaicism can be explained by nondisjunction during the cleavage or later stages of development (Fig. 9–24).

The Blakeslee group (Avery, et al., 1959) described combinations of trisomy, translocation, and other anomalies in the Jimson weed. These combinations led to interesting and, for a while, puzzling chromosomal configurations during meiotic prophase. For example, they discovered an extra chromosome, both arms of which were homologous to each other. If the system of identifying chromosomes by numbering their arms is used (Fig. 9–21), this kind of anomaly can be represented as 1–2, 1–2, 2–2, and would often appear at diakinesis as shown in Fig. 9–25. Another kind of translocation trisomy could be represented as 1–2, 1–2, 2–3, 3–4, 3–4, in which chromosome 2–3 is the translocate as well as the trisomic. During diakinesis, this group of chromosomes could form either a looping catenation of five chromosomes in telosynapsis or, rarely, a configuration that resembles a pair of eyeglasses (Fig. 9–25).

Euploidy.   It is among the euploids that the most economically significant chromosomal anomalies are found. Euploids are cells or organisms in which the number of sets of chromosomes occur in regular multiples. The specific number of multiples is implied in the names of euploids in which the coefficient is reasonably small, such as in *monoploids* (1n), *triploids* (3n), and *tetraploids* (4n). Although the terms *octoploid* (8n) and *dodecaploid* (12n) exist, it is more common practice to use the general term *polyploid* when the coefficient is larger than four.

Euploid cells have no difficulty during mitosis

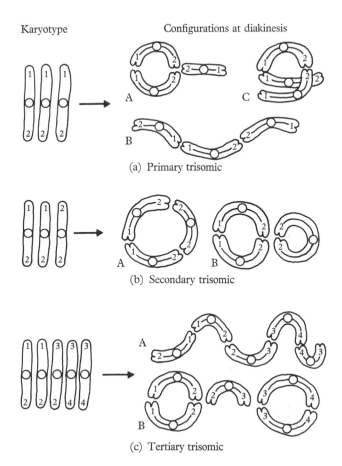

Karyotype   Configurations at diakinesis

(a) Primary trisomic

(b) Secondary trisomic

(c) Tertiary trisomic

Fig. 9–25.   Chromosomal configurations that may be encountered at diakinesis in (a) primary, (b) secondary, and (c) tertiary trisomics. Alternative configurations for each group are indicated by capital letters.

(unless the coefficient of "ploidy" is enormous), because in mitosis each chromosome behaves independently of all others. Considerable difficulty is encountered during meiotic cell division, however, due to the processes of synapsis and recombination. Because most animal species reproduce sexually and therefore are obligated to insert meiosis between generations of individuals, euploidy (other than diploidy) is extremely rare among them. In fact, many cases of presumed euploidy in animals are explainable in terms of transverse centromeric fission (Astaurov, 1969), as discussed later in this chapter. Asexual reproduction is commonplace in plants, however, and accordingly this method of propagation is often called "vegetative." Many wild species of plants reproduce both sexually and asexually, and there are many that employ only the latter method. Examples of methods of vegetative reproduction are depicted in Fig. 9–26. Herbaceous plants such as geraniums and coleus lend themselves readily to propagation by "slipping," whereas woody plants such as fruit trees and ornamental shrubs must be grafted if they are to be propagated asexually.

Euploidy in plants has been very beneficial to mankind. Euploids are often larger and hardier, and they often have larger fruit and larger and more complex blossoms than their diploid counterparts. The most familiar examples cited for comparison are diploid versus polyploid apples, and diploid versus polyploid roses. The fruits of horticultural varieties of euploid apples such as Macintosh, Gravenstein, and Baldwin, which are so popular in America, are far bigger, sweeter, juicier, and longer-lasting than the fruit of the diploid "crabapple." Euploid roses have several rows of large petals that make their wild diploid counterpart, however lovely, appear as country cousins. The list of comparisons is seemingly endless.

The reasons for the frequent euploid exaggeration of the quantitative characters of a species are not well understood. Often the nuclei are larger, doubtless because of the great increase in DNA, and the cells are correspondingly larger. This is because cells tend to maintain what is technically known as a constant *nucleoplasmic ratio* (the ratio of the volume of the nucleus to the volume of the cytoplasm). Although this may lead to larger organism size in some cases, in others there is little difference because the organism compensates for the larger cell size by producing fewer cells during its development. Other reasons for exaggeration may have the same basis as quantitative or *polygenic* inheritance, as discussed in Chapter 6, although this would be difficult to reconcile with the principle of genic balance.

Although extremely rare, euploidy in animals is not unknown. It has been studied in *Drosophila* and in certain amphibian species such as the newt *Triturus* and a few higher vertebrates (Abdel-Hameed, 1972). In man and other mammals, euploidy appears to be lethal or sublethal according to Carr (1971), who completed an extensive study of spontaneous abortion in human beings. He described thirty abortuses that were either triploid or triploid plus, and seven that were tetraploid. He estimated that at least one percent of all conceptuses in man are triploid, and almost all of these abort by the end of the third month of pregnancy.

Autoploidy and Alloploidy.   The euploids that have been discussed up to this point belong to a class called *autoploids.* Their supernumerary set or sets of chromosomes may be a consequence of one or more aborted mitotic divisions, resulting in chromosome doubling. This mechanism is believed to account for certain spontaneous and most of the artificially induced polyploids in plants, as described below. Another mechanism leading to euploidy is believed to be nondisjunction of the entire genome during meiosis, leading to diploid gametes. In either case, the supernumerary set or sets of chromosomes are duplications of the chromosomes that are already present in and characteristic of the particular individual or race in question. It is for this reason that the term autoploidy (*auto* = self) is used.

Less common, but possibly of greater evolutionary significance, is the form of euploidy known as *alloploidy*. The classic example of alloploidy is the case of *Raphanobrassica* which was reported by Karpechenko in 1927. Karpechenko produced a hybrid by cross-pollinating radishes (*Raphanus sativus*) and cabbages (*Brassica oleracea*). Each of these species is characterized by a diploid

Natural   Artificial

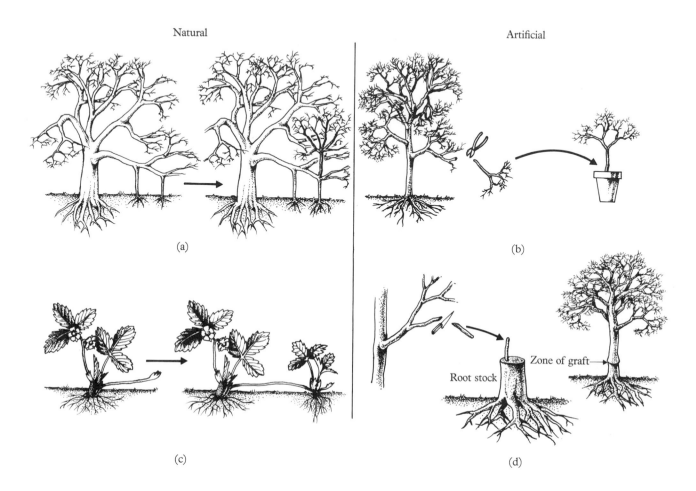

(a)   (b)

(c)   (d)

Fig. 9–26.   Some methods of asexual reproduction in plants. (a) A branch that touches the ground may sprout adventitious roots, and a new plant may then grow from this. The banyan tree is an example of a tree that grows roots from its branches, which penetrate the earth and enable the tree to spread radially to form a small forest of one tree. (b) Called "slipping," a branch or shoot is cut from the main plant and rooted either in water or, better yet, in coarse sand and water. Rooting hormone may be added for plants that resist this method of propagation. (c) A "runner" is formed by some species of plants. This may run above ground, as in strawberries, or underground, as in many grasses. The tip of the runner sprouts roots and a new plant develops. (d) Grafting is done by cutting a small branch or "scion" from a plant and affixing it to another compatible plant with a good root system. This method is used to propagate varieties that either cannot easily be induced to form roots by slipping or do not form vigorous roots.

number of 18 chromosomes. The hybrid between them also has 18 chromosomes, but 9 of them are *Brassica* chromosomes and 9 are *Raphanus* chromosomes, which are so dissimilar from each other that synapsis does not occur between them during meiotic prophase. Consequently, the hybrid was considered a double monoploid and, accordingly, was highly sterile. Quite fortuitously, Karpechenko discovered a hybrid that was not sterile, but sexually reproduced other individuals similar to itself. This hybrid was found to have 36 chromosomes, and at meiosis normal pairing was observed. The plant proved to be a double diploid (*amphidiploid*), possessing two sets of radish (18R) chromosomes and two sets of cabbage (18C) chromosomes. Because this anomalous plant is able to reproduce itself and is sexually isolated from both parent species (since it cannot produce fertile hybrids with them), it represents a new species and has been named *Raphanobrassica* (Fig. 9–27).

Many species of plants and a few of animals are suspected to have arisen as allopolyploids like *Raphanobrassica*, except to have done so spontaneously in nature. One of these appears with little doubt to have arisen within the last two hundred years. This is a species of cord grass, *Spartina townsendii*, which is found in abundance on the coast of southern England, northwestern France, and the Netherlands. The first specimen of *S. townsendii* was discovered in 1870 in a highly circumscribed area of southern England in which the long-known European species, *S. stricta*, overlapped the range of an introduced American species, *S. alterniflora*. Not only does *S. townsendii* combine many of the physical characteristics of *S. stricta* and *S. alterniflora*, but its diploid number of chromosomes (2n = 126) equals the sum of *S. stricta*'s (2n = 56) and *S. alterniflora*'s (2n = 70) diploid numbers (Huskins, 1930).

The well-known golden or Syrian hamster *Mesocricetus auratus* (2n = 44), was believed at one time to be an alloploid derived from natural hybridization between the European hamster, *Cricetus cricetus* (2n = 22), and one of the Asiatic species

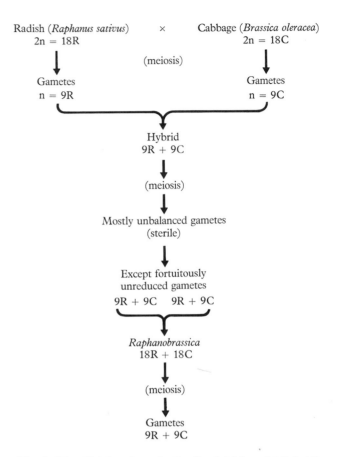

Fig. 9–27.   Origin of synthetic alloploid (amphidiploid) *Raphanobrassica*.

such as *Cricetulus griseus* (2n = 24). Further studies of the chromosomes of these species favors an alternative origin of the *M. auratus* karyotype, namely, by centromeric fission and fusion. This mechanism was first alluded to by Robertson in 1916. Robertson proposed that many variations in chromosome number are a result of complicated processes of redistribution of chromosome segments by reciprocal translocation, inversion, and fragmentation, while the total amount of genetic material remains nearly constant. Figure 9–28 shows how karyotypes can be altered by either transverse fission or fusion of centromeres. The difference in karyotypes between several species of

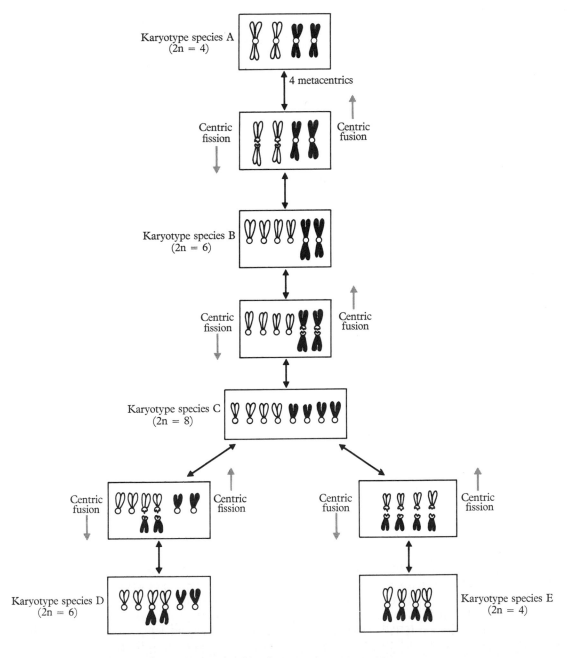

Fig. 9–28. Some variations in chromosome number that may occur as a consequence of chromosome fragmentation involving the centromere (centric fission) and translocation (accompanied by centric fusion). For changes in karyotype from centric fission, read down; for changes from centric fusion, read up. The chromosome pairs in the original (?) species are distinguished by representing one pair white and the other pair shaded.

closely related animals can be accounted for by these kinds of mechanisms (Astaurov, 1969; Todd, 1970).

Attempts to discover or retrace the origin of two important agricultural crops make interesting stories. One of these is American cultivated cotton, which figured so prominently in the economy of the South and led to the widespread sociological problems that the nation has had ever since the Civil War, and the other is the high-glutin bread wheats. American cultivated cotton, *Gossypium hirsutum*, is a hardy plant that produces large bolls of long staple cotton fibers. It has 26 pairs of chromosomes, of which half are long and half are short. American wild cotton, *G. thurberi*, has 13 pairs of small chromosomes; Asiatic cotton, *G. arboreum*, has 13 pairs of long chromosomes. If the hybrid produced by crossing wild American cotton with Asiatic cotton is induced by the drug colchicine (see p. 241) to double its chromosome number, an amphidiploid is produced that is fully fertile and closely resembles American cultivated cotton. Like American cultivated cotton, it has 13 pairs of long chromosomes and 13 pairs of short chromosomes. These results suggest that American cultivated cotton developed when Asiatic cotton was introduced to America and hybridized with the native species, followed by chromosome doubling to produce an amphidiploid.

The case for the origin of modern bread wheats such as *Triticum estivatum* and *T. spelta* is even more persuasive. These wheats have 21 pairs of chromosomes. Two other cultivated wheats, *T. monococcum* and *T. dicoccum*, have 7 and 14 pairs of chromosomes, respectively. At first it was thought that the bread wheats arose as amphidiploids of the latter two species, but attempts to produce them synthetically by means of colchicine and hybridization of these species failed. Then a hybrid was derived by crossing a variety of *T. dicoccum* with the goat grass *Aegilops squarrosa*, after both had been rendered tetraploid by colchicine treatment. The goat grass had 7 pairs of chromosomes.

The amphidiploid hybrid produced in this way had 21 pairs of chromosomes. Like the bread wheats it was fully fertile, and when it was crossed with *T. spelta*, a fertile hybrid resulted that showed normal meiosis. Finally, this synthetic amphidiploid had the characteristics of a primitive form of bread wheat. There is little doubt, then, that the economically important bread wheats represent allopolyploids between one or more primitive wheat species and a species like common goat grass.

MECHANISMS OF MUTATION

Some consideration has already been given for mechanisms whereby mutations arise at the molecular level. Specifically, alterations in the base composition of DNA may arise spontaneously as a consequence of tautomeric shifts, base ionization, or accidental copy errors during replication. Similar alterations may be induced by the use of chemical mutagens such as base analogs, acridine dyes, hydroxylamines, and alkylating agents (Table 9–1), and in microorganisms by ultraviolet irradiation. Similarly, some of the mechanisms by which internal rearrangements of chromosome structure, aneuploidy, and euploidy arise spontaneously have also been described. These include unequal or diagonal crossing-over and chromosome non-disjunction (cf. pp. 220 and 232, and Figs. 9–15 and 9–24). The latter has not been difficult to understand since it was first postulated by Bridges, but mechanisms that imply breakage of chromosomes have presented many problems. Duplications, deletions, inversions, and translocations all imply chromosome breakage or fragmentation at some time. Chromosomes are complex structures, however, evidently consisting of DNA complexed, at least in part, with protein, RNA, lipids, calcium, and small quantities of various other substances of low molecular weight (Ris, 1957). How such a complex structure can be broken and repaired by such unrelated mechanisms as irradiation and chemical mutagens, as well as during the course of

Table 9–1.   The principal chemical mutagens that directly affect DNA, and their specific effects. (Key references are given for the benefit of individuals who wish to investigate each category more thoroughly.)

Mutagen	Mutant effect	References

Base analogs
  5-bromouracil
  2-aminopurine
  2,6-aminopurine
	Base-pair substitutions
	  (transitions)
		Litman and Pardee (1956)
		Benzer and Freese (1958)
		Freese (1959, 1968).
		Brenner, et al. (1961)

Nitrous acid (HNO$_2$)	Base-pair substitutions (transitions) Deletions	Tessman (1962) Yanofsky, et al. (1966) Beckwith, et al. (1966)
Hydroxylamines	Base-pair substitutions (transitions)	Freese, et al. (1961) Philips and Brown (1967)
Mustards	Base-pair substitutions (transitions) Frameshift	Auerbach and Robson (1946) Loveless (1966) Malling and de Serres (1968)

Nitrosoguanidine (NG)	Base-pair substitutions
	  (transitions and
	  transversions)
	Deletions
		Mandell and Greenberg (1961)
		Whitfield, et al. (1966)
		Langridge and Campbell (1969)

normal crossing-over, is difficult to reconcile and has been the object of much study and speculation.

The Hit and Breakage-First Hypotheses

Following the pioneering experiments of Muller (1927) and Stadler (1928) with X-rays, it became evident that a large proportion of induced mutations involved breakage or fragmentation of chromosomes. The relationship between dosage and mutation frequency was found to be linear for gene ("point") mutations and exponential for macrolesions. These relationships led to the formulation of the *hit hypothesis*, which in essence proposed that ionizing radiations such as X-rays bring about physical breaks in chromosomes at the points of impact (Timoféef-Ressovsky, 1934). This led in turn to formulation of the *breakage-first hypothesis* which proposed that, just as in normal crossing-over (Chapter 4), the primary event in

radiation-induced mutations is a chromosome break, following which the broken ends may either remain open or rejoin. If they rejoin, the original configuration is restored. If they remain open, the chromosome fragments that lack centromeric connection will be lost during subsequent cell division, and thereby be deleted. On the other hand, if multiple breaks occur relatively close to one another in place and time, inversions, translocations, ring chromosomes, and other such abnormalities are possible. The breakage-first hypothesis fails to explain how the breaks occur in the first place, however. The effect of mutagenic radiation is not a physical one, as the hit hypothesis implies, like a bullet breaking chemical bonds simply by striking them. It is a chemical effect brought about by ionization of the atoms through which the rays pass. How such a nonspecific effect and how the many relatively unreactive chemical mutagens can disrupt the enormous number of covalent bonds that theoretically must be broken in order to fragment so complex a structure as a chromosome is very difficult to comprehend. Understanding becomes even more difficult when one considers that the effect is often delayed, in some cases by a matter of hours, and that some mutagens are oxygen-dependent, others are affected by respiratory inhibitors, and still others are independent of these factors.

The Exchange Hypothesis

The *exchange hypothesis* was conceived and developed by S. H. Revell (1959) as a result of detailed and extensive studies of chromosome breakage induced in the broad bean *Vicia faba* by means of X-rays. According to this hypothesis, chromosome breakage is a symptom of some other kind of disturbance in the cell and/or chromosome that promotes chromosomal exchange. This disturbance is similar to the normal processes of crossing-over except that nonhomologous chromosomes may be involved in the exchange. The exchange may be consummated or not, depending upon other considerations in the disturbance. Revell proposed that chromosomes are so constructed as to undergo segmental exchange under certain conditions of proximity and the cell cycle. These conditions are normally met as a consequence of the unique phenomenon of synapsis during the pachytene stage of meiosis. In this situation, exchanges take place without requiring that previous breakage be induced by such drastic agents as ionizing radiations or radiomimetic chemical mutagens (mutagens that produce the same effects as radiation). Furthermore, in natural meiosis errors occasionally occur, resulting in the appearance of the *same kinds of aberrations as one finds produced by radiation or radiomimetic mutagens*. In addition, the various kinds of induced aberrations occur with the relative frequencies that characterize spontaneous aberrations, and thus with frequencies that are predictable by the exchange hypothesis (Kihlman, 1961).

That the kinds of damage caused by ionizing radiations is basically chemical in nature was implied in the results obtained by Thoday and Read (1947). They found that X-ray-induced chromosomal aberrations were greatly influenced by oxygen tension in the cells' environment at the time of treatment. Their studies have been extended, and the relationships between oxygen dependency and the kinds of chromosome-breaking agents used are shown in Table 9–2 (Kihlman, 1961). According to this table, chromosome-breaking agents can be classified in three groups: (1) agents that are inhibited by reduced oxygen concentration or by the use of respiratory inhibitors, (2) agents that are dependent directly upon oxygen availability and are not influenced by respiratory inhibitors, and (3) agents that produce an effect independent of oxygen concentration.

The induction of polyploidy is an entirely different matter because it does not involve any direct action on the chromosomes or the DNA. Nevertheless, euploidy can be induced by a variety of means, all of which interfere with one or more of the normal processes of cell division. Plant tissues are generally more susceptible to these methods than animal tissues are. Traumatizing growing tissues by cutting or bruising can cause cells undergoing mitosis to return to interphase without completing division of the cytoplasm, although the chromosomes have already doubled. This phenomenon is occasionally encountered as a result of cutting or pinching off the apical bud of a plant. Frostbite, especially in the spring when the cells of buds are dividing rapidly, may cause a similar effect, as has been demonstrated experimentally. Elevated temperatures may also be effective in this way.

The mitotic spindle is sensitive to a number of chemical substances including such chemically unrelated molecules as sulfanilimide, chloral hydrate, colchicine, and podophyllotoxin (Dawson, 1962). Colchicine and its derivatives have proved to be extremely effective and efficient, and have been used experimentally for many years (Fig. 9–29).

These chemicals prevent the assembly of the protein subunits that comprise the microtubules of the mitotic spindle, thereby preventing the formation of this structure. They do not interfere appreciably with any of the other cell components or processes in the concentrations used, and so the effect is expressed in a straightforward way as an inability of the treated cell to complete cell division, with the consequent doubling of chromosome number. Continued or repeated treatment of plant cells with colchicine can lead to increases in chromosome numbers to more than a thousand in a single cell.

Colchicine is obtained from the autumn crocus *Colchicum*. Because it is more toxic to animal tissues than to plant tissues, its less toxic chemical relative, colcemid (Fig. 9–29c), is often used in experimental work with animal cells. One very important use of colchicine or colcemid is in cytological studies with tissue culture. Cells are raised by tissue culture until a population of approximately the desired number of cells is obtained, and then they are treated with one of these inhibitors. As each cultured cell enters mitosis, it progresses as far as metaphase and then stops. In a rapidly growing culture, it does not take long for the major part of the cell population to reach this stage. The investigator can then harvest the cells and find that a large proportion of them are at the stage that is ideal for studying their chromosomes.

Table 9–2. The principal chromosome-breaking agents categorized according to the influence of oxygen concentration on their effect.

Type of effect (delayed or not delayed)	I. The effect dependent on oxygen concentration			II. The effect independent of oxygen concentration
	A. The effect inhibited by respiratory inhibitors		B. The effect not inhibited by respiratory inhibitors (oxygen can usually be replaced by nitric oxide)	
	1. The effect inhibited by uncoupling agents	2. The effect not inhibited by uncoupling agents		
Delayed	Maleic hydrazide, ethyl alcohol	N-methyl-phenyl nitrosamine	Visible light, acridine orange, potassium cyanide	Alkylating agents (nitrogen mustard, diepoxypropyl ether, N-nitroso-N-methyl urethan, Myleran, β-propiolactone)
Not delayed	N-methylated oxypurines (e.g., 8-ethoxycaffeine, 1,3,7,9-tetramethyluric acid)		X-rays	Alpha rays, ultraviolet light

[From B. A. Kihlman (1961), "Biochemical aspects of chromosome breakage." *Adv. in Genetics* **10**:41. By permission.]

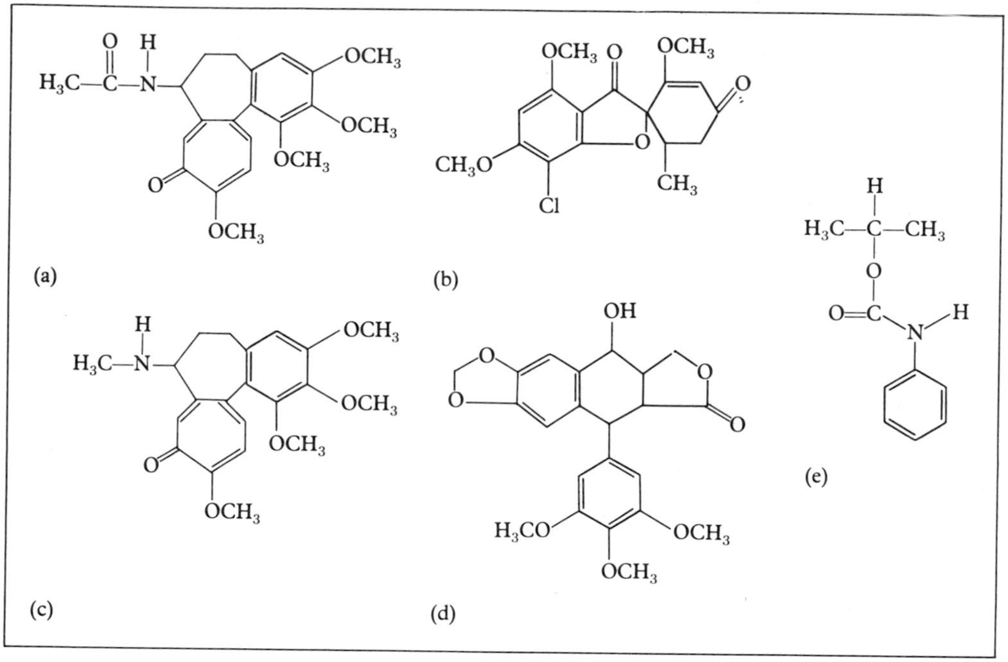

Fig. 9–29. A variety of mitotic spindle inhibitors: (a) colchicine, (b) griseoful-vin, (c) colcemid, (d) podophyllotoxin, and (e) isopropyl-n-phenyl carbamate (IPC).

Without the use of an aid such as this, the investigator would have to comb through hundreds or even thousands of cells to find one in mitotic metaphase.

The microtubule inhibitor griseofulvin (Fig. 9–29b) has been found to be very effective as an inhibitor of growth in fungi, and is frequently used by veterinarians to combat fungus infections in domestic animals. It sometimes has unfortunate side effects on myeloid tissues, so it is usually not recommended for human therapy.

HOW MUTATIONS ARE DETECTED

Until relatively recently, the discovery of mutations was a matter of luck. Most mutations that have been incorporated into domestic animals and plants were originally discovered quite by acci-dent, and then propagated by selective breeding by farmers or animal husbandmen who were shrewd enough to recognize the advantages of them. One example is the short-legged breed of Ancon sheep, a breed that is descended from a mutant ram discovered by Seth Wright, a sharp-eyed Yankee farmer, in Dover, Massachusetts, in 1791. The short-legged breed eliminated the need for high fences in areas lacking predators. A more recent example of an accidentally discovered mutation in sheep is the polled (hornless) condition in the popular Dorset breed, discovered in 1951 at North Carolina State College. It has been propagated ever since, and now it is a well-known breed in its own right.

The "attached-X" mutation in *Drosophila* represents a relatively recently discovered mutant that proved to be of great value as a tool in genetics. The wife of the famous geneticist T. H. Morgan, herself a geneticist, discovered in 1922 an unusual female in her stock of *Drosophila*. It escaped, but luckily was found on a window in the laboratory and was bred. It was found that the two X chromosomes of this female were joined by centric fusion (Fig. 9–28), which resulted thereafter in permanent nondisjunction of these chromosomes.

Since the discovery of mutagenic agents, it has been possible to induce mutations at will, although the mutations are still random and must be sifted in order to be characterized. Ingenious methods of detecting mutations have been devised, whereby even "invisible" mutations can be uncovered in many cases. The most commonly used methods for each of the principal kinds of genetically useful organisms are described below.

Bacteriophage Mutations.   Bacterial cultures are prepared that consist of a uniform carpet of bacteria growing on agar plates. A diluted suspension of bacteriophage is then either sprayed or flooded onto the surface of this carpet, and the preparation is incubated. Each virus infects the bacterium with which it comes in contact and reproduces itself a hundredfold or more within a matter of 20 to 40 minutes. Then the cell is lysed. The released particles invade adjacent bacterial cells, and thus the infection spreads. Within a few hours, each original infection has spread to produce a spot of lysed bacteria that is large enough to be visible with a low-powered magnifying glass or even the unaided eye. Such spots are called "plaques." Mutant phage have been discovered that differ in terms of the so-called "morphology" of the plaques they produce. The *r*-mutants of T4 phage, which have been extensively investigated by Seymour Benzer (1961), are expressed by the large, sharply demarked plaques they make when grown on strain B of *E. coli* (Fig. 9–30). A large number of mutations that affect plaque morphology have been discovered in phage by this kind of technique.

Bacterial Mutations.   Penicillin kills bacteria by interfering with their ability to form cell walls after fission. Hence penicillin kills only those cells that

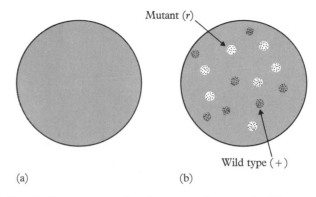

Mutant (*r*)

Wild type (+)

(a)   (b)

Fig. 9–30.   A comparison between plaques of wild type and *r*-mutants of T4 bacteriophage on a plate of *E. coli*. (a) The appearance of a uniform plate culture of *E. coli* before infection with phage particles. (b) The appearance of viral plaques, which represent areas of infection in which the bacterial cells have been lysed. Each plaque statistically represents the action of the progeny of a single virus particle. The large, distinct plaques characterize the *r*-mutants.

are growing. This phenomenon has been exploited by geneticists as one way of separating certain kinds of mutants from bacterial cultures. These mutants are the so-called "biochemical" mutants, because they all represent deficiencies in the organism's ability to synthesize one or more of the kinds of enzymes it requires for normal growth on minimum medium, as described in Chapter 7. For example, if wild-type *E. coli*, growing on minimum medium, is treated with a mutagen such as ultraviolet light and then exposed to penicillin, all the growing cells will be killed. Because they have biochemical mutations that make them incapable of growth on minimum medium, the cells that are not growing are not killed by the penicillin. The penicillin is then removed either by washing or by exposure to the enzyme penicillinase, and the suspended cells are transferred to complete medium. The biochemical mutants will grow on or in complete medium, but they will be a mixed lot, and now must be separated. This is achieved by the ingenious method of "replica plating" devised by Lederberg and Lederberg (1952).

The Lederberg Technique of Replica Plating.   An agar plate of complete medium is inoculated with a diluted suspension of mixed mutants. Each cell or viable unit will produce a colony of identical progeny on this agar plate. A swatch of sterilized velvet is then pressed onto the agar plate, just as a rubber stamp is pressed onto an ink pad. The velvet is then pressed in succession onto freshly prepared agar plates, each of which consists of minimum medium supplemented by one of the known biochemical requirements of *E. coli*. These are the components of complete medium which wild-type organisms can synthesize on their own. Each time the velvet is pressed onto a plate, a few organisms are transferred to locations corresponding to the positions occupied by the parent colonies on the plate of complete medium (Fig. 9–31). The biochemical requirement of the various mutant types can thereby be determined and correlated with the parent colony. From the parent colony, a pure stock of mutant cells can then be obtained. The method of replica plating has been modified in several minor ways, and has been applied to fungi and other microorganisms.

Neurospora Mutations.   The life cycle and the techniques of culturing the breadmold *Neurospora crassa* have already been described (pp. 61 and 170, and Figs. 4–8 and 8–8).

Following the techniques of Beadle and Tatum (1941), Woodward and his collaborators (1945) devised a method for isolating mutants of *Neurospora* on a large scale. They cultured irradiated spores in batches in liquid minimum medium. Spores whose nutritional requirements had not been affected by the radiation (i.e., the nutritional wild types) grew and produced hyphae. The culture solution was then strained through a filter that would trap these hyphae, but would permit the ungerminated spores—the radiation-affected spores—to pass through. These spores were then transferred to supplemented agar culture tubes and the various mutant types were sorted out (Fig. 9–32).

Drosophila Mutations.   The first techniques ever devised specifically for identifying mutations were devised for *Drosophila* by H. J. Muller (1928). One of these, the so-called *ClB* or "grandsonless-lineage" technique, was designed specifically to detect the occurrence of recessive lethal mutations on the X chromosome. He mated X-rayed males with females that carried a recessive lethal gene (*l*), the genetic marker bar (*B*), and an inversion on one of the two X chromosomes (Fig. 9–33). The inversion prevented crossing-over between the lethal gene and the bar gene, which would result in viable gametes (Fig. 9–19). In effect, the inversion served as a crossover suppressor and is designated by the symbol *C*. Thus this X chromosome is identified as *ClB*. The other X chromosome of the females was normal. The daughters produced by mating the *ClB* females with X-rayed males would be of two classes—half would be bar-eyed, because they inherited the *ClB* chromosome from their mother, and half would be wild type. Muller selected out the bar-eyed daughters and mated them with normal males. One would expect that the grandsons produced by this mating would be of two classes—those that inherit the *ClB* chromosome from their mother and those that inherit the other X chromosome. In this case, the "other" X chromosome was passed down from the X-rayed grandfather. Actually, any males that receive a *ClB* chromosome would be inviable and would not be found, due to the *l* allele in hemizygous condition. If a recessive lethal had been induced on the X chromosome that the grandfather passed on to the daughter in this lineage, the grandsons that would receive it would also be inviable. Consequently, Muller was able to quantify the frequency of recessive lethals induced on the X chromosome via X-rays by counting the number of bar-eyed daughters that failed to produce any grandsons (Fig. 9–33).

Another technique developed by Muller (1928) made use of a strain of flies he developed that included an inversion and a recessive lethal on

Fig. 9–31.   The Lederberg technique of replica plating. (a) Bacteria that have been exposed to a mutagenic agent such as ultraviolet light are plated on complete medium and incubated until clearly discernible colonies form. Mutated cells that will not grow on the components of complete medium are automatically eliminated. This diagram assumes that the viable organisms form a convenient pattern of colonies. (b) A swatch of sterilized velvet, stretched under a slightly smaller petri dish, is pressed onto the agar plate and then withdrawn, with some cells from each colony adhering to it. (c) The swatch of velvet is then pressed successively onto a plate of minimum medium, plus several plates of minimum

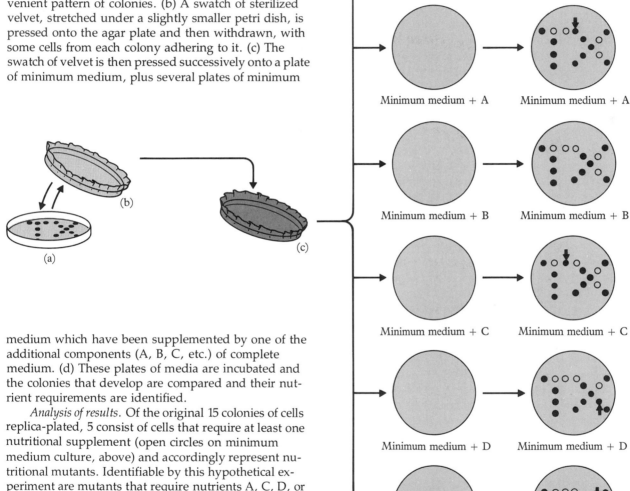

medium which have been supplemented by one of the additional components (A, B, C, etc.) of complete medium. (d) These plates of media are incubated and the colonies that develop are compared and their nutrient requirements are identified.

Analysis of results. Of the original 15 colonies of cells replica-plated, 5 consist of cells that require at least one nutritional supplement (open circles on minimum medium culture, above) and accordingly represent nutritional mutants. Identifiable by this hypothetical experiment are mutants that require nutrients A, C, D, or E (arrows). One mutant, which occupies the center-top position of the "T," does not grow on any of the plates shown. It may be a double mutant that requires two nutritional supplements, which could be determined by another set of plates containing minimum medium plus various combinations of supplements, two at a time.

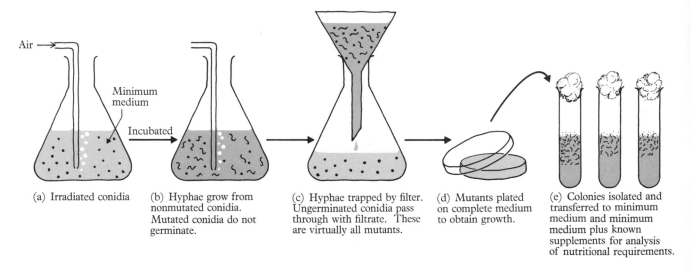

Fig. 9–32.   Schematic representation of a method for isolating large numbers of mutant spores from irradiated conidia of *Neurospora crassa*. Irradiated conidia are suspended in aerated liquid minimum medium (a) and cultured (b) to permit germination and growth of hyphae by spores not nutritionally affected by the radiation. The entire culture is then passed through a filter (c) that will hold back all the hyphae, permitting only ungerminated spores to pass. The filtrate is then plated on complete medium (d) in order to obtain germination and growth of these spores, most of which represent nutritional mutants. Colonies of hyphae are then isolated and transferred to culture tubes of minimum medium and minimum medium plus known supplements for analysis (e).

one of the autosomes, and a nonallelic recessive lethal on its homologue. This is viable because the two lethals, although on each of the two homologues, are not allelic and therefore not expressed. The homologous chromosome also included several recessive genes for visible traits, and it represented the "test" chromosome. A strain constituted in this way represents a balanced lethal, as described earlier, and breeds true as a heterozygote. Homozygosis is not possible because of the recessive lethals and because of the crossover-suppressing effect of the inversion. Muller X-rayed such strains over a number of generations, with the result that a number of recessive mutations could accumulate in the test chromo-some. Then he outcrossed individuals from these strains and interbred their progeny. The recessive markers on the test chromosome should show in one-quarter of the F2 produced by this interbreeding. Failure of a marker to be present implied that it was linked to an induced recessive lethal. The recessive lethal originally incorporated on the test chromosome did not influence the results because it was located far enough away from the genetic markers to behave as if it assorted independently, as described in Chapter 4. Modifications of these techniques developed by Muller (Mukai, 1964) have been used successfully for detecting the induction of mutations in *Drosophila* and other dipterans.

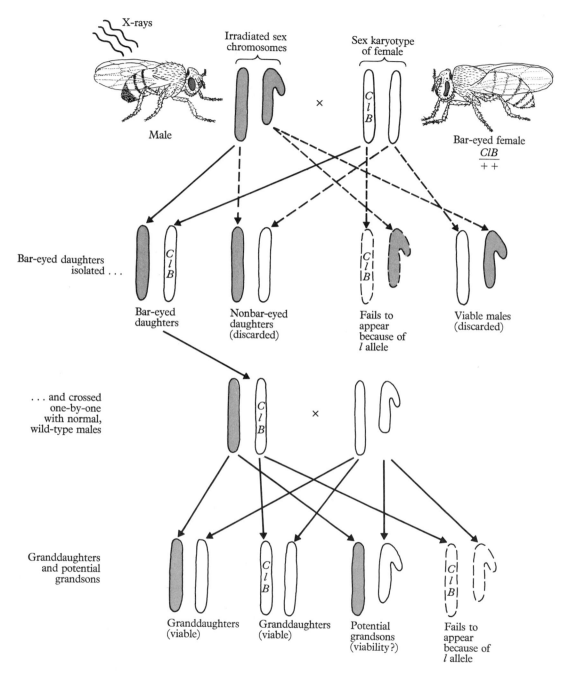

Fig. 9–33. Muller's *ClB* or "grandsonless" method for detecting induced, sex-linked, lethal mutations in *Drosophila melanogaster*. (See text for explanation.) Daughters carrying a paternal X chromosome in which a lethal has been induced fail to produce male offspring, and are identified accordingly.

MUTATION RATES

Mutation rates can be expressed in various ways. Muller (1928) calculated both spontaneous and X-ray-induced mutations in terms of percent lethal mutations per chromosome per generation. Timoféef-Ressovsky (1934), also working with *Drosophila,* found that different genes and even different alleles of the same gene differed in their mutation rates. Because of this, mutation rates are usually calculated for specific genes and are often expressed as number per hundred-thousand (10^5) gametes. In bacteria, a high degree of accuracy or "resolution" can be obtained because of the enormous numbers of progeny that are produced in a short time, and their mutation rates are often expressed as numbers of mutations per 10^5 or 10^6 viable units (for example, 1.4×10^{-6}).

Representative mutation rates for specific loci of various organisms are shown in Table 9–3. The great difference in mutation rates between microorganisms and the larger, multicellular organisms can be attributed to differences in generation time. Each cell division represents a generation in bacteria, whereas in *Drosophila* and other higher organisms, each gamete is the product of perhaps hundreds or even thousands of cell divisions.

Mutation rates can be modified in various ways. As already described and discussed, they can be increased by the use of mutagenic agents. For example, Muller (1928) discovered that the relationship between mutation frequency and X-ray dose is a direct one—that is, doubling the dose resulted in doubling the frequency of lethal mutations in the chromosomes he studied. The effect appears to be independent of either the dose rate, and therefore dose intensity, or dose wavelength. Subsequent work by Muller and later by Stadler showed that the rate of mutations induced by X-rays *per se* is also independent of temperature. This is not so for spontaneous mutations, however—Muller had shown earlier that mutation frequency increases with increased temperature. The concentration of oxygen can often affect the rate at which mutations are induced, as first demonstrated by Thoday and Read (1947) and shown in Table 9–2.

Table 9–3. Spontaneous mutation rates at particular genetic loci in a variety of organisms.*

Organism	Trait	Mutation per 100,000 gametes
Bacteria (from many sources)		
E. coli (K 12)	To streptomycin resistance	0.00004
	To phage T1 resistance	0.003
	To leucine independence	0.00007
	To arginine independence	0.0004
	To tryptophan independence	0.006
Salmonella typhimurium	To threonine resistance	0.41
	To tryptophan independence	0.005
Diplococcus pneumoniae	To penicillin resistance	0.01
Neurospora crassa (from Kolmark and Westergaard; Giles)	To adenine independence	0.0008–0.029
	To inositol independence	0.001–0.010
	(one inos allele, JH5202)	(1.5)

*Mutations to independence for nutritional substances are from the auxotrophic condition (e.g., *leu⁻*) to the prototrophic condition (e.g., *leu⁺*).

Drosophila melanogaster males		
(from Glass and Ritterhoff)	y^+ to *yellow*	12
	bw^+ to *brown*	3
	e^+ to *ebony*	2
	ey^+ to *eyeless*	6
Corn (from Stadler)	Wx to *waxy*	0.00
	Sh to *shrunken*	0.12
	C to *colorless*	0.23
	Su to *sugary*	0.24
	Pr to *purple*	1.10
	I to i	10.60
	R^r to r^r	49.20
Mouse (from Carter, Lyon,		
and Phillips)	b^+ to *brown*	0.85
	p^+ to *pink eye*	0.85
	s^+ to *piebald*	1.70
	d^+ to *dilute*	3.40
Man (from Neel, 1962)	Epiloia	
	England (Gunther and Penrose; Penrose)	0.4–0.8
	Retinoblastoma	
	England (Philip and Sorsby)	1.2
	USA, Michigan (Neel and Falls)	2.3
	USA, Ohio (Macklin)	1.8
	Germany (Vogel)	1.7
	Switzerland (Böhringer)	2.1
	Japan (Matsunaga and Ogyu)	2.1
	Aniridia	
	Denmark (Mollenbach)	0.5
	USA, Michigan (Shaw, Falls, and Neel)	0.5
	Achondroplasia (chondrodystrophy)	
	Denmark (Mørch)	4.2
	North Ireland (Stevenson)	14.3
	Sweden (Böök)	7.0
	Japan (Neel, Schull, and Takeshima)	12.2
	Partial albinism with deafness	
	Holland (Waardenburg)	0.4
	Pelger's anomaly	
	Germany (Nachtsheim)	2.7
	Japan (Handa)	1.7
	Neurofibromatosis	
	USA, Michigan (Crowe, Schull, and Neel)	13.0–25.0
	Microphthalmos-anophthalmos	
	Sweden (Sjögren and Larsson)	0.5
	Huntington's chorea	
	USA, Michigan (Reed and Neel)	0.5

Sufficient variability in mutation rates has been encountered in various strains of the same species, suggesting that spontaneous mutation rates may be to some extent under genetic control. This has been confirmed by the demonstration of so-called "mutator" genes. For example, Rhoades (1941) described a dominant mutator gene in maize that elevated the mutation rate of genes at other loci. More recently, Koch (1971) demonstrated that changing a base pair in T4 bacteriophage increased the rate of transitions at the adjacent base pair up to 23-fold. On the other hand, Johnson and Bach (1966) showed that polyamines such as spermine suppress the mutagenic action both of mutator genes and of the mutagen 2-aminopurine in E. coli. The mechanisms by which mutation rate modifiers act are still poorly understood, however. For further information and bibliography, the reader is referred to Drake (1970).

QUESTIONS

1. A strand of DNA,

G-T-A-C
C-A-T-G

is treated with a mutagenic amount of each of the following substances:

a) 5-bromouracil
b) Nitrous acid
c) Hydroxylamine

Outline the changes in DNA that could result from the use of each of these mutagens.

2. During the evolution of photosynthesis, those prokaryotes that were unable to tolerate oxygen were forced to seek shelter from an atmosphere that was becoming increasingly rich in oxygen. These organisms are known as strict anaerobes. Other organisms evolved mechanisms for coping with oxygen—they became aerobic. What relationship might have existed between the divergent evolution of anaerobes and aerobes and the fact that at least two separate mechanisms exist for DNA repair?

3. Sketch the synaptic figures that would result from the following pairs of homologous chromosomes:

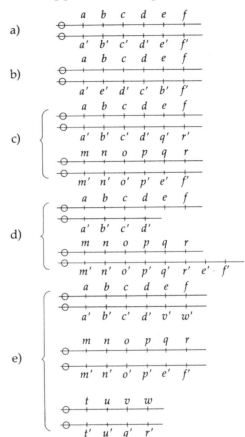

4. The term semisterility is used when 50 percent of an organism's gametes are genetically nonfunctional. What kind of macrolesions can lead to semisterility? Document your answer with diagrams.

5. If you were presented with a *Raphanobrassica* plant and asked to prove that it was an amphidiploid, how would you go about doing so?

6. Two parental organisms have the following pairs of homologous chromosomes:

♀   ♂

1 2 3 4 5 6   7 8   1 2 5 4 3 6   7 8
1 2 3 4 5 6   7 8   1 2 5 4 3 6   7 8

Sketch the synaptic figures that would appear during gamete formation in the F_1. Sketch the figures that would result from a crossover between genes 3 and 4 (in the F_1 during gamete formation). Repeat for a crossover between genes 2 and 5.

7. Given the following reciprocal translocation in *Drosophila*:

$$
\begin{array}{ll}
\text{a b c} & \text{d e f g h} \\
\hline
\text{1 2 3} & \text{4 5 6 7 8} \\
\text{a b c} & \text{d 5 6 7 8} \\
\hline
\text{1 2 3} & \text{4 e f g h}
\end{array}
$$

sketch the meiotic figure that would appear at metaphase I. Repeat, but assume a crossover between genes 5 and 6.

8. Why is it easier to establish pure-breeding strains of tetrasomics than trisomics?

9. Why do you think that aneuploidy is more common with the sex chromosomes than the autosomes?

10. Which of the common chromosomal anomalies are conducive to speciation? Why?

11. In a mating between two individuals, both of whom are trisomic for chromosome 21 (Down's syndrome), what proportion of the offspring would be expected to have the disease?

12. Why do you think that the relationship between X-ray dosage and point mutations is linear, whereas the relationship between X-ray dosage and macrolesions is exponential?

13. Why do you think that the effect of X-rays is cumulative relative to point mutations (i.e., independent of the dose rate), whereas for macrolesions there is

a definite correlation between dose rate and frequency of macrolesions produced?

14. What is the rationale for the medical practice of administering relatively large doses of penicillin for a prolonged period of time? Would penicillin be effective as a preventive drug?

15. If crossing-over were not suppressed in *ClB* females, what results might be expected?

16. Speculate on the ways in which a mutator gene might work.

17. Why would you expect the mutation rate to be greater in prokaryotes than in eukaryotes? What about the rate per generation?

18. How might it be possible for two particular genes to be linked in one individual, but be on nonhomologous chromosomes in another member of the same species?

19. Can an egg exist with a *ClB* genotype? Why?

20. What is the maximum ratio of males to females that can appear in the F_2 generation of a *ClB* cross? Assume that there are no induced mutations in the male.

21. How could you account for the viability of a zygote formed by a *ClB* egg that is fertilized by a sperm with a lethal mutation induced on its X chromosome?

22. Consider the following methods for detecting mutations: attached X analysis, examination of plaque morphology, penicillin treatment on minimum medium, replica plating, *ClB* cross, and balanced lethal system. Which of these methods is effective to do each of the following jobs? Explain.

   a) Determination of mutation rate.

   b) Determination of kind of mutation.

   c) Determination of lethal mutation rate.

   d) Analysis of nutritional mutations.

   e) Quantitative recovery of particular mutants.

10

Gene Structure: Recombination in Prokaryotes

It is apparent from the preceding chapter that most mutations occur in a random sort of way, with respect to both kind of mutation and location within the genome. If we accept the principle that all chromosomes arise from preexisting chromosomes and that mutation frequencies have probably always been at least as high as those observed today, then it follows that every chromosome existing today must have acquired innumerable mutations during its passage through the vast reaches of time. Since the overwhelming majority of mutations, irrespective of their cause, are deleterious to one degree or another, we are forced to the paradoxical conclusion that all chromosomes that exist today are incapable of existence. Obviously there must be some mechanism or mechanisms whereby deleterious mutations may be removed from chromosomes. One mechanism is natural selection, by which defective chromosomes are eliminated *in toto* from the gene pool. Another mechanism is crossing-over, which is manifested as genetic recombination.

Although the phenomenon of genetic recombination has been known for many years, it has been regarded primarily as a means of introducing new and "advantageous" mutations into an already normal genome, thereby raising the selective value of the genome even higher. This represents what we in this technological age call "progress," and is entirely consistent with the way in which we have become accustomed to looking at nature and evolution. Although this interpretation may not be entirely wrong, it may represent a distortion of reality. It is questionable that the intricate mechanism of genetic recombination evolved as a means of taking advantage of the highly improbable event of a beneficial mutation. This implies a kind of entelechy and clairvoyance that cells and chromosomes probably do not possess. It becomes even more questionable when one considers that the pressures for mere survival are and have been so great that every mechanism of living matter appears to be geared to this one purpose. The appearance of genetic combinations that exceed what is needed for survival alone is therefore probably quite fortuitous. It is much more realistic, therefore, to regard crossing-over and genetic recombination as mechanisms that have evolved primarily for replacing defective parts of the genome (Fig. 10–1). Thus the concept that the driving force of evolution is survival, as opposed to progress and transcendency, becomes as applicable to genes and chromosomes *per se* as to organisms.

How do chromosomes identify defective genes preparatory to replacing them with non-defective ones? One possibility is that they don't; that crossing-over is both a frequent phenomenon and a random one. If this is the case, gene replacement is a matter of chance. Another possibility is that as a consequence of homologous pairing (synapsis), areas of mismatch are detected and serve as the stimulus for initiating a sequence of events that culminates in recombination. This postulate helps to explain the observation that chromosome breakage is most frequent under conditions in which the chromosomes undergo synapsis (Revell, 1959).

When a molecule of DNA breaks preparatory to recombination, where is the break located in relation to the nucleotide components of the genes? In other words, are breaks restricted to points between the last base pair of one gene and the first base pair of the next? Do they violate the integrity of the base-pair triplets by occurring at either of the two points within them? Or are their locations completely random? The answers to these questions are not entirely known, but it appears that the location is not limited to any particular pattern of base pairs. Thus the break can and does occur at points other than the phosphodiester bonds between adjacent genes. As long as the base pairs are comparable in the two molecules at and adjacent to the point of breakage, it should not matter where the break occurs. If the break takes

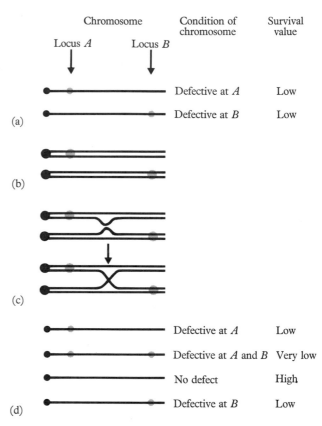

Fig. 10–1. The removal of genetic defects from a pair of chromosomes by crossing-over and recombination. (a) Mutational defects are represented in both members of a pair of chromosomes. (b) Replication occurs to produce pairs of chromatids, duplicating the defects as well. (c) Synapsis and crossing-over occurs. (d) Segregation of chromatids with the appearance of one chromosome with neither defect. This chromosome has a higher probability of survival than any of the others.

place between molecules at a position in which the base pairs do not correspond, however, both products of the recombination would be abnormal (Fig. 10–2), and therefore the position of the break does become important. It is difficult to test whether or not the site of break is preferentially outside the limits of a gene mutation, however,

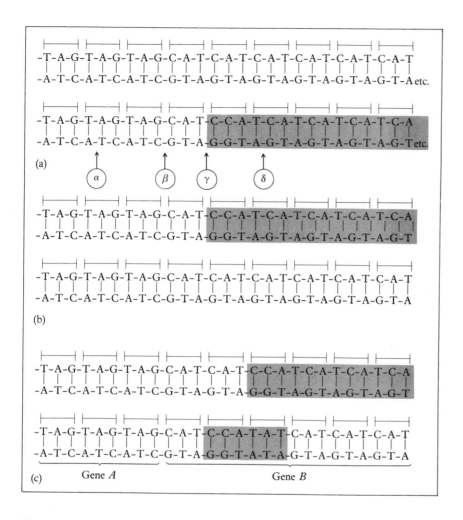

Fig. 10–2. Relationship between the position of break and the products of exchange. (a) Hypothetical base-pair sequence in a segment of DNA at the junction of two genes *A* and *B*. The upper segment assumes a "normal" sequence of triplets. The lower segment has a frameshift mutation in gene *B* (shaded area). (b) No matter where the point of break and exchange takes place *outside* the limits of the defect, the products of exchange will be unaltered. For example, the break can occur at point α, which lies between base pairs within a triplet, or at point β, which lies between the adjoining triplets of the two genes, or even at point γ, which lies within the unmutated part of gene *B*. (c) If the point of breakage lies *within* the limits of the defect, such as at point δ, *both* products contain abnormal triplets in patterns that differ from the original mutation. Since the frequency with which new mutant forms emerge from heterozygotes does not exceed that emerging from homozygotes, it appears that the site of breakage must preferentially be outside the limits of the genetic defect.

because these limits are usually so small that the chance of recovering breaks within them is correspondingly small. Moreover, what might be discovered in this connection in viruses or bacteria, which lend themselves to resolving highly infrequent events, may be of little or no relevance to sexually reproducing eukaryotes.

That DNA breakage and segmental exchange do take place at other points has been demonstrated innumerable times, and has provided many insights into gene structure. Although recombination is a regular occurrence in sexual reproduction and, therefore, a phenomenon that one generally would associate with eukaryotes, it is in bacteria and viruses that it has been most extensively investigated and analyzed. There are good reasons for this. The principal reason is the fact that enormously large numbers of these organisms can be generated and evaluated in a very short time. Since breakage and exchange can occur at any point on the chromosome or DNA molecule with essentially equal probabilities for all points, the probability of their taking place at a specified point is very small. The smaller the probability, the rarer the event, and accordingly, the larger the population of individuals required in order to detect and quantitate that event.

RECOMBINATION IN VIRUSES

Recombination is observed in its simplest and possibly most primitive form in the bacterial viruses. As described in greater detail in Chapter 7, the infection of bacteria by phage begins with introducing the phage's nucleic acid into the bacterial cell. Replication of the nucleic acid then occurs, which is followed or accompanied by synthesis of the specific phage proteins, and then by assembly of these proteins and the nucleic acid replicas into phage progeny particles. It is during the so-called vegetative phase, when the phage nucleic acid is replicating, that recombination is believed to take place. But there is some evidence that in phage, the replication process *per se* is not a necessary condition for recombination. If only one infecting particle is involved, recombination has no meaning, because all replicas are identical. If, however, two or more particles infect and these particles represent at least two genetically different classes (mixed infection), then recombination does have significance and, if it occurs, can be detected and its frequency measured.

Host-Range Mutants

Bacteriophage T2 will infect and lyse most cells of *E. coli* strain B. The cells of this strain that have proved to be resistant to T2 infection have been isolated and cultured as a separate substrain, designated B/2. The range of hosts that wild-type T2 can infect is therefore limited. The mutant phage host-range (h) extends this range to include B/2, and accordingly can infect all members of strain B.

Plates of *E. coli* B that have been inoculated with either wild-type (h^+) or mutant (h) phage will develop clear plaques, as described in Chapter 9. Plates of *E. coli* B/2 that have been inoculated with h^+ phage will show no plaques, but those inoculated with h will. Since h phage produces clear plaques on either strain, it will produce them on plates that contain a mixture of both strains of cells. Wild-type T2, on the other hand, will produce cloudy or *turbid* plaques on a mixed culture of B and B/2, because only the B cells will be lysed (Fig. 10–3a).

Rapid-Lysis Mutants

If an infected cell is reinfected (superinfection) by additional phage before the lysozyme concentration has reached the critical level necessary for lysis, further lysozyme synthesis is inhibited and lysis is delayed. During plaque formation by wild-type phage, more and more infective particles are produced and superinfection becomes increasingly frequent. The rapid-lysis mutants (r) are bacteriophage whose lytic activities are unaffected by superinfection, as their wild-type counterparts (r^+) are. As a consequence, the plaques produced by r-mutants are sharply defined and show much more rapid growth. These plaques, therefore, are

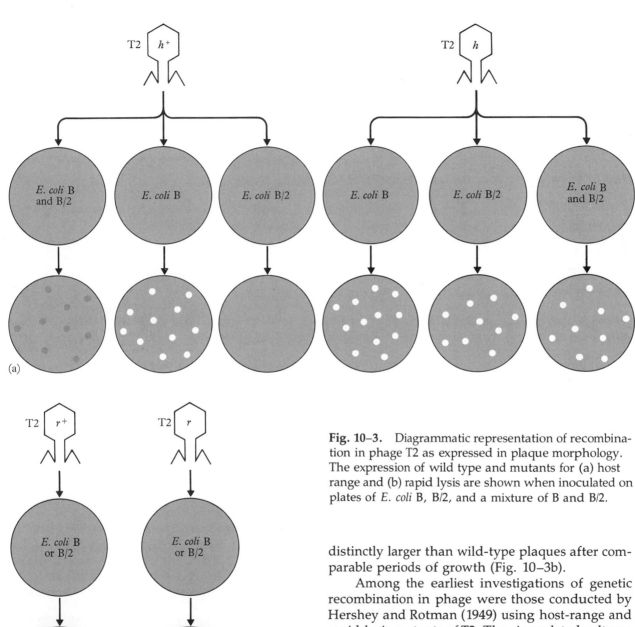

(a)

(b)

Fig. 10–3. Diagrammatic representation of recombination in phage T2 as expressed in plaque morphology. The expression of wild type and mutants for (a) host range and (b) rapid lysis are shown when inoculated on plates of *E. coli* B, B/2, and a mixture of B and B/2.

distinctly larger than wild-type plaques after comparable periods of growth (Fig. 10–3b).

Among the earliest investigations of genetic recombination in phage were those conducted by Hershey and Rotman (1949) using host-range and rapid-lysis mutants of T2. They inoculated cultures of *E. coli* B with large enough numbers of both these mutants (hr^+ and h^+r) to assure mixed infection of all cells. They waited for lysis to occur and then harvested the phage progeny and used them to inoculate mixed plates of *E. coli* B and B/2. Four kinds of plaques appeared on these plates: (1) small clear, (2) large turbid, (3) small turbid, and (4) large clear (Fig. 10–4). The genotypes that correspond to these four plaques are hr^+, h^+r, h^+r^+, and hr, respectively. The last two classes are recombinants, and must have arisen by recombination in the mixedly infected cultures of *E. coli* B (Fig. 10–4).

The mechanism by which recombination is effected in bacterial viruses is not known. It occurs in mixed infections even when DNA synthesis, and hence replication, is suppressed by means of potassium cyanide (Tomizawa and Anraku, 1964). Therefore it appears that it is the products rather than the process of replication that are required. In Chapter 4 it was shown that recombination in eukaryotic organisms is best explained by some sort of "breakage and reunion" hypothesis, which proposes essentially the following sequence of events: (1) synapsis of homologous chromatids, (2) breakage of chromatids at identical points, and (3) cross-reunion of chromatids to effect recombination and chiasma formation. In these cases, the recombination classes must be equal (with exceptions recorded in fungi—cf. p. 72). That is, given the cross $ab \times a^+b^+$ in which a and b are on the same chromosome, the recombinant classes ab^+ and a^+b are equal in number. Hershey and Rotman (1949) found that this was not the case in phage if lysis were induced early. On the contrary, they found no correspondence at all between the frequencies of the two classes. The later lysis occurred after infection, however, the closer the recombinant classes did correspond.

According to Visconti and Delbruck (1953), who conducted a statistical analysis of recombination in phage, the phage DNA molecules undergo a series of "matings" (close associations and recombinations) within the host cell between the time they are replicated and the time they are removed from the DNA "pool" for packaging in their protein coats. Early lysis releases particles that have undergone possibly only one mating, whereas later lysis includes particles that have undergone perhaps five matings. Accordingly, any

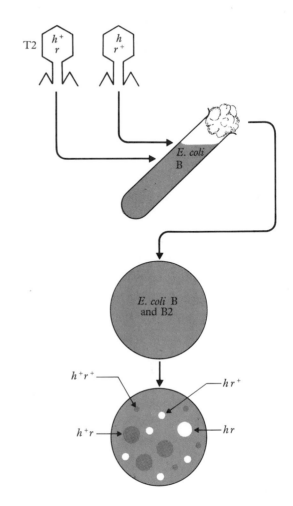

Fig. 10–4.   A broth culture of *E. coli* B is mixedly infected with both host-range and rapid-lysis mutants. After lysis, the progeny are harvested and inoculated on plates containing a mixed culture of B and B/2. (See text for explanation of results.)

inequality between recombinants resulting from one mating would tend to be canceled out by reciprocal inequalities in a subsequent mating. Therefore, the results obtained by early lysis probably reflect more accurately what is going on during recombination.

These results favor an old theory of Belling (1931), later revived, refurbished, and renamed by Lederberg (1955), which proposed that recombination occurs during replication as a consequence of the growing DNA molecule's copying first one parent molecule and then another, as illustrated in Fig. 4–13. If three or more parent molecules are involved in the matings occurring in the DNA pool, then nonreciprocal recombination could occur in this way. However, doubt on this explanation is cast by the findings of Tomizawa and Anraku (1964), cited above, and those of Meselson and Weigle (1961). By using the heavy isotopes of nitrogen (^{15}N) and carbon (^{13}C) to label phage DNA, Meselson and Weigle demonstrated very clearly that recombination does involve breakage and reunion, and that it occurs after rather than during replication (Fig. 10–5). The results obtained by Hershey and Rotman (1949), therefore, must have some explanation other than the copy-choice mechanism of Belling and Lederberg. For example, it has been proposed that any DNA molecule emerging from a mating has a 50 percent chance of remaining in the pool for a subsequent mating, and correspondingly a 50 percent chance of being isolated in a protein coat. This could explain the inequality of recombinants in a small sample—one becomes incorporated in a particle and is identified, while the other remains in the pool and is not (Hayes, 1968).

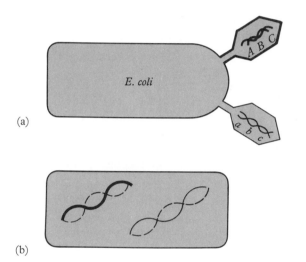

(a)

(b)

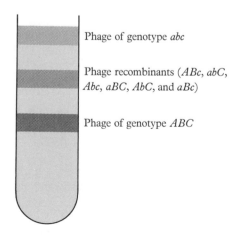

Phage of genotype *abc*

Phage recombinants (*ABc*, *abC*, *Abc*, *aBC*, *AbC*, and *aBc*)

Phage of genotype *ABC*

(c)

Fig. 10–5. Schematic representation of the experimental evidence of Meselson and Weigle (1961) supporting the theory that recombination involves breakage of DNA after DNA replication has occurred. (a) Bacteria growing in a normal medium are infected with two strains of phage. One strain, indicated by heavy lines, has the heavy isotopes ^{15}N and ^{13}C incorporated into its DNA. The other kind of phage contains light DNA. These phage also differ in genotype. The heavy phage is of genotype *ABC* whereas the light phage is of genotype *abc*. (b) The phage DNA then replicates, forming either all light or all hybrid molecules. New DNA is here represented by dashes and is of the light variety. (c) The phage progeny are recovered and subjected to prolonged centrifugation in a CsCl solution. The light phage (*abc*, parental type) remain in the region of lowest density, whereas the hybrid phage (*ABC*, parental type) band in an area of greater density. Between these two bands are the recombinant phage particles.

Dissecting Viral Genes

Using recombination data, Benzer (1955) demonstrated both the spatial limits of a gene and the location of mutation sites within it. Hundreds of r-mutations have arisen independently of one another and have been tested in T4 bacteriophage. Although they involve different sites in the DNA molecule, they appear to be functionally alike, all producing large, sharply defined plaques on plates of sensitive bacteria. But some of them fail to produce plaques with the strain of E. coli known as strain K. They infect the cells of this strain and kill them, but they cannot lyse them, and so plaques do not form. These are called rII mutants in order to distinguish them from the other r-mutants which can produce plaques in plates of strain K bacteria. Benzer showed further that the rII mutants consisted of two groups which he called rIIA and rIIB (Fig. 10–6). He showed this by a genetic test known as the *complementation test* described below.

The rationale of the genetic complementation test is similar to the following analogy. Many car owners have been frustrated trying to remove license plates attached by a pair of nuts and machine screws, unless they use both a screwdriver and a wrench. Using a screwdriver alone succeeds merely in turning the screw while the nut goes around with it, still firmly attached. Using the wrench alone has the same effect. Both tools must be used together, the action of one complementing the action of the other. Likewise, assume that virus gene products A and B are necessary to cause lysis (removing the screw); that strain *a* lacks the ability to produce A but can produce B (the screwdriver); and that strain *b* cannot produce B but can produce A (the wrench). Between them, both gene products are provided and lysis occurs. In order to complement each other in this way, the precise nature of the respective functions must be different, although they may contribute toward the same end. Thus the license plates are removed by the screwdriver gripping the screw and the wrench gripping the nut.

Benzer mixedly infected E. coli K with rII mutants in pairs (e.g., r101 + r102, r102 + r103, etc.) and found that in some cases lysis resulted (complementation) and in others it did not. He was able to assemble the mutants into two groups, A and B, such that any combination of an A and a B mutant showed complementation, while any two within the same group did not. Since the functions of rIIA and rIIB complement each other, they must be different genes, even though their end result (phenotypic effect) is apparently the same. If they are different genes, then they should show not only complementation, but recombination as well. They do—Benzer recovered from the progeny produced by mixed infections of rIIA and rIIB mutants in E. coli B phage particles that would both lyse cells of strain K and produce wild-type plaques. For example, if the rIIA mutant r101 and the rIIB mutant r51 are used together to infect strain B bacteria, the genotypes of particles that can be recovered will be r101 (parental), r51 (parental), r101+r51+ (wild type), and r101 r51 (double mutant).

Benzer (1961) found recombination between some of the r-mutants of the same gene. He mixedly infected strain B bacteria with phage carrying mutations in the rIIA gene, two at a time. He tested their progeny on plates of strain B, since both mutants and wild type grow well in this strain and are clearly distinguishable from each other, and on plates of strain K. By appropriate dilution methods, he was able to measure the frequency of recombination between pairs of mutants and to determine the relative distances between them. Essential to this procedure was measuring the frequency of reversion from mutant to wild type, which is readily done in controls. Because of the enormous numbers of phage particles that could be produced and analyzed, Benzer could have detected recombination frequencies as low as 0.0001 percent. The lowest frequencies he found, however, were about 0.02 percent. Tessman (1965) induced mutations in the rII region, thereby

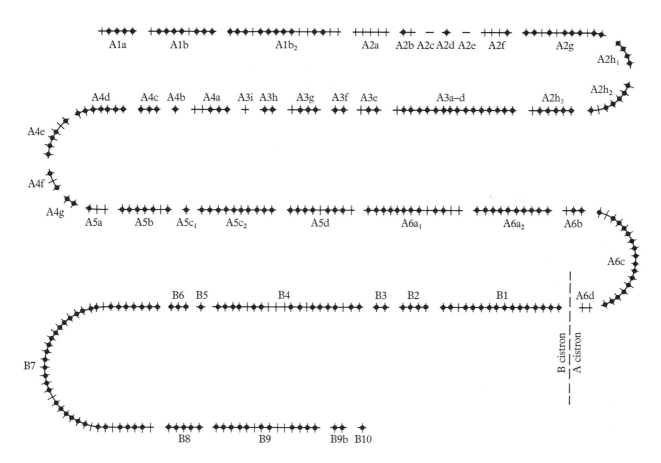

Fig. 10–6. The *r*II mutations in phage T4. The *r*II mutations occur in two groups (*r*IIA and *r*IIB) that are distinguished by complementation testing. [From S. Benzer (1959), *Proc. Nat. Acad. Sci.* **45**:1616–1617. By permission.]

greatly increasing the number available for study and, correspondingly, the chance of obtaining mutations very close together, and obtained frequencies of recombination as low as 0.00001 percent.

On the assumption that the frequency of recombination between mutant sites is a direct function of the distances between them (cf. Chapter 4), Benzer constructed a map of the *r*II region (Fig. 10–6) that shows the sites of the various spontaneous mutations with which he worked and the relative lengths of the A and B genes. As a result of

Benzer's studies, the following statements can be made about T4 genes. (1) Two genes having the same function (phenotypic expression) can be distinguished as separate genes if they complement each other. That is, normal function is obtained by combining two mutants if the mutations are in different genes. (2) Mutations that do not show complementation are mutations of the same gene. (3) A gene may have more than one mutational site, as evidenced by recombination between mutations at those sites.

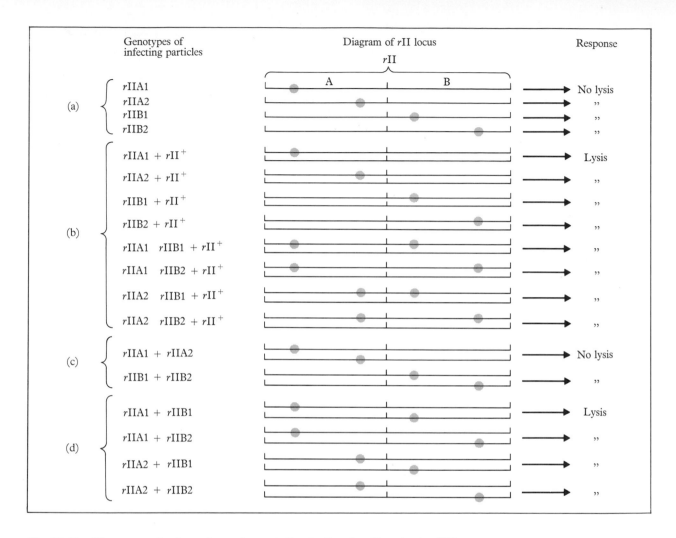

Fig. 10–7. Diagrammatic view of complementation testing for *r*II mutants of T4 bacteriophage in strain K of *E. coli*. (See text for explanation.) (a) Virus particles carrying mutations in four different regions of the *r*II locus effect no lysis when inoculated singly. (b) Mixed infections by any of the four mutants with wild-type particles produce lysis. (c) Mixed infections by two different single mutants in the same cistron fail to effect lysis. (d) Mixed infections by two single mutants in different cistrons effect lysis (complementation). Shaded areas indicate relative location of mutations.

The Cistron

Benzer's findings are in agreement with or have been confirmed by genetic studies of other organisms, principally bacteria, fungi, and *Drosophila*. When Benzer made his discoveries, a gene was considered to be a unit of function. Accordingly, the *r*II region of the T4 DNA would constitute a gene. Because there was no distinct difference in function between divisions *r*IIA and *r*IIB despite the fact that they show complementation, Benzer recognized the need to clarify genetic nomenclature. He introduced the term "cistron" to apply to the genetic unit within which complementation does not occur, but recombination may. He adopted the term from the *cis-trans* effect, which was first discovered and described in *Drosophila* (see Chapter 4). In diploid organisms, linked mutations may be on the same chromosome of a homologous pair (*cis* position) or on different members (*trans* position) of that pair. If recessive mutations involve the same gene, heterozygotes that have them in the *cis* position will be wild type, whereas heterozygotes that have them in the *trans* position will be mutant. If the mutations involve different genes, there will be no phenotypic difference between *cis* and *trans* heterozygotes. In the cases in which he mixedly infected strain K cells with T4 phage, Benzer observed similar differences in response, as diagrammed in Fig. 10–7. Hence mutations that showed a positive *cis-trans* effect (wild type when *cis*, mutant when *trans*) were members of the same recombining group, which he called a *cistron*. Subsequently, ideas about the nature of the gene have changed, so that what is meant by "gene" in molecular terms is clearly understood, and the structural term "cistron," which fulfilled an important need for a time, is falling into disuse. Benzer speculated that the *r*IIA and *r*IIB cistrons probably code different polypeptides, which secondarily become complexed to form a protein that determines the phenotype. On this basis, his term cistron is quite synonymous with the term gene as it is used today.

RECOMBINATION IN BACTERIA

Genetic recombination in bacteria may be facilitated in one of three ways—by conjugation, transformation, or transduction. Each of these ways enables the genomes or parts of genomes of two individuals to come in contact, which is a necessary condition for recombination. How recombination is finally effected between DNA molecules is conjectural (see Chapter 4), but there is no evidence to indicate that the mechanism is any different in bacteria and viruses, for in neither one is the DNA packaged in a complex chromosomal structure as it is in eukaryotes. Nevertheless, studies of genetic recombination in bacteria have shed considerable light on gene structure and the function of genetic material.

Conjugation

Until the discovery of genetic recombination in bacteria by Lederberg and Tatum in 1946, bacterial reproduction was believed by nearly all microbiologists to be an exclusively asexual process. Lederberg and Tatum grew two triple nutritional mutant strains of *E. coli* K in mixed culture in yeast-beef broth (complete medium), and then plated the progeny on agar medium to which various supplements had been added. The supplements corresponded to each of the nutritional deficiencies that characterized the mutant strains. They found a variety of recombinants, including wide-type K, that could not be explained by reversion or by transformation as plausibly as by sexual union. Subsequently their hypothesis was confirmed by microscopic demonstration of conjugation in strain K organisms (Fig. 10–8). It is an infrequent event, for only one cell in a million, according to Lederberg and Tatum's estimates, could be classified as a recombinational type in the cultures they investigated.

Bacterial conjugation is regulated by an infectious sex factor, F, which is present in some individuals but lacking in most. Factor F (described more fully in Chapter 11) is an episome which induces its host cell to establish cytoplasmic connection by way of a bridge or *sex fimbria* (Fig. 10–8) with a cell that lacks the F-factor. The basic mechanism responsible for this behavioral process is evidently related to the fact that bacteria harboring the F-factor (such bacteria are designated by the symbol F$^+$) synthesize a different surface component. This component, which can be detected immunologically as a specific surface antigen, is believed to lower the electrical charge at the host cell's surface and thereby facilitate adhesion to any F$^-$ cell (one lacking the F-factor) with which it comes in contact during random collisions (Hayes, 1968).

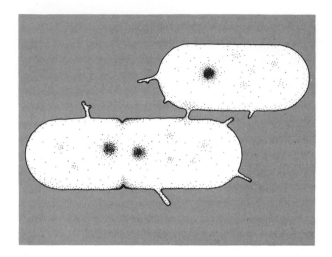

Fig. 10–8. Schematic representation of conjugation in *E. coli.* The upper cell is the donor and is designated Hfr. The lower cell is the F⁻ recipient.

The F-factor is a DNA molecule of about 250,000 nucleotide pairs which replicates autonomously in the host cell and is structurally and spatially independent of that cell's DNA. However, the factor may become inserted (*integrated*) in the host cell's DNA (discussed in Chapter 11), in which case it replicates as a part of that DNA. It is in bacterial strains with integrated F-factor that genetic recombination occurs with high frequency (more than one in 10^3 cells, as compared with less than one in 10^6 described by Lederberg and Tatum). These strains are designated as Hfr (*h*igh *f*requency *r*ecombination) and have been used extensively in recombination studies in bacteria.

The reason that Hfr strains show a high frequency of recombinants appears to be due not to any increase in conjugation tendency, as one might infer, but to the way in which the integrated factor affects the host's DNA. As deduced by Wollman and Jacob (1955, 1958) from their studies of conjugation, and demonstrated microscopically by Cairns (1963), the entire genome of *E. coli* is represented by a single molecule of DNA, and this molecule exists as a closed circle. The F-factor, which is also a circular molecule when not integrated, is opened and inserted into the host's DNA molecule in its integrated condition (Fig. 10–9). Replication of the integrated factor is synchronized with replication of the host's DNA, and thus the factor loses the autonomy that characterizes it in F⁺ strains. The integrated factor has polarity, as one might expect of DNA, and this polarity is expressed at the time of conjugation. After or while completing a round of replication, one of the daughter rings of DNA breaks at one of the two points of junction of the bacterial component and the F-factor. Which of the two points is the point of breakage is determined by the polarity of the F-factor—that is, if the polarity is signified by representing the F-factor as an arrow, the break always occurs at the tail of the arrow. The free end of the bacterial DNA, called the "origin" (O), then leads the way as the molecule migrates from the Hfr cell, through the sex fimbria, into the recipient F⁻ cell (Fig. 10–10).

The site of insertion and polar orientation of the F-factor is probably random, or nearly so, but once the factor has become integrated, all descendants of that Hfr cell will have the factor integrated at the same site and with the same polarity, and these characteristics become stable characteristics of the strain. Wollman and Jacob (1955), in a series of ingenious experiments, mixed a strain of Hfr cells that was sensitive to streptomycin with a strain of F⁻ cells that was resistant to streptomycin and incorporated a number of selected mutant genes (genetic markers). After eight minutes incubation, which had been demonstrated to be sufficient time to permit cell-to-cell adhesion to occur and conjugation to be initiated, they started removing samples from the culture at regular time intervals. Each sample removed was subjected to strong shearing forces in a food blender in order to separate Hfr and F⁻ cells—that is, to interrupt

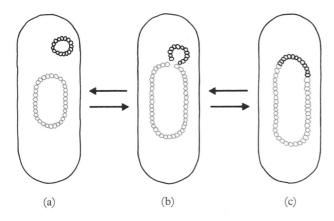

Fig. 10–9. The transition from the F$^+$ to the Hfr state in *E. coli*. (a) The F-factor in an F$^+$ cell exists as an autonomous, circular unit as indicated by the small circle. (b and c) Integration of the F-factor, converting the cell to the Hfr condition.

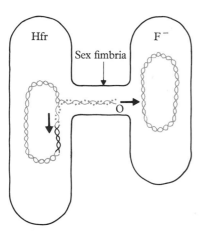

Fig. 10–10. The replication and transfer of DNA from an Hfr donor to an F$^-$ recipient. Newly synthesized DNA strands are indicated by broken lines; the integrated F is in black. The origin (O) leads the way through the sex fimbria.

mating—and then was treated with streptomycin in order to inhibit growth of the Hfr cells. The sample was then distributed over a range of plates of supplemented agar medium in order to identify the various genetic markers that characterized the initial Hfr and F$^-$ strains. The results, depicted schematically in Fig. 10–11, showed that the genetic markers are arranged linearly in a fixed order, and that the distances between markers, expressed in units of time, are constant. This finding confirmed previous deductions regarding the linear arrangement of genes and represents another method for mapping bacterial DNA.

Different strains of Hfr bacteria were found to be characterized by different genes adjacent to the origin and/or by a reverse order of gene transfer during conjugation, as shown in Fig. 10–12. These findings served as the basis for the postulates that the bacterial DNA is a circular molecule and that the F-factor has polarity. As stated earlier, the former postulate has been verified microscopically by Cairns (1963).

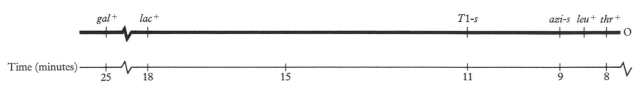

Fig. 10–11. A partial genetic map of *E. coli* DNA with time in minutes as the unit of distance. (See text for explanation.)

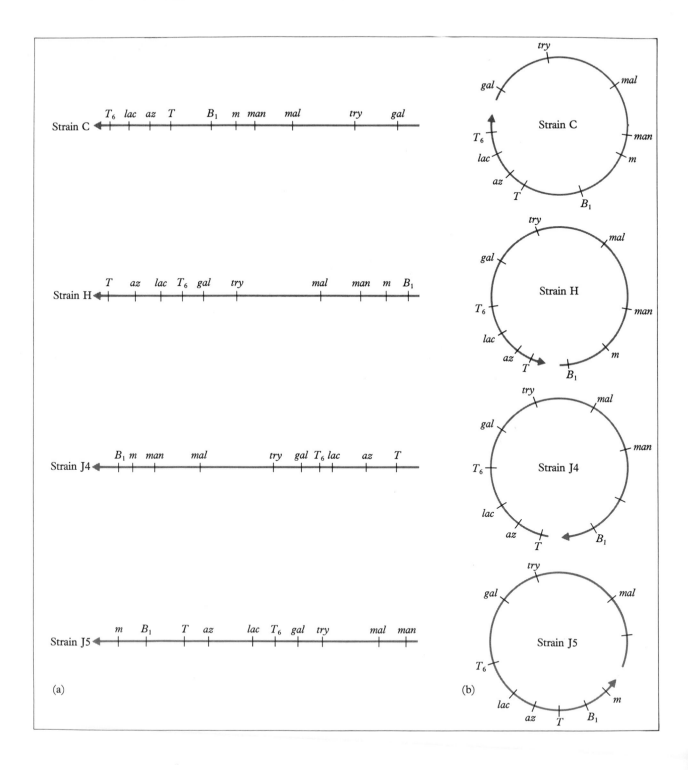

(a)

(b)

◀ **Fig. 10–12.** (a) Differences in point of origin during transfer in four Hfr strains of *E. coli.* (b) These results are explained by postulating differences in the site of integration of the F-factor in the circular chromosome. [Data from Hayes, Jacob, and Wollman (1963).]

Subsequent work has culminated in a detailed circular map of the DNA of *E. coli* K (Fig. 10–13) constructed in time units from zero to ninety minutes, assuming that the rate of transfer for the entire molecule at 37°C is uniform and is completed in ninety minutes (Taylor and Thoman, 1964; Hayes, 1968). The last part of the molecule to be transferred is the F-factor. The position and orientation of the F-factor in some of the Hfr strains described and investigated are indicated in the inner part of the circular map (Fig. 10–13).

One of the important points implicit in the experiments described above is that, in order for the Hfr genes to be identified in the F⁻ cells, they must have been incorporated into the genome of these cells. This is so because they are identified in terms of the phenotypes of colonies, which represent clones of millions of individual cells that have been produced by multiple division generations of the exconjugants (the F⁻ cells isolated after conjugation). The presence of the Hfr alleles in the exconjugant genomes can be accounted for only by genetic recombination. It follows, therefore, that recombination does not require the presence of two entire genomes, but rather that fragments of genomes can be exchanged. It is interesting to note that when a donated fragment is incorporated into a complete, recipient DNA molecule, it almost invariably occupies the same position or locus as it did in its parent molecule. This implies the existence of some mechanism whereby genetic segments are recognized and placed in their proper position before recombination occurs. The mechanism may be nothing more than similarity of base sequence, and thus may resemble the mechanism that is the basis of *in vitro* hybridization studies with DNA and RNA (cf. Chapters 8 and 11).

Results such as the foregoing and those that follow raise the question of what becomes of the DNA segment for which the recombined segment was exchanged. It may be carried along for several cell generations, presumably without replicating. Therefore, being progressively diluted, it may eventually be lost.

Transformation

As mentioned in previous chapters, bacterial transformation was first described by Griffith in 1928 in *Diplococcus pneumoniae* and further explained by Avery, MacLeod, and McCarty in 1944. In brief, they demonstrated that DNA extracted from one strain of *D. pneumoniae* (also referred to as *Pneumococcus*) could be taken up by intact cells of another strain and alter its genotype with respect to certain characters to that of the donor strain. This phenomenon strongly suggests that genetic recombination is involved somewhere in the process, for the transformation is permanent and transmissible. Although it is beginning to become apparent that conjugation, transformation, transduction, and sexual reproduction may be viewed as alternative ways of bringing homologous strands of DNA together so that genetic recombination can take place, bacterial transformation has something special to offer in terms of casting light on the mechanism of recombination itself. In addition, transformation has proved to be a useful way to dissect the bacterial genome and to construct genetic linkage maps like those already described. Since these maps conform to what has already been derived and described by other methods, they will not be described here. However, it is sufficient for our purposes to make the observation because it is an important one, and then move on to other aspects of this fascinating phenomenon, such as mechanisms of entry of transforming DNA and the theories of how it is incorporated into the recipient's genome.

The DNA that can transform must exceed a certain minimum size, which depends upon the species of the recipient cell. This reflects the fact

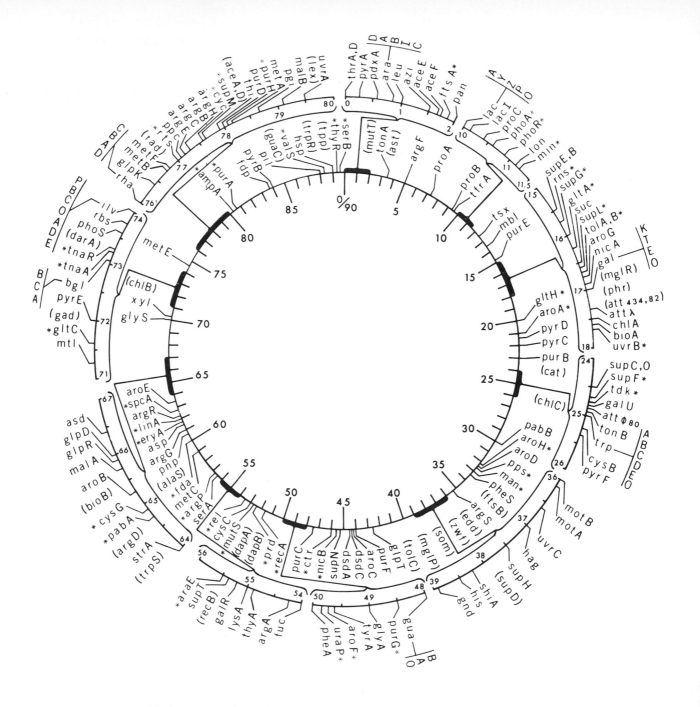

Fig. 10–13. Revised linkage map of *E. coli*. The outer circle represents the genetic map of strain K12. The inner circle shows the position of insertion and orientation of the F-factor in some of the strains that have been investigated. [Courtesy A. L. Taylor and C. D. Trotter (1967), *Bact. Rev.* **31**:337. By permission.]

that an important limiting factor in transformation is DNA uptake, which appears to involve both specific receptor sites on the surface of the recipient cell and enzymes within the cell. In *Pneumococcus*, for example, DNA segments with molecular weights less then 5×10^5 are not taken up (Rosenberg, et al., 1959), whereas in *Bacillus subtilis* the minimum size is about 10^7. No upper limit in size has yet been found, however.

Although the kind of DNA that will bring about transformation in bacteria evidently must be derived from the same or a very closely related species as the cell being transformed, there is considerable variation in the degree of relationship that is required simply for DNA uptake. According to Schaeffer (1964), uptake in *Haemophilus influenzae* appears to be restricted to DNA from closely related species. *Diplococcus* and *Streptococcus*, on the other hand, are nonspecific, for they seem capable of accepting any double-stranded molecule of DNA of proper molecular weight regardless of its species of origin. For example, Hotchkiss (1954) found that DNA from calf thymus inhibits the uptake of pneumococcal transforming DNA which, together with other evidence, suggests that the calf DNA is taken into the cell. On theoretical grounds, it appears that the major factor preventing nonspecific DNA from transforming is the recombination process itself—the fact that the DNA must be sufficiently homologous with some segment of the recipient cell's DNA to be recognized by it and substituted for it. However, the entry process is evidently specific as well, because bacterial cells have so-called "restriction" enzymes that cleave and render ineffective some varieties of entering DNA. These cells also possess enzymes called modification methylases. The methylases act on specific kinds of entering DNA too, but their action is counter to that of the restriction enzymes. Presumably, a methylase enzyme methylates and thereby blocks those sites on entering DNA that are attacked by the restriction enzymes (Marx, 1973).

The variety of cells that are capable of being transformed is not well known. In bacteria, proved transformation has been described in *Bacillus, Diplococcus, Haemophilus, Neisseria, Rhizobium,* and *Streptococcus* (Schaeffer, 1964). Aaronson and Todaro (1969) reported transformation of human cells in tissue culture by DNA extracted from simian virus. Whether this constitutes the same kind of mechanism as that described above for bacteria is doubtful, because the introduction of viral DNA into the DNA of an animal cell would most probably represent insertion, as opposed to recombination, and more closely resemble the mechanism by which the F-factor (or viral DNA) is integrated into a bacterial chromosome, as discussed in Chapter 12.

The fact that small fragments of DNA and single-stranded DNA are not taken up by cells competent for transformation suggests that uptake is not dependent on pores or other perforations in the cell wall. Moreover, the number of DNA molecules that are taken up is highly limited (about ten per cell), and the uptake is energy-dependent (Stuy and Stern, 1964). The ability of bacteria to be transformed depends upon their physiological state, which can be correlated in turn with the phases of the life cycle. This state of transformational competence lasts about fifteen minutes in typical cultures of *Pneumococcus*, and it corresponds to the phase during which the cell membrane is invaginating and the membranous organelles known as mesosomes move out from the cytoplasm (Miller and Landman, 1966). It is now believed that competent bacteria have acceptor sites for transforming DNA, and that the mesosomes may be the sites of the enzymes involved in DNA uptake (Wolstenholme, et al., 1966).

Lacks (1962), using ^{32}P as a radioactive tracer, found that transforming DNA in *Pneumococcus* is broken down into single strands as it enters competent cells, and one of the strands is further degraded to acid-soluble fragments. The intact strand is associated with protein and is inactive until it becomes incorporated into the recipient cell's genome. Lacks proposed that the protein involved is an exonuclease, which separates the strands as it draws them into the cell and breaks one of them down.

According to Fox and Allen (1964), the transforming DNA is inserted into the host's genome while it is still a single strand. They propose that when it comes into position at its homologous site on the host's DNA, the single strand of transforming DNA undergoes pairing with its complementary strand at this site. An exonuclease then excises the unpaired strand and insertion of the transforming strand takes place. At least until the next round of DNA replication, this area of the genome is hybrid.

The relationship of transforming DNA insertion to recombination in higher forms is highly suggestive because, as described in Chapter 4, it is becoming increasingly clear that recombination in diploid species is a complex, enzymatically regulated process. Furthermore, the exquisiteness of the processes that come into play suggests that bacterial transformation is not an accidental phenomenon or a biological curiosity, but a highly evolved mechanism by which microorganisms take advantage of genetic diversity in the population and thereby adapt to changing demands in the environment (cf. Chapter 12). Hence, although the mechanisms by which genetic material is exchanged between microorganisms may differ from those in diploid species, the mechanisms of insertion or incorporation of this material into the genome doubtless have many elements of similarity including evolutionary significance.

Transduction

As described briefly in Chapter 7, bacterial transduction is the phenomenon by which an infecting virus transfers genetic information from one bacterium to another. The basic nature of the process was first explained by Zinder and Lederberg in 1952. They discovered that filtrates of cultures of one strain of *Salmonella* could, under appropriate experimental conditions, "transduce" genetic characteristics from that strain to another strain, one at a time. The strains they used differed in that one (designated LA22) was lysogenic and the other (designated L2) was not. A lysogenic strain is one that has a virus integrated in its genome—in this case phage P22. For genetic markers, Zinder and Lederberg used a variety of nutritional mutants and streptomycin resistance. The explanation of their findings is represented diagrammatically in Fig. 10–14.

As vectors for transferring genetic information from one cell to another, viruses are unquestionably effective but not very efficient. The size of the DNA fragment they can transduce is limited by the internal capacity of the protein capsule, which is designed to accommodate the phage DNA. This has a molecular weight on the order of 6×10^7 (phage P1) as compared to about 2.8×10^9 molecular weight for the *E. coli* genome—that is, the phage DNA is about one-fiftieth the length of that of *E. coli* (Ikeda and Tomizawa, 1965a). Empirically, only the amount of bacterial DNA that corresponds to one or two genes is transduced at a time. Ikeda and Tomizawa (1965b), by use of ^{32}P- and ^{3}H-labeled thymidine in phage P1, showed that transducing particles carry only bacterial DNA fragments—no phage DNA—and that such particles represent only about 0.3 percent of the total. On the basis of this and other considerations such as recognition by appropriate modification methylases, it is estimated that the probability of transduction for any given genetic marker is between 10^{-5} and 10^{-4} (Ikeda and Tomizawa, 1965a). It is therefore not the efficiency of transduction that is impressive, but that it happens at all. More impressive than this are the biological implications of this kind of phenomenon (discussed in Chapter 12) and the possible use of virus vectors in genetic "engineering."

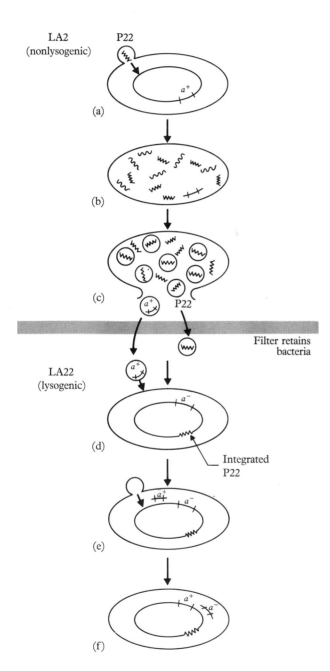

LA2
(nonlysogenic)

P22

(a)

(b)

(c)

P22

Filter retains
bacteria

LA22
(lysogenic)

(d)

Integrated
P22

(e)

(f)

Fig. 10–14. Postulated mechanism of transduction in *Salmonella*. Strain LA22 is a lysogenic strain that has integrated into its genome the DNA of phage P22, and so it is not ordinarily lysed by P22. Strain LA2 is non-lysogenic and is lysed by infections of P22. If cultures of LA2 carrying the wild-type allele a^+ are infected with P22 (a), reproduction of the phage occurs (b), followed by formation of progeny particles and lysis of the host cell (c). Segments of the host's genome occasionally are included within a phage protein capsule, as shown in (c). Virus progeny are passed through a bacterial filter into a culture of LA22 that is deficient for a (a^-). In the diagram, a phage particle that includes the a^+ allele from the LA2 strain becomes adsorbed on a cell of LA22 (d). The bacterial DNA segment is injected into the host cell (e), and by genetic recombination replaces the gene a^- (f). The progeny of this cell can be isolated in medium that will not support the growth of a^- mutants. [Adapted from Zinder and Lederberg (1952).]

QUESTIONS

1. Why are most mutations harmful to the organism?

2. If you wished to change a nutritionally mutant strain of bacteria to the wild type, how would you do it using each of the following techniques?

   a) Transformation

   b) Conjugation

   c) Transduction

3. What is the minimum number of physical sites of crossing-over that must occur during a single recombination in *E. coli* F$^-$?

4. Why do you think that F$^+$ cells yield a high frequency of episome transfer and a low frequency of chromosome transfer, whereas Hfr cells show the opposite effect?

5. Why do you think that so many different mutational sites are known for the *r*II region in T4 phage but so few are known for any gene in man?

6. Conjugation, transformation, and transduction are all methods of bringing together different strands of DNA. However, once the "new" DNA gets inside the recipient cell, it is more likely to integrate and recombine if it got there via conjugation or transformation than by transduction. Why might this be so?

7. A variety of Hfr strains are isolated from an F$^+$ strain of *E. coli*. Each of the Hfr strains are mixed with F$^-$ cells and the technique of interrupted mating is followed. Each strain transfers its genes in a unique order as follows:

Strain	Order of transfer
1	A B C D E F G
2	G A B C D E F
3	D E F G A B C

Draw a map of the F$^+$ genome from which the Hfr strains were derived.

8. Eight *r*II deletion mutations were tested in all possible combinations for recombination. The following data represent complementation tests—+ indicating the ability to yield r^+ recombinants, and 0 indicating no r^+ recombinants were formed.

	1	2	3	4	5	6	7	8
1	0	0	0	0	0	0	0	0
2		0	+	+	+	+	0	0
3			0	0	+	0	+	0
4				0	0	0	+	+
5					0	0	+	+
6						0	+	0
7							0	0
8								0

Draw a map showing the order of the mutations.

9. Does the fact of transduction throw any light on the evolutionary origin of viruses? Explain your answer.

10. Explain how mutation, selection, and recombination could work together to form virus particles that are resistant to antibodies A, B, and C?

11. If no recombination occurs, what is the appearance of plaques formed by a mixed infection of hr^+ and h^+r T2 phage on a mixed plate of *E. coli* B and B/2? What about a mixed infection of hr^+ and h^+rII on a mixed plate of *E. coli* B, B/2, and K?

   If recombination took place in each of these situations, what kind of plaques would appear? How could you isolate any one particular kind of viral progeny?

12. Reconcile the following statements:

   a) *r*IIA mutants will only complement *r*IIB mutants, and

   b) some pairs of *r*IIA mutants show recombination.

13. Define a gene in structural and functional terms.

14. During mating of an F$^+$ cell and an F$^-$ cell, a replica of the sex factor enters the F$^-$ cell, changing it to F$^+$. How is it that any F$^-$ cells still exist?

15. In the design of an experiment involving conjugation or transformation, what is the value of using cells that are streptomycin-sensitive?

16. If bacterial transformation is a process that occurs in nature and if some bacteria are relatively nonspecific with respect to transforming DNA, speculate on the integrity of bacterial species.

11

Genes in Development

Although it is well understood and accepted that the fertilized egg contains in its DNA all the information that the adult organism will require, it is not *known* that such is the case. For example, under no conditions so far discovered will a zygote produce myosin, or hemoglobin, or pancreatic lipase, or any of a large number of kinds of substances that characterize the cells and tissues of adult organisms. We *infer* that the zygote has the specific information required for the synthesis of these substances tucked away somewhere in its DNA, because this is the only inference that makes sense. If it were otherwise, we could not explain the transmission of adult characters from one generation to the next. However, this inference implies the existence of some mechanism that prevents certain genes from expressing themselves in certain situations, in this instance the situation of "zygoteness." Such a mechanism must exist. Otherwise the zygote would be doing everything that all the cells of the adult organism do—that is, it would essentially be a single cell organism such as an amoeba or a bacterium. We must therefore accept the notion of differential gene expression—that is, the existence of a mechanism or mechanisms by which certain genes may be repressed at certain times and activated to express themselves at other times. Accordingly, in the zygote we would find repression of all genes that are not concerned with "zygoteness" and activation of those that are.

If the notion of gene activation/repression is acceptable for the zygote (and for prokaryotes, Chapter 8), why should it not be equally acceptable for the cells of the adult eukaryote? Why should it not be that the genes for myosin synthesis are repressed in red blood cells, that the genes for hemoglobin synthesis are repressed in muscle cells, and so on, while the genes appropriate to their normal function are activated? An alternative to this hypothesis would be the existence of a mechanism whereby red blood cells during differentiation receive only those genes from the total genome that are concerned with red blood cell function, and whereby muscle cells receive only "muscle DNA," and so on. This idea is not acceptable, as is explained below. A strong argument against it is the exquisiteness of mitosis, by which all cells of the adult are produced from the zygote. Its very elegance and precision is reconcilable only as a mechanism for assuring that *all* of the genome of the parent cell is transmitted without loss, duplication, or alteration to each of the daughter cells.

THE VARIABLE GENE ACTIVITY THEORY

If the idea that certain genes are activated in the zygote and others are activated in the adult cells is acceptable, then there should be no difficulty in accepting the idea of a sort of series of switchings on and off of genes in sequence as the zygote undergoes the structural changes that characterize growth and development. This idea, which is known as the *Variable Gene Activity Theory of Cell Differentiation*, is widely held by geneticists and developmental biologists today.

Each type of somatic cell of an adult organism is very different both morphologically and physiologically from the original zygote. Bridging the differences is a series of cell generations that to varying degrees are transitional between the "generalized" zygote and the differentiated cells of the adult. If these cells are transitional, then they are identical to neither the zygote nor the adult cell, and therefore are different from both. The differences of the transitional cells might reasonably be attributed to a difference in the spectrum of genes that are activated in them in contrast to the ones that are activated in the zygote on the one hand, and in adult cells on the other. That there is a single transitional cell between the zygote and a given adult cell is possible, but in most cases it appears unlikely. It is more likely in these cases that there is a number of transitional cells that constitute a series of intermediate, progressive types, each differing from the other in terms of which

genes are activated and which are repressed. According to this view, the specialized functions of cells are acquired stepwise during the course of development. There are many examples that illustrate this principle—for example, the red blood cells of vertebrates show a number of intermediate types during their histogenesis (cf. Wilt, 1965; Fantoni, et al., 1968, Godet, et al., 1970). The cells that synthesize lactic dehydrogenase isoenzymes (cf. Whitt, 1970) and skeletal muscle cells (Richler and Yaffe, 1970) show transitional types. The process by which cells change from embryonic type (without specialized function) to adult type (with specialized function) is called *differentiation*.

The Theory of Variable Gene Activity depends upon the premise that there is a molecular relationship between DNA and the structure of cellular proteins, such as described by the so-called "central dogma" of molecular biology. The evidence for this relationship has been amply reviewed in preceding chapters of this book. The Theory of Variable Gene Activity also assumes that every cell of a multicellular organism contains the full complement of genetic DNA that characterized the original zygote. As mentioned above, the mitotic process in all its detail is reconcilable only if it assures just this. For many years it has been known that genetic DNA is carried by the chromosomes, that the number of chromosomes is constant for a given species, and is, with a few explainable exceptions, the same for all somatic cells of any individual. Direct evidence for the constancy of genetic DNA has come from chemical determinations of the total DNA content of various cells (Mirsky and Ris, 1949; Boivin, et al., 1948; Inui and Takayama, 1970), which have shown that all somatic cells of a given adult individual contain the same amount of DNA, and that this amount is twice that of the germ cells. In 1964, McCarthy and Hoyer concluded from their investigations that not only is the total amount of DNA the same, but all of the polynucleotide sequences are present and appear in the same relative proportions.

GENE AMPLIFICATION

In the oocytes of some species and in certain embryonic cells, the phenomenon of *gene amplification* has been described (Macgregor and Kezer, 1970). In these situations, certain portions of the genome are replicated in *excess* of the normal DNA complement. The best established example of this occurs in amphibian oocytes (Gall, 1968), in which the nucleolar organizer regions of the chromosomes have proliferated hundreds of DNA copies of their nucleotide sequences, and these copies become incorporated in as many minute nucleoli. As

will be seen below, this represents a mechanism for enabling the oocyte to synthesize vast quantities of ribosomal RNA in a relatively short period of time. Other perhaps more celebrated though less well-understood examples of gene amplification occur in the chromomeres of the giant salivary gland chromosomes of *Sciara, Chironomus,* and *Drosophila* (Ashburner, 1970).

Interest in the giant chromosomes, already great because of their unusual size and their being useful landmarks for genetic mapping, was heightened by the reports of Beermann in 1952 and by Breuer and Pavan in 1955. They suggested that the swelling of the bands (the chromomeres) is evidence of gene activity. They interpreted the swollen puffs as localized increases in DNA and RNA and they described changes in puffing pattern with developmental time (Fig. 11–1). Subsequent work (Keyl, 1965) indicated that the chromomeres

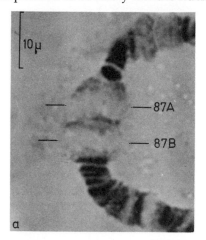

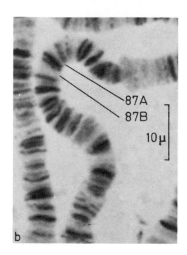

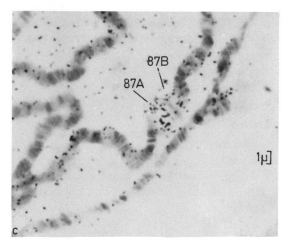

Fig. 11–1. (a) The appearance of puffs in the 87A and 87B regions of a polytenic chromosome of *Drosophila melanogaster,* compared to (b) the same regions without puffs. (c) The concentration of radioactive RNA (^{3}H-uridine) in puffs induced by temperature shock. [Courtesy of E. G. Ellgaard and U. Clever (1971), *Chromosoma* (Berlin) **36:**60–78.]

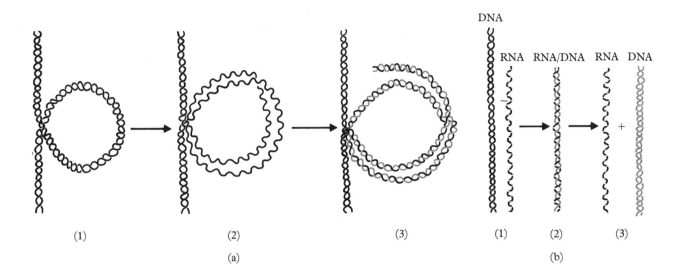

Fig. 11–2. Two models of mechanisms for gene amplification. (a) Replication of a specific DNA segment. (1) The DNA strand forms a loop, as postulated for chromomeres, puffs, and lampbrush chromosomes. (2) Component strands are separated enzymatically, and (3) replication of the looped segment takes place. Newly synthesized strands are shown shaded. One of the loops is excised, and the other remains to repeat the process. (b) Synthesis of DNA by transcription from RNA. (1) A molecule of RNA is transcribed on a segment of DNA. (2) The RNA moves away and by "reverse transcriptase" serves as a template for synthesis of a single strand of DNA. The DNA strand separates from the RNA template, a complementary strand of DNA is synthesized, and the RNA is free to transcribe another molecule of DNA. [(a) after Kehl (1965); (b) after Temin (1971).]

are sites of gene amplification. The mechanism of the amplification is not understood (Wallace, et al., 1971). Figure 11–2 represents suggested models. Other studies of the puffing itself (Beermann, 1966; Ellgaard and Clever, 1971) showed that there is a direct correlation between the degree of puffing of individual chromomeres and their uptake of uri-

dine. Moreover, it has been shown that puffing is under genetic control (Beermann and Clever, 1964) and is correlated with variable gene activity, as is discussed below.

Considering the wide range of functions that the genome of a cell must make provision for, one would predict that in any given differentiated cell of specialized and correspondingly limited breadth of function, only a small part of the total genome would be active. Stated another way, a lymphocyte would not use the genes that it has for functions that are irrelevant to lymphocyte function, and so it would use only a fraction of its total genome. In calf thymus, the amount of DNA available for transcription of RNA is no more than 5 to 10 percent of the total DNA—that is, 90 to 95 percent of the DNA is "masked" (Paul and Gilmour, 1966, 1968).

Retention of Genetic Potential

If all somatic cells, however highly specialized, do indeed have the full genome characteristic of the original zygote, then they should have the same potential as that zygote—namely, the genetic potential to generate an entirely new and complete individual. One would need merely to find a way to unmask or activate the appropriate genes in the nucleus and possibly, but more difficult, to render the cytoplasm favorable for a renewed burst of mitotic activity. The cells of lower organisms can approximate these requirements more readily and show well-developed regenerative powers accordingly. The well-known flatworm *Planaria* can regenerate an entire individual from either the anterior or posterior half of its body, or even less. The higher one ascends the phylogenetic tree, evidently the more limited are the regenerative powers. Crustaceans can regenerate appendages or parts of some organs, but nothing more. Some species of salamanders can regenerate digits or even an entire leg, but anurans (frogs and toads) cannot. Some species of reptiles can regenerate a tail, but the regenerative powers of other terrestrial vertebrates are quite limited to certain kinds of cells and tissues to the extent that some kinds, such as neurons and skeletal muscle, cannot even regenerate themselves. However, it has been known for many years (Bloom, 1937) that mammalian cells that are not highly differentiated in structure or function, such as histiocytes, endothelial cells, and lymphocytes, can be trans-

formed into other types of cells. Petrakis and others (1961) described the transformation of monocytes in tissue culture to macrophages and then to fibroblasts that subsequently synthesized collagen (cf. also Z. Cohn, 1968). In 1950, Stone induced regeneration of neural retina from pigmented retina, although cells of the latter are totally different in both structure and function from the former and are normally not antecedent to them. The climax to all such studies was reached by Gurdon (1962), who transplanted nuclei from intestinal epithelial cells of amphibian tadpoles into enucleated eggs of the same species and derived adult individuals from them, some of which subsequently mated and reproduced (Gurdon and Uehlinger, 1966). In 1970, Gurdon and Laskey reported derivation of embryo frogs from tadpole epithelial cells that had been propagated in tissue culture prior to implantation into enucleated eggs of the same species. By transplanting the nuclei obtained from these embryos when they were in the blastula stage to new enucleated eggs, Gurdon and Laskey were able to obtain several clones of individuals that appeared normal and underwent metamorphosis to adult form. By this technique of serial transplantation, they were able to obtain a greater number of successes than with a single passage, which implies that the longer a differentiated cell nucleus can reside in an undifferentiated cell, the greater the likelihood that it will dedifferentiate. This also implies that the composition of the cytoplasm can influence the activities of the nucleus.

Somatic Cell Hybridization

Another experimental approach to the genetic analysis of differentiation is the technique of *somatic cell hybridization*. While culturing somatic cells of mice *in vitro*, Barski, Sorieul, and Cornefert (1960) discovered that under certain conditions, pairs of cells would fuse and their chromosomes would become incorporated within the same nucleus. The kinds of cells that do this spontaneously in tissue culture are those that can be maintained successfully in tissue culture indefinitely without showing senescence (retardation and/or cessation of growth), as well as many kinds of tumor cells. Tissue culture cell fusion can also be induced by "transforming" viruses, such as the simian virus SV40 or the Sendai virus. That fusion has occurred can be detected readily if the "parent" cells differ in some details of their karyotype.

Following the fundamental discovery of Barski, Sorieul, and Cornefert, somatic cell hybridization has been achieved not only between different strains of the same species, but between different species. In 1965 Harris and Watkins, exposing mixed cultures of mouse and human cells to ultraviolet-inactivated Sendai virus, obtained multinucleate heterokaryons. Weiss and Green (1967) observed mouse-human hybrids with the chromosomes of both species located in the same nucleus.

Davidson (1973) reported a series of experiments with a line of differentiated mouse kidney cells. These cells, originally derived from a tumor, synthesized a tissue-specific enzyme called ES-2. When the kidney cells were hybridized with mouse fibroblasts, which do not synthesize ES-2, synthesis of the enzyme ceased. It was inferred that the ES-2 portion of the kidney-cell genome was suppressed. When the kidney cells were hybridized with human fibroblasts, ES-2 activity was detected in some clones of cells but not in others. This difference could be correlated with a karyotype difference between the clones. In those hybrids that had lost a particular human chromosome (chromosome 10), ES-2 activity was detectable. In those hybrid clones that had retained chromosome 10, ES-2 activity was suppressed. It was inferred from these results that the human fibroblast produced a diffusible regulator substance that suppresses the synthesis of ES-2, and that the regulator gene or genes involved are located on chromosome 10. Support was given to this conclusion when it was found that ES-2 activity was regained in subsequent generations of hybrid cells once chromosome 10 was lost. (It should be noted that the loss of chromosomes from hybridized cells is not an unusual occurrence. In fact, combinations of cells from mice and primates are generally characterized by a rapid loss of the primate chromosomes.)

In general, it has been found that when differentiated and undifferentiated cells are fused, the differentiated cell function is lost or reduced. This is consistent with the Variable Gene Activity Theory which assumes both gene activation and gene repression.

THEORIES OF EMBRYONIC DEVELOPMENT

For many years, theories of preformation were in vogue, presumably because they did not conflict with theological beliefs of the times. According to the preformation theory held during the eighteenth and nineteenth centuries, an individual exists preformed in its entirety within the ovum; development is therefore little more than a matter of growth in size. Preformationism was even extended to the cellular level after the nucleolus was discovered, in the belief that the nucleolus is an incipient nucleus; that is, the nucleolus grows to become a nucleus, while the nucleus enlarges to become a cell. This idea assumed the existence of a series of ever-smaller particles within the nucleolus, each destined ultimately to become a cell. One wonders if the emergence of the last preformed particle was predetermined to coincide with Armageddon!

As microscopy improved, it appeared that animal eggs were rather generalized cells with no evidence of structural preformation. The theory of development by epigenesis, which was initially proposed by Aristotle, regained favor and sparked the emergence of the science of embryology by capturing the imaginations of some of the world's most able scientists during the late nineteenth and early twentieth centuries. According to the theory of epigenesis, development of the individual (ontogeny) consists of a series of causal interactions between the various parts of the embryo as it grows. This theory has the happy advantage of admitting environmental influences, but happier yet of setting the stage for the Variable Gene Activity Theory as a mechanism for epigenesis.

The preformationists were not easily put down, however. With the discoveries of Balfour, Needham, and many others, it became evident that the fertilized egg is characterized by a kind of "chemical preformation." Support for this idea came from the once-celebrated experiments of Vogt, Mangold, and Ethel Harvey, who in various ways showed that specific organs of the embryo have cytoplasmic counterparts, as it were, in specific localities in the egg.

It is generally considered today that the egg does have three-dimensional organization of some of its chemical components, and that doubtless this does play a prominent role in providing the kind of situation that epigenesis requires in order to begin. Obviously there could be no "causal interactions between the various parts" if there were no various parts. The matter of point of view appears to be the crucial element in such arguments, as is evident when one considers that one prominent developmental biologist (Raven, 1961) defends the view of preformationists. He asserts that all the information needed by the adult is preformed in terms of genetic DNA within the egg, and that epigenesis elicits no change in this specific information; rather the environment provides substrate and energy essential to the multiplication of this information.

In attempting to explain cellular differentiation and embryonic development by variable gene activity and epigenesis, another major premise is needed. There must exist some mechanism for signaling the turning-on and turning-off of the appropriate genes at the appropriate times for a cell as it differentiates and ultimately acquires its adult or specialized function. The *Theory of Embryonic Induction* provides for such a mechanism.

The Theory of Embryonic Induction is attributed to Hans Spemann (1936), who demonstrated that during gastrulation in amphibian eggs, cells that constitute the so-called dorsal lip of the blastopore differentiate into chorda-mesoderm and, while so doing, they acquire the ability to "organize" the developing embryo in terms of its anterior-posterior and left-right axes. An important feature of the mechanism by which it organizes the embryo is by "inducing" the overlying ectoderm to develop into a platelike structure, the *neural plate*, which later differentiates into the brain and spinal cord.

Subsequent work by many experimental embryologists has confirmed and extended Spemann's original thesis. For example, it was soon found that the optic cup, from which the eye subsequently develops, induces the ectoderm opposite it to differentiate into the crystalline lens. The notochord and neural tube of vertebrate embryos induces adjacent mesodermal cells to differentiate into components of the axial skeleton and trunk musculature (Holtzer and Detweiler, 1953; Watterson, et al., 1954; Strudel, 1955). Schowing (1961) showed that various parts of the skull are induced by particular regions of the brain and notochord. The development of the wings and legs in birds has been shown to involve a series of reciprocal inductions (Hampé, 1960), one aspect of which is illustrated in Fig. 11–3. Development of the metanephric kidney in birds and mammals depends upon a similar mutual feedback, namely, induction action between the developing ureter and the adjacent mesenchyme (Grobstein, 1955). These are but a few of the clear-cut examples of embryonic induction that have been described since Spemann's fundamental discoveries.

How does induction relate to variable gene activity? This has not been definitively demonstrated yet, but the operon model of Jacob and Monod (cf. Chapter 8) suggests a mechanism, and attempts to isolate inducer molecules (Yamada, 1961; Saxén, et al., 1968) have yielded results that are not inconsistent with the more basic features of this model. In brief, as illustrated in Fig. 11–4, one can conceive that the inducing cell releases an inducer substance which acts upon a susceptible cell, called a

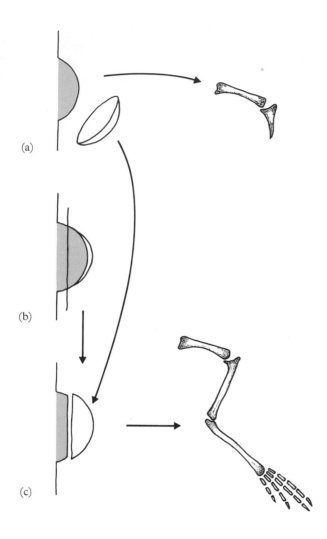

Fig. 11–4. ▶ **Fig. 11–4.**   Diagram representing two theories of inductor action. (a) Short-circuit induction, in which the inductor substance operates directly on the cytoplasmic metabolic machinery of the competent cell to influence its differentiation. (b) Long-circuit induction in which the inductor exerts its influence via the genes of the competent cell. [From M. E. Wolff (1966), *Cell Differentiation and Morphogenesis* (Beermann, et al., eds.). Amsterdam: North Holland. By permission.]

"competent" cell, in such a way as to affect either genetic transcription or mRNA translation. In substance, this means that the inducer material, like the repressor or corepressor substances of Jacob and Monod, either turns genes on, turns them off, or affects their translation through mRNA. That inducer substances are released has been demonstrated many times (Saxén, et al., 1968). Their chemical nature has not yet been described, but it appears that inducers may include a very broad range of molecules including steroids, nucleic acids, polypeptides, and proteins (Saxén, et al., 1968). The specificity of inducer action is a function of the inducer itself, its concentration, the competence of the "target" cell, the presence and/or amounts of other inducer substances, and the concentration gradients of the various inducers in the induction field (Toivonen, et al., 1961). The mechanism of induction is evidently as complex as differentiation itself, which is hardly surprising.

Many questions remain. According to Brachet (1944), the pioneer in modern developmental biology, the kinds of questions that will elicit the most useful answers pertain to the molecular aspects of development. For example, what is the molecular or biochemical nature of "competence"? What determines competence, and why are some cells limited in competence while others possess a broad range of potential? When is differentiation a function of gene transcription and when is it a function of mRNA translation? What are the roles of the various species of RNA in differentiation and development? Some of these questions are beyond the scope of this book, but others are clearly in the province of genetics and will now be considered.

Fig. 11–3.   Induction of limb development in the chick. (a) If the apical cap of ectoderm is removed from the mesodermal component of the limb bud, only the femur and a portion of the tibia differentiate. (b) If the apical cap and the superficial mesenchyme of the limb bud are removed, no limb development ensues. (c) If the superficial mesenchyme is removed, but the apical cap is grafted onto the basal mesenchyme, all components of the limb develop. The apical cap itself, apart from its inductive action, contributes only to the covering (skin) of the limb. [From A. Hampé (1960), *J. Embryol. Exptl. Morph.* **8**:247. By permission.]

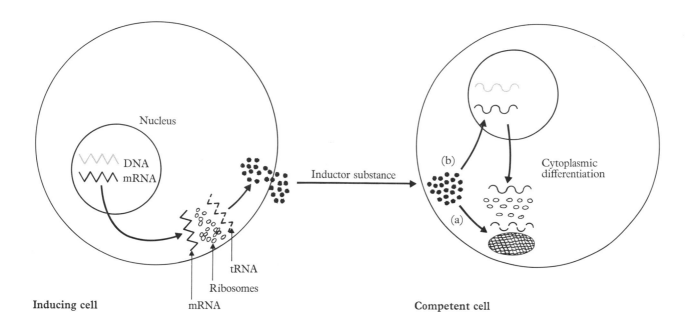

MOLECULAR ASPECTS OF DEVELOPMENT

Oocytes and Oogenesis.   One can start by considering the fertilized egg, for it shows the maximum range of competence, namely *totipotence* —the ability to give rise to a total and complete individual organism. It must be borne in mind, however, that the fertilized egg is preceded by a history of development that in certain respects is complex (Grant, 1965), but about which much remains obscure. In some species, the events of oogenesis can be patently bizarre (Raven, 1961), so much so that they suggest that oogenesis involves something much more than producing a haploid but otherwise generalized cell. Whereas biologists tend to think of oocytes as generalized cells of minimal specialization, some developmental biologists regard them as highly specialized cells (Beams, 1964). All of the cytoplasmic organelles are well developed in oocytes, which is indicative of high metabolic and biosynthetic rates. Rather peculiar

to oocytes are the vitelline membrane, numerous and often very well-differentiated nuclear pores, and a specialized kind of endoplasmic reticulum (ER) called *annulate lamellae*, which is evidently produced as a proliferation of the outer nuclear membrane (Beams, 1964). But of more interest to geneticists are the chromosomes and nucleic acids of oocytes.

In the roundworm *Ascaris megalocephala*, large amounts of chromosomal material are eliminated during oocyte formation (Boveri, 1887). In the insects *Sciara* and *Miastor*, and in the copepod *Cyclops furcifer*, the germ cells are products of cells that have a greater number of chromosomes than somatic cells (Beermann, 1966). For example in *Miastor*, 36 of the 48 chromosomes of the zygote are eliminated from the nuclei of somatic cells, but the full complement is retained in the germ cell line (White, 1946). In *Cyclops furcifer*, on the other hand, interstitial components of the chromosomes are discarded without any reduction in the number of chromosomes (Beermann, 1966). In the water beetle *Dityscus*, unequal division during oogenesis results in the oocytes receiving extra chromatin material. Raven (1961) proposed that oocytes contain genetic information in excess of that represented by their chromosomes, and he felt that this information is in the egg cytoplasm in the form of redundant nucleic acid. Significant quantities of cytoplasmic DNA have been found in the oocytes of frogs (Haggis, 1964) and mice (Reamer and Levy, 1964). Dawid (1965) showed that frog eggs have 300 to 500 times as much DNA as is found in somatic cells, and he presented evidence that the excess is indeed in the cytoplasm but that it is largely, if not entirely, mitochondrial DNA. Because of the extremely large size of frog eggs, they contain enormous numbers of mitochondria. In contrast to these findings is the view of Elsasser (1958) that the informational content of the adult is far greater than could be stored in the fertilized egg in its primary sequence of macromolecular subunits. He feels that cells gain new information as they develop.

The oocytes of many animal species show a kind of chromosomal morphology that has attracted considerable attention in recent years, although this peculiarity was described many years ago by Rückert (1892), Marechal (1907), and others. This morphological variant is called *lampbrush chromosomes*, because the chromosomes appear with lateral filamentous projections like those of old-fashioned brushes that were used to clean out the soot from kerosene lamps (Fig. 11–5). The investigations of Gall (1954), Callan (1963), Izawa, et al. (1963), and others have shown that the lampbrush effect is attributable to multiple strands of polynucleotides that occur at certain points along the chromosome. That these are RNA strands that are polymerized on DNA templates has been proved by showing that when DNA transcription is inhibited by treatment with actinomycin-D (Izawa, et al., 1963), the lampbrush filaments do not appear. Hence these filaments of RNA are DNA-dependent.

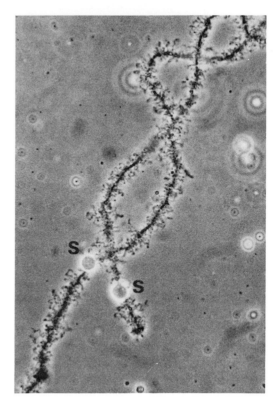

Gall (1968) showed in the clawed toad *Xenopus laevis* that the nucleolar organizer region of the chromosomes is multiplied many times to produce hundreds of extrachromosomal nucleoli within each nucleus. Miller and Beatty (1969) isolated such nucleoli and subjected them to electron microscopic and autoradiographic analysis (Fig. 11–6). They found that each nucleolus contained a strand of DNA, presumably a copy of the corresponding chromosomal segment of DNA, and this strand had the characteristic structure of the lampbrush chromosomes. Their analysis showed that the DNA is a single, double-helical molecule that, together with protein, forms an axial filament. At more or less regular intervals along this axial filament are the lampbrush clusters or "matrices" of RNA filaments. Each matrix covers about 2 to 3 microns of axial filament length (Fig. 11–6), which corresponds roughly to the length of DNA

◀ **Fig. 11–5.** Lampbrush chromosomes from an oocyte of a newt (*Triturus vulgaris meridionalis*) showing the sphere-organizing regions (S) characteristic of most amphibian oocytes. [Courtesy of G. Mancino, et al. (1970), "Spontaneous aberrations in lampbrush chromosome XI from a specimen of *Triturus vulgaris meridionalis* (Amphibia, Urodela)," *Cytogenetics* **9:**260–271. Karger, Basel. By permission.]

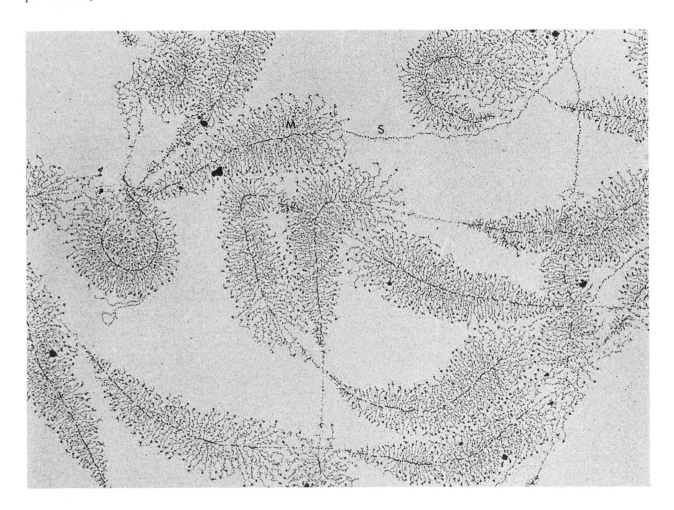

Fig. 11–6. Portion of a nucleolar core isolated from an oocyte of the newt *Triturus viridescens* showing matrix units (M) separated by matrix-free segments (S). The axial fiber can be broken by deoxyribonuclease. The fibrils of the matrix can be removed by ribonuclease, trypsin, or pepsin. Magnification X25,000. [Courtesy of O. L. Miller, Jr., and B. R. Beatty (1969), *Science* **164:**955–957.] Copyright © 1969 by the American Association for the Advancement of Science. By permission.]

needed to code a molecule of precursor rRNA. Precursor rRNA has been shown to be a rather long molecule, about 40S, which later is cleaved to 18S and 28S rRNA. Inasmuch as the time required for incorporation of tritiated uridine into the matrical RNA corresponds with the time for the appearance of labeled precursor rRNA, Miller and Beatty concluded that the axial strand of DNA that is covered with matrix represents a gene that is coding for multiple copies of precursor rRNA. If they are correct, then their figures may represent the first

established pictures of a gene. The functional significance of lampbrush chromosomes, therefore, becomes clear—they represent chromosomes that are actively engaged in mass production of RNA. Brown, et al. (1972) extended the molecular analysis of amplified ribosomal DNA and found that each molecule consists of repeating units, as shown in Fig. 11–7.

Oocytes of echinoderms, fishes, amphibians, birds, and mammals synthesize large quantities of RNA, which is transferred to the cytoplasm. Al-

Fig. 11–7. (a) Electron micrograph of a single molecule of amplified ribosomal deoxyribonucleic acid (rDNA). This molecule contains eight complete repeats. (b) A tracing of the molecule designates the two general regions within each repeat. The numbers refer to the lengths in microns of the DNA between arrows. [Courtesy of D. D. Brown, P. C. Wensink, and E. Jordan (1972), *J. Mol. Biol.* **63**:57–73.]

(a)

though some of it is doubtless used for metabolic requirements of oocyte maintenance, most of it appears to be masked or in stored form (Davidson, 1968). Spirin (1966) proposed that this RNA is combined with masking protein in the form of cytoplasmic particles, which he called "informosomes." Although all kinds of RNA are synthesized by oocytes, they are done so at different times and rates. Ribosomal RNA and 4S and 5S RNA are synthesized during the early stages of meiosis of the oocytes. The rRNA is synthesized at an extremely rapid rate during diplotene, which is the stage of lampbrush-chromosome formation in amphibians. For example, over 95 percent of the RNA synthesized per unit time during the lampbrush stage in *Xenopus* is ribosomal (Davidson, 1968). Messenger RNA, on the other hand, is not synthesized in large amounts until the terminal stages of oogenesis (Piatogorsky and Tyler, 1967), and tRNA appears to form at about the time of fertilization from low molecular weight RNA which was synthesized earlier (namely, the 4S and 5S RNA) (Brown and Littna, 1966; Brown, 1967).

The Fertilized Egg.   Experiments with radioactive labeling and with administration of actinomycin-D and puromycin have shown that the large stores of cytoplasmic RNA built up by the oocytes is utilized during cleavage after fertilization has occurred. The unmasking of the RNA is effected by fertilization itself, although the mechanism by which this is accomplished is not understood. As long ago as 1915, it had been shown in fishes (Newman, 1915) that cleavage and blastulation were under maternal control—that is, in modern terminology, they were under the control of the genome that existed in the oocyte *before* fertilization. In recent years, evidence has accumulated from a variety of sources and by a variety of techniques substantiating this early finding, showing its applicability to insects, echinoderms, and other vertebrates. Considerable evidence for this has come from hybridization experiments (Chen, 1967) that have shown consistently maternal patterns of development up to the stage of gastrulation. Development during cleavage and blastulation, then, depends upon the translation of stored informational RNA.

What happens at fertilization to trigger the mechanism involved is not well understood. It cannot be the introduction of the male genome, *per se*, or its centrioles, because merely pricking the egg membrane or using a variety of chemical agents can often be as effective in initiating cell cleavage as sperm penetration. In fact, the importance of sperm penetration as such has been challenged and relegated to the role of a nonspecific stimulus

(b)

for cleavage. The more important change in mature oocytes, namely, activation of the cytoplasmic RNA and protein synthesis, is induced hormonally, at least in the various species that have been studied (Smith, et al., 1966). Gurdon (1967) suggests that the factor stimulating the DNA synthesis necessary for cleavage is cytoplasmic, and that it is induced late in oogenesis by luteinizing hormone (LH) from the pituitary. This factor does not act until fertilization or shortly thereafter, however. It appears, then, that fertilization is at first no more than a stimulus for the release of Gurdon's cytoplasmic factor. According to other investigators (De Terra, 1969), all cytoplasmic signals that regulate nuclear behavior must originate at the cell surface, and the surface-membrane effect of sperm penetration could constitute such a signal. Metafora, et al. (1971) extracted from ribosomes of unfertilized sea urchin eggs a protein that inhibits ribosome function by repressing the binding of mRNA and amino-acyl-tRNA to the ribosomes. They proposed that this protein is removed from the ribosomes as a consequence of fertilization.

Cleavage and Blastulation.   As mentioned above, pregastrular development is controlled by the maternal genome in both its morphological and biochemical aspects (Tennent, 1914; Chen, 1967). This has been shown by hybridization experiments in which the morphogenetic characteristics of the hybrid are compared with those of the parent species (Moore, 1941; Greg and Løvtrop, 1960). It has also been shown by observing the effects of blocking RNA transcription during fertilization and cleavage (Gross and Cousineau, 1964). In these cases, cleavage and DNA synthesis proceed and a blastula is formed. Gastrulation and cellular differentiation do not occur, however. If actinomycin-D, which is used as the inhibitor of transcription, is administered in pulses at various times and for various durations after fertilization, it is found that the informational RNA necessary for gastrulation is synthesized during the middle and late blastula stages (Giudice, et al., 1968).

Striking differences are found in the electrophoretic banding patterns of proteins produced by eggs during cleavage, blastulation, and gastrulation in sea urchins (Termin and Gross, 1965; Gross, 1967), with a pronounced shift occurring during late blastula. The diversity of proteins as determined by column chromatography in sea urchins starts to increase during the middle-to-late blastula stage, and then becomes more pronounced thereafter (Ellis, 1966). Similar results were obtained by means of immunological methods (Westin, et al., 1967). Moreover, when the RNA of developing embryos is labeled with ^{32}P and analyzed by sucrose gradient techniques, it is found that informational RNA is greater in the blastula stage than during cleavage, but that the amount of rRNA remains steady until after blastulation (Brown and Littna, 1964).

Cell division during cleavage is at first synchronous, and the daughter cells are equivalent in size and cytoplasmic content. In the large, yolky eggs characteristic of most nonmammalian vertebrates, only the first three divisions are synchronous. The third, which is horizontal or meridional, results in four small, relatively yolk-free cells plus four large, yolky cells. The former divide subsequently at a faster rate than the latter, both because of their small size and, according to Balfour, because of lack of the retarding action of yolk particles. Each meridional division of the yolky cells results in cells of varying yolk content and size, because of the gradient of yolk concentration that existed in the oocyte and zygote. The consequence of this is a corresponding mitotic gradient along a central axis that runs through the center of maximum concentration of yolk and the center of the original zygote. This is the so-called "animal-vegetal axis," a term used by descriptive embryologists who defined the yolk-free cytoplasm as constituting a zone of the egg called the *animal pole*, and the dense deposits of yolk constituting a zone called the *vegetal pole*. Zygotes that are virtually yolk-free are nevertheless characterized by a slightly eccentric nuclear position, so that a gradient of cell size, although slight, is established by the meridional cleavage planes, and the effect of establishing a mitotic gradient is still achieved. The first expression of a mitotic gradient is asynchronous cleavage, thereby establishing variation in the cell-to-cell environmental relationships which, according to some theorists, is sufficient to set the stage for epigenesis. But, as is described below, there is evidence that the stage for epigenesis is set during oogenesis.

Gastrulation and After. The gastrula stage is the first stage of cellular differentiation in higher organisms. Up until this time, every cell has been essentially totipotent, although there exist many species in which this is not the case, as evidenced by transplantation experiments. During gastrulation, the three primary germ layers, ectoderm, endoderm, and mesoderm/mesenchyme, are segregated. In animals such as the chordates, the mesoderm/mesenchyme component is known as the *chorda-mesoderm*, which proliferates cranially from the region of the dorsal lip of the blastopore and represents the first definitive embryonic inductor.

The signals for gastrulation and all it implies are called by the embryo genome. As stated earlier, gene transcription of the embryo genome starts in the blastula stage, as demonstrated by studies with actinomycin-D (Giudice, et al., 1968), electrophoresis (Gross, 1967), amino acid incorporation (Ellis, 1966), ^{14}C uptake (Kafiani, 1970), and ^{3}H-uridine uptake (Monesi and Salfi, 1967). If transcription is prevented during this time, gastrulation either is prevented or is highly abnormal, and further development fails to occur. The burst of mRNA and tRNA synthesis that characterizes the blastula stage is triggered by some cytoplasmic factor or factors (Gurdon and Brown, 1965; Gurdon, 1969). The nature of this factor is uncertain, however. Gurdon (1969) believes it is a protein, for he has demonstrated by nuclear transplantation that cytoplasmic proteins migrate into the nucleus and that this migration is accompanied or fol-

lowed by DNA synthesis. Kafiani (1970) presents evidence that RNA polymerase and the DNA template for mRNA transcription exist in the nuclei of the cells during cleavage, but transcription is suppressed by a cytoplasmic factor until blastulation or even later.

Denis (1968) and Gurdon (1969) proposed the existence of different categories or sets of genes in multicellular organisms. One set is concerned with the functions that are basic to all cells, and so are comparable to the genome of a bacterial cell, for instance. These genes are under continuous regulation by the concentration of the direct or indirect products of their activity, as is the case in bacteria and is the basis of the theories of Jacob and Monod. The other set of genes is concerned with specialized cell functions (hemoglobin or myosin synthesis, for example), and these genes, according to Gurdon, are not subjected to continuous control, but only during times of mitosis when the chromosomes come into direct physical contact with the cytoplasm. This is consistent with the observations that in many situations, major changes in cell differentiation are associated with or immediately preceded by mitosis. Hence at mitosis, the daughter cells would be "reprogrammed" like the parent cell except when developmentally important agents, such as unequal distribution of components of egg cytoplasm, hormones, inducers, etc., should bring about program changes that lead toward specialization of function.

It has been possible to establish the times at which program changes are expressed as either transcriptional or translational, as shown above. The two occurrences of major program changes discussed so far have been during oogenesis, when evidently the cell is programmed for cleavage and blastulation by the oocyte genome, and during blastulation, when the embryo is programmed for gastrulation. From this point onward, programming differs much more markedly from one cell to another, both in terms of specificity of program and time of programming, for it is now that epigenetic processes move into "high gear." However, it has not been so easy to establish either the times at which significant changes in the cytoplasm occur to initiate the changes in programming or the chemical nature of the cytoplasmic changes.

As mentioned earlier, the dorsal lip of the blastopore of vertebrate eggs includes cells (Spemann's organizer) that determine the axial polarity of the embryo. But Curtis (1962) and Harris (1964) have shown that the cytoplasmic factors, or their precursors, that are responsible for this organizing effect can be traced to a particular locality in the cortex of the fertilized egg. If this area is removed from the fertilized egg, gastrulation and concomitant organizer formation do not occur.

Yet if it is transplanted from a fertilized egg to a cleaving embryo, it has no influence. On the other hand, if this area is traced into the eight-cell stage and transplanted from there to an abnormal location in a fertilized egg, a secondary embryonic axis develops after gastrulation (Fig. 11–8). Among other things, these observations show that epigenesis really starts much earlier than during the gastrula stage. It can, in fact, be traced back to the oocyte in which the cytoplasmic determiner identified in the fertilized egg was undoubtedly formed and spatially located. Davidson (1968) was led to state, " . . . it would appear that storage in the egg cytoplasm of molecules whose function is

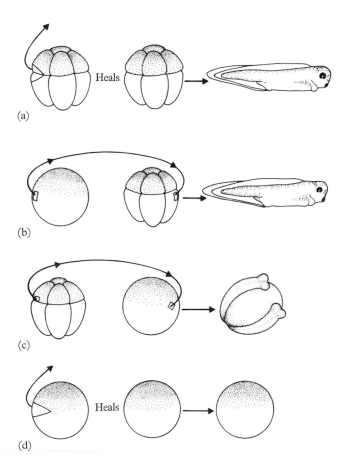

(a)

(b)

(c)

(d)

Fig. 11–8. Localization of cytoplasmic factors responsible for the primary organizing effect in amphibian development. (a) Removal of the gray crescent cortex from an egg in early cleavage does not prevent normal development. (b) Grafting the gray crescent cortex from an uncleaved zygote to the ventral margin of an egg in early cleavage similarly results in normal development. (c) Grafting the gray crescent cortex from an egg in early cleavage to the ventral margin of an uncleaved zygote induces secondary embryonic axes. (d) Removal of the gray crescent cortex from an uncleaved zygote prevents morphogenesis, although cleavage does occur (compare with a). [From A. S. G. Curtis (1962), *J. Embryol. Exptl. Morph.* **10**:416. By permission.]

the selective specification of embryo gene activity is an extremely general and probably universal mechanism in animal development."

The identity of these cytoplasmic molecules has not been established. It is likely that they are quite varied. As far as the earliest stages of embryogenesis are concerned, some of these molecules are doubtless protein (Gurdon, 1969); in other cases certain neurohumors (serotonin, acetylcholine, adrenalin, noradrenalin) have been implicated (Buznikov, et al., 1970); and in still others the proportion of potassium ions to sodium ions appears to play a very significant role (Kafiani, 1970). In later stages of embryogenesis, specific exogenous inducer substances are involved, as discussed earlier, and these may include a wide variety of chemical substances (Tiedemann, 1967). Whether or not they are absorbed by competent cells and become the regulating cytoplasmic molecules, as implied schematically in Fig. 11–4, is not known, but it appears to be unlikely.

Primary Site of Inducer Action

The primary site of metabolites that influence gene regulation in bacteria is the cytoplasm. It has been shown that the molecule that mediates gene regulation in carbohydrate metabolism in bacteria is cyclic adenosine monophosphate (cAMP). Cyclic AMP is synthesized in the cytoplasm from ATP under the influence of the cytoplasmic enzyme adenyl cyclase, and it is degraded by another cytoplasmic enzyme, a phosphodiesterase. The kinetics of ATP$\rightleftharpoons$cAMP is influenced by the concentration of glucose in the cytoplasm, so that as the concentration of glucose increases, the concentration of cyclic AMP decreases. According to one view (Chambers and Zubay, 1969; Eron, et al., 1971), when the concentration of cyclic AMP rises, it facilitates the attachment of the protein CAP (cf. p. 204) and RNA polymerase to the *lac* operon—if cAMP is not present, this attachment will not occur even though the lactose concentration is sufficient to inactivate the repressor (cf. Chapter 8).

According to another view (Aboud and Burger, 1970), cAMP operates at the translational level by releasing a translational block. Cyclic AMP also plays a role in the synthesis of the enzyme tryptophanase in *E. coli*, and here again, it is believed to act at the translational level (Pastan and Perlman, 1969).

In his studies on the cellular slime mold *Dictyostelium discoidum*, Bonner (1969) discovered a diffusible substance, which he called *acrasin*, that was released by the individual amoeboid cells of this species when their food supply was cut off for one reason or another. This substance causes the separate cells to aggregate and form a syncytial mass called a *pseudoplasmodium*, which then undergoes morphogenetic changes that culminate in the production of a stalked fruiting body from which reproductive spores are released. Spores that are transported to a favorable environment germinate and give rise to a new generation of amoeboid cells. Acrasin has been shown to be cAMP. However, cyclic AMP is ineffective in inducing aggregation of the vegetative amoebocytes if the nutritive requirements are met.

The effect of epinephrin on liver cells is glycolysis, whereby glycogen is phosphorylated to glucose phosphate. This effect is mediated by cAMP. Moreover, a number of other hormones whose actions and target cells are specific (such as ACTH, glucagon, and calcitonin) act through cAMP. In these cases, it has been demonstrated that the direct site of action of the particular hormone is the cell membrane of the target cell, and that the action is the release or activation of adenyl cyclase located in the cell membrane (Rasmussen, 1970). It has been proposed that the specificity of hormone action in these cases resides in the target cells themselves, both with respect to recognition of the hormones through specific binding sites on their surfaces and the effect of the action mediated by cAMP. The possibility that the specificity of developmental inducers is also a function of specific binding sites on the surfaces of competent cells (Steinberg, 1964), as well as the more general postulate that the cell membrane is the primary site of action of inducer substances, has been suggested in the past (Weiss, 1950) and reaffirmed quite recently (De Terra, 1969; Sunshine, et al., 1971).

When the insect hormone *ecdysone* is injected into dipteran larvae, puffing of certain chromomeres of the giant salivary gland chromosomes is induced (Beermann, 1966) (Fig. 11–1). Measurements of changes in electrical potential across the cell and nuclear membranes of these cells, both before and after ecdysone administration (Kroeger, 1968), revealed a pronounced increase starting within a minute after injection and lasting approximately twelve minutes. Within fifteen minutes after injection, puffs started to appear. The same effect was achieved in the absence of ecdysone by increasing either the intracytoplasmic or intranuclear concentration of potassium ions. It was concluded from this work that the primary site of ecdysone action is the cell membrane, altering its ion exchange characteristics so that there is a rise in the K^+:Na^+ ratio, resulting in the induction of puffing as a direct consequence of this rise. The importance of this ionic ratio in early development has already been mentioned (Kafiani, 1970).

For many years, developmental biologists have identified, the cell membrane as being of primary importance in morphogenesis (Moscona, 1962). It has been suggested that the specificity of aggregation and adhesion that characterize histogenesis and organogenesis is attributable to specific mucoprotein matrices coded by the genome and localized on the cell surface (Moscona, 1962, 1968). It is well established that polysaccharides, probably in the form of mucoproteins, are found on the surfaces of almost all cells (Revel and Ito, 1967). Based on the studies of cellular aggregation in sponges, it has been proposed that a "glycocalyx" 100 to 300 A in diameter, made up of mucoprotein particles, exists on cell surfaces, and that these particles endow cells with their morphogenetic proclivities (Humphreys, 1967), although high-resolution electron microscopy has failed to provide morphological evidence of their existence.

Gene Activator-Repressor Mechanisms

The mechanism or mechanisms by which the appropriate genes are selected for transcription during morphogenesis are poorly understood. The Jacob-Monod model stimulated considerable research and speculation, and is still widely considered to provide the greatest insight into the way in which genes can be selectively activated. Davidson (1968), however, believes that the Jacob-Monod model has limited applicability to multicellular organisms, particularly animals, because it depends upon the existence of polycistronic operons, which at best are rare in higher organisms. There is little doubt that histones mask DNA in animal cells (Georgiev, 1969), but it is not known how they can do so selectively, because only about thirty different kinds of histones have been described in animal and plant cells, which is insufficient for the variety of genes to be repressed. Davidson proposes that histone repression is selectively mediated by other molecules that do have the capacity for gene recognition. Other authorities feel that the histones do have the potential even though the variety of histones is small, because these histones have theoretically the potential to multiply this number sufficiently to provide the variety required for gene selectivity. Acetylation of histones, for example, which is known to occur in living nuclei, would greatly increase the variety (Goodwin, 1966). That acetylated histones do not bind strongly to DNA presents a problem, however. In fact, it has been proposed that acetylation is a mechanism for *unbinding* the histones from DNA as a condition of gene activation. More recently it has been proposed that the histones can theoretically exist in a variety of stereoisomers and, hence, can provide patterns of hydrogen bonding that are varied enough to meet the requirements of gene selectivity (Lewin, 1970).

If an intermediary molecule is involved, as Davidson (1968) and Britten and Davidson (1969) suggest, this molecule must meet certain requirements. Proteins are logical candidates for polynucleotide recognition and binding because they do this with rRNA (rRNA methylating enzymes) and because both the lambda-phage and *lac*-operon repressors have been shown to be proteins (Ptashne, 1967a,b; Riggs and Bourgeois, 1968). But RNA is also a logical candidate because it could recognize and bind with any DNA that has a complementary base sequence. Moreover, there are massive quantities of RNA in the nucleus of developing cells that have not been accounted for. These are large molecules—up to 90S—that never leave the nucleus unless they have been completely degraded, and so they do not represent cytoplasmic RNA precursors. They hybridize with five times more DNA than does cytoplasmic RNA (Shearer and McCarthy, 1967), they are polymerized very rapidly, and they are short-lived. These characteristics make this heavy RNA a very attractive candidate for the job as the regulatory intermediary in histone repression. What DNA, then, serves as the template for the polymerization of heavy RNA? It is unlikely that it is the same DNA sequence that is going to be repressed, because this DNA must serve as the template for mRNA. Not only is heavy RNA different from mRNA in size, it must have a different function, and these factors are incompatible with template identity. The amount of DNA in eukaryotic cells is far greater than that in prokaryotic cells, even though the number of enzymes needed is not greatly different. Vogel (1964) calculated that human cells contain approximately a hundred times more DNA than can be accounted for in terms of estimated structural gene content. Davidson (1968) proposes that this "nonstructural" DNA content could be the template for heavy RNA. Considering that *Salmonella* has five regulatory genes involved in the pathway of histidine synthesis, which is a relatively straightforward and uncomplicated pathway, one would expect that the number of genes required for regulation of an animal cell's genome during as complex a process as differentiation must be enormous. If gene activity is regulated by DNA/RNA intermediaries, then Vogel's findings should not be surprising. The DNA content of a cell may mimic the epitomy of bureaucracy wherein more DNA is needed to control itself than to contribute to the enzymatic complement of the cell. Finally, considerable evidence is accumulating in support of the theory that nonhistone proteins selectively remove histone from specific gene sites when and as the products of these sites are needed (G. S. Stein et al., 1975).

QUESTIONS

1. Outline a kind of experimental procedure that you think would provide direct proof that a zygote contains all the genetic potential of the adult.

2. Can you suggest any reason why the ability to regenerate lost parts has decreased with evolutionary progression?

3. Given the hypothesis (which may not be true) that the repressed DNA in *Planaria* cells is derepressed by a different mechanism than in mammalian cells, how would you go about proving this hypothesis?

4. Reconcile the existence of embryonic inducer substances to the operon model and to current knowledge of the mechanism of protein synthesis using each of the following alternatives:
   a) Inducers function at the level of mRNA transcription.
   b) Inducers function at the level of translation.
   c) Inducers function by directly affecting enzymes.

5. In view of the uniqueness of lampbrush chromosomes and the strange events of oogenesis in many species, how can the simple statement, Somatic cells are diploid; gametes are haploid, be supported?

6. Discuss this statement: The functional importance of an egg cell lies not only in the genes that it carries, but also in its cytoplasmic organization.

7. Comment on the following: If the events of blastulation and cleavage are under the control of maternal DNA, an early embryo is no different genetically than any other growth on or in the maternal organism. Furthermore, once the secrets of variable gene activity are discovered, any nucleated, differentiated cell could be returned to a zygote-like state. Therefore, any surgery resulting in the removal of maternal cells constitutes an abortion.

12

Delayed Mendelian and Non-Mendelian Inheritance

In 1905 Bateson, Saunders, and Punnett reported on the inheritance of various characteristics in the sweet pea (*Lathyrus odoratus*). In addition to being among the first to observe partial linkage, they came across another feature of inheritance that did not comply with Mendelian rules. If a pure-breeding strain of sweet pea with long pollen grains was fertilized by pollen from a pure-breeding round-grain variety, the F_1 plants produced all long pollen grains and the F_2 segregated into an approximate ratio of three long to one round. These results would be entirely unremarkable if the trait concerned any other part of the plant. We would say that this characteristic is acting in typically Mendelian fashion, long being dominant to round. However, segregation of genes occurs *before* the formation of pollen grains, which are therefore haploid. It is not possible to have a genotypically heterozygous pollen grain. What appears as a "normal" series of crosses is really not normal at all. This is especially obvious when contrasted with other studies involving pollen-grain characteristics. For instance, the heterozygote formed in rice by crossing a starchy pollen strain with a waxy one produces starchy grains and waxy grains in equal number (Parnell, 1921). These results are more nearly typical and certainly more "normal" than the apparent lack of segregation in the sweet pea.

In 1909, Correns reported on the inheritance of leaf coloration in a number of plants. Most of his work indicated typically Mendelian inheritance, the variants usually being inherited as recessives. However, one of his strains of four o'clock (*Mirabilis jalapa*) did not behave as expected. Though most of the shoots in this variety had their yellowish-green leaves striped with white to varying extents, a few shoots on a normally variegated plant had solid green leaves and a few had white leaves. Flowers on the green shoots yielded green progeny even if the pollen had come from a white or variegated flower. Similarly, pollen from green, variegated, or white shoots, when used to fertilize flowers on white shoots, had white progeny. These quickly died due to lack of photosynthesis. When fertilized with the three types of pollen, flowers in the variegated shoots yielded green, white, and variegated offspring in no consistent ratio and in apparent independence of either the pollen source or degree of variegation of the "mother" flower.

Total disregard of the rules of segregation and a tendency to look like the mother are effects of two different hereditary processes: delayed Mendelian inheritance and nonchromosomal inheritance. However, differences in the effects of these types of heredity exist, and it is on the basis of such differences that they are distinguishable.

DELAYED MENDELIAN INHERITANCE

As the name implies, traits displaying delayed inheritance conform to the principles of chromosomal genetics, but are sidetracked just a bit by ties to the parent. The unusual results of the pollen grain studies of Bateson, Saunders, and Punnett fall into this category. Evidently the pollen mother cell produces something that is carried over to the pollen grains and determines their shape. In most cases of delayed inheritance, the "something" that is carried over is unknown. It may be long-lived mRNA, a complex of enzymes, or even particular structural or enzymatic peculiarities. In any event, the phenotype of the pollen grain reflects the genotype of its parent cell. A plant that is heterozygous or homozygous for the long pollen grain gene will produce long pollen grains even if the pollen grain itself carries only the recessive gene for roundness.

In the case of sexually reproducing animals, the ties are usually between the maternal parent and offspring. Such "maternal inheritance" results from the two important features that are characteristic of the egg but not the sperm: (a) the orientation of the mitotic spindle axis, and (b) a relatively high cytoplasmic continuity between oocyte and egg with little or no contribution from the sperm. As explained in Chapter 11, maternal cyto-

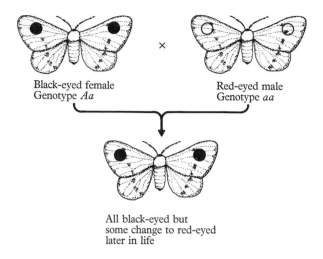

Black-eyed female
Genotype *Aa*   ×   Red-eyed male
Genotype *aa*

All black-eyed but
some change to red-eyed
later in life

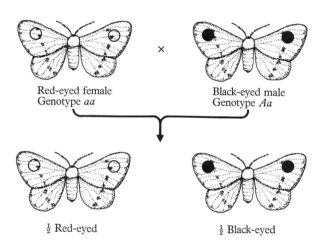

Red-eyed female
Genotype *aa*   ×   Black-eyed male
Genotype *Aa*

½ Red-eyed   ½ Black-eyed

plasmic components shape the early development of a number of organisms. In this section, the concern is with maternal effects that persist at least until sprouting, hatching, or birth, thus influencing the phenotype of the offspring.

The normal eye color of the moth *Ephestia kuhniella* is black, with a mutation for red inherited as a recessive. Kuhn (1937) found that the phenotype of the testcross offspring of heterozygotes differed depending upon the sex of the heterozygote (Fig. 12–1). The dominant gene *A* results in the production of kynurenin, a precursor

◀ **Fig. 12–1.**   Inheritance of eye color in the moth *Ephestia kuhniella*. The normal color is black; the "red" mutation is inherited as a recessive. In these reciprocal crosses, maternal inheritance is apparent though temporary. *A* signifies the wild-type allele; *a* stands for the recessive red allele. (Eye color is symbolized as a solid circle (black) or open circle (red) on the wings for convenience.)

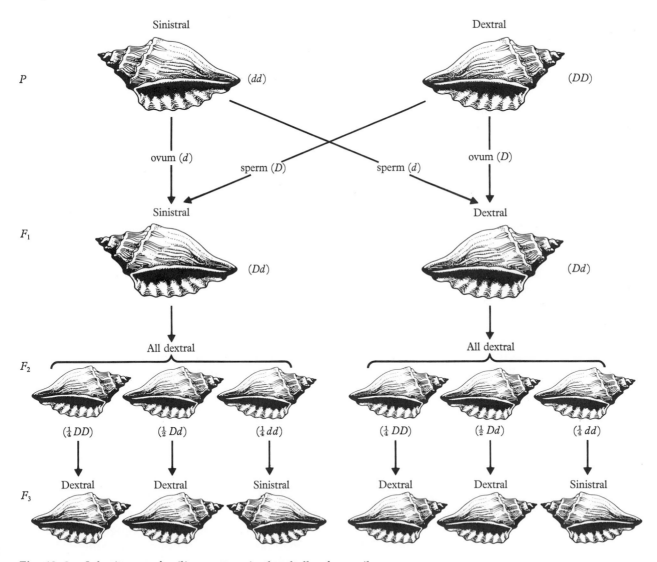

Fig. 12–2.   Inheritance of coiling pattern in the shells of a snail. Dextral coiling (*D*) is dominant to sinistral coiling (*d*). This set of reciprocal crosses clearly shows permanent maternal inheritance. These snails can reproduce either by self-fertilization or by mating with another snail.

of black eye pigment. This material is a cytoplasmic constituent of eggs produced by *AA* or *Aa* females. As a consequence, *aa* larvae produced by heterozygous females will initially have black pigment. The continuation of pigment production depends on the genotype of the larva itself. If its genotype is *aa,* it will have the black pigment in its early stages, but later in life the eye color will change to red.

A comparable situation exists in snails, except that the effect induced by maternal genes is permanent throughout the life of the offspring. The shell and body organization of some snails coil to the right (dextrally); other snails of the same species coil to the left (sinistrally). Pure-breeding lines of each kind can be established either by self-fertilization of these hermaphroditic animals or by cross-fertilization. The results of a set of reciprocal crosses are shown in Fig. 12–2.

A 3:1 ratio occurs, with dextral (*D*) dominant to sinistral (*d*), but it is delayed one generation. Of equal significance in solving this puzzle is the fact that offspring of pure-breeding parents show maternal inheritance irrespective of the paternal genotype or phenotype. Earlier work by embryologists (Conklin, 1903) demonstrated that spiral cleavage in molluscs is dependent on the orientation of the spindle during the first few cell divisions—if the spindle is tipped to the right of the midline, coiling to the right will occur, and vice versa. Furthermore, the orientation of the spindle in the early zygote is maternally set during oogenesis, prior to fertilization by the sperm.

Developmental consequences of delayed gene action can be detrimental, as in the case of the "grandchildless" mutation in *Drosophila.* Females homozygous for this allele, which acts as a recessive, are fertile, although not fully so, but the sex organs of their offspring fail to mature, so that they are fully sterile, irrespective of the genotype of the male parent. Presumably the effect is transmitted via the egg cytoplasm where it acts to prevent normal reproductive development.

NON-MENDELIAN INHERITANCE

In contrast to delayed gene action, non-Mendelian or nonchromosomal heredity refers to the replication and transmission in eukaryotic organisms of DNA that is extrachromosomal. The general term *plasmid* has been given to these nonnuclear, self-reproducing entities. For the sake of simplicity, we can divide non-Mendelian traits into two categories. The first category includes traits that are probably dependent, or partially dependent, upon the DNA of certain cytoplasmic organelles, such as chloroplasts and mitochondria. The second category includes traits that are determined by other DNA components. In some cases, these components may be "foreign" substances in the sense of having an extracellular origin. In either case, the site of the DNA concerned is primarily the cell's cytoplasm rather than the nucleus.

Inheritance via cytoplasmic DNA tends to be maternal because most of the zygote's cytoplasm is derived from the egg. Therefore, reciprocal crosses give different results, a situation similar to that of delayed Mendelian inheritance. However, the unusual phenotypic ratios do not disappear after one generation, as is normally the case with delayed gene action. Plasmid inheritance implies perpetuation through DNA replication and, hence, is a second system for the transmission of traits. Nevertheless, in most cases the nuclear and nonnuclear pathways interact and a dependence exists between them. The total phenotype of a cell or organism, then, is a reflection of the interplay of nuclear genes, nonnuclear genes, and the environment.

Plasmids of Cytoplasmic Origin

Chloroplast Inheritance. The study of Correns cited at the beginning of this chapter was the first recorded observation of nonnuclear inheritance of chloroplast characteristics. The white regions of a variegated leaf are white because the chloroplasts in that region have lost their pigmentation.

Chloroplasts are present in the cytoplasm of the egg and they represent the plastids in the shoot involved. Consequently, flowers on the white-leaved shoots (which obtain their nourishment from colored regions of the plant) will produce female gametes containing colorless plastids. Female gametes produced by green regions will contain functional chloroplasts. Those from variegated sections can have green, white, or a mixture of both plastid types. The cytoplasmic contribution of the male gamete, which is contained within the pollen grain, is minimal and is usually devoid of plastids.

Many characteristics of chloroplasts are totally regulated by nuclear genes. Other traits seem to be controlled by nonnuclear means alone or in combination with the nucleus. One of the best known of the latter concerns color in corn. If a plant is homozygous for the recessive nuclear gene *iojap* (from Iowa, the source of the strain, and japonica), some of the chloroplasts are evidently caused to mutate to the colorless state, resulting in striped or

variegated leaves. A set of reciprocal crosses for this strain is shown in Fig. 12–3.

The deviant chloroplast trait is believed to be initiated by the mutant nuclear condition *ij/ij*, and if striped parents are always male, the trait behaves in a Mendelian manner (Fig. 12–3a). However, if the female parent is striped, the plastid characteristic is perpetuated and transmitted independently of the nucleus. By chance, a striped plant will produce some "green" eggs, some "white" ones, and some containing both green and colorless plastids. On this basis, and because of the linear nature of cell division during monocot development, the offspring can be green, white, or striped in varying degrees. No repeatable ratios are obtained and there is no correspondence between nuclear genotype and plastid phenotype. Furthermore, an *ij/ij* plant *does have* green sections; it is capable of making chlorophyll. Therefore, the nuclear condition cannot be one of inhibiting or blocking synthetic pathways of chlorophyll synthesis.

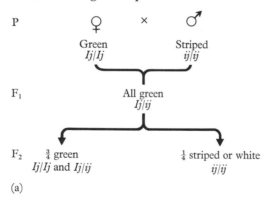

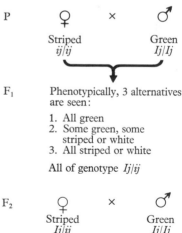

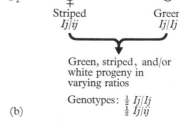

Fig. 12–3.   Inheritance of the striped *iojap* trait in corn, *Zea mays*. (a) When the female parent is normal, Mendelian inheritance is exhibited. (b) When the female parent is striped, the typical Mendelian ratios are not observed. Apparently the *ij/ij* genotype causes a mutation in the chloroplast, resulting in the striped or white condition. Once initiated, the *iojap* trait is inherited cytoplasmically. (See text for further explanation.)

Three of the criteria for identifying non-chromosomal heredity are met by the *iojap* example: (1) a difference in reciprocal crosses, (2) continuing maternal inheritance irrespective of genotype, and (3) non-Mendelian segregation. In distinguishing cytoplasmic heredity from delayed Mendelian action, the continuation of the trait for more than one generation is crucial. This is a reflection of the difference between the presence of an acquired substance that is reproductively passive and a separate genetic system that is reproductively active and self-perpetuating.

Indirect evidence has accumulated that points to semiautonomous control of such traits as plastid pigmentation, multiplication, and differentiation (Wilkie, 1970). Chloroplasts contain up to 10^{-15}g DNA—roughly enough for 100 genes that are each 1000 nucleotide pairs in length. About 15 percent of this DNA is similar to nuclear DNA in weight and base composition, and the two will hybridize. The remainder of the chloroplast DNA differs from nuclear DNA, and annealing will not occur between them. Moreover, it has long been recognized that plastids have a complex development life cycle of their own. They are self-replicating structures and their DNA can, in fact, replicate at a rate independent of the nuclear rate. A particular kind of ribosome is present in chloroplasts, as is RNA and the enzymes that are required for protein synthesis (Scott and Smillie, 1967; Smillie, et al., 1967). In a manner similar to that of nuclear DNA, replication of chloroplast DNA is semi-conservative, synthesis of its RNA can be inhibited by DNase or actinomycin-D, and RNA-protein translation ceases upon treatment with RNase or puromycin. Finally, two physical properties of nuclear DNA can also be shown to occur independently in plastids: (1) mutation caused by ultraviolet light of 260-mμ wavelength, and (2) photoreactivation with visible light (Granick and Giber, 1967).

Mitochondrial Characteristics. In his work with baker's yeast *(Saccheromyces cerevisiae)*, Ephrussi

(1953) discovered that one or two colonies per thousand grew to only one-third to one-half the normal size when cultured on standard medium. As in bacteria, each colony represents the asexually produced progeny of a single cell. The production of a few such "petite" colonies from normal cells was a regular and repeatable phenomenon, but if cells from the petite colonies were replated, they always bred true and formed petites. The change apparently was permanent. Treatment of yeast with acridine dyes induced normal cells to mutate to petites that were indistinguishable from those that arose spontaneously, but fully as significant as this was the fact that this mutation occurred in 100 percent of the treated cells. This finding is quite inconsistent with the random nature of mutations that characterize nuclear genes, but it is apparently characteristic of nonnuclear genes.

When it was noticed that the petite character could be mimicked by growing normal cells in the absence of oxygen, the physiological difference between the two growth states was discovered. Petite cells lack one or more of the respiratory enzymes needed for aerobic respiration. The amount of energy obtained by anaerobic respiration of foodstuffs is a mere fraction of the total that can be obtained if the anaerobic glycolytic reactions are followed by the aerobic Krebs cycle and electron transport chain. Hence the less energy obtained, the slower the growth rate. Additionally, the petite cells cannot normally produce spores, development of which requires a sexual process that seems to depend upon the energy derived from aerobic respiration. They can, however, grow vegetatively and produce new cells.

Haploid yeast cells can be mated with others of opposite mating type. If diploid zygotes obtained in this way are plated on standard medium, they will reproduce sexually to form diploid colonies. If the zygotes are plated on a medium containing only sodium acetate and raffinose as the carbon source, meiosis may occur to yield four haploid spores per cell (Fig. 12–4). These then can germinate to form haploid colonies. Under these

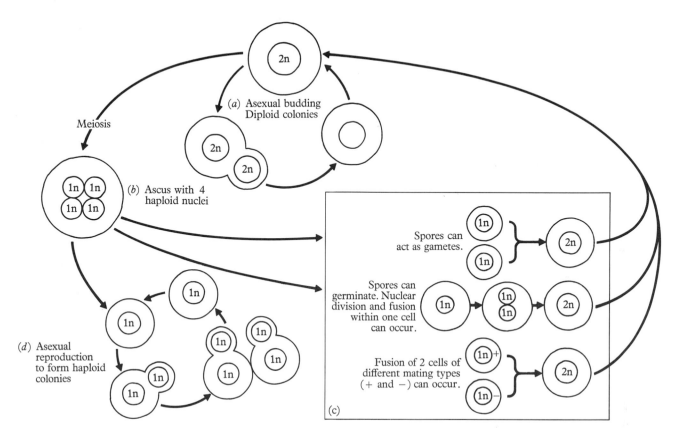

Fig. 12–4. The life cycles of yeast. (a) Vegetative reproduction by budding generally occurs when conditions are favorable. (b) When living conditions become unfavorable, meiosis may occur leading to the production of haploid spores. (c) Various ways of reconstituting the diploid state. (d) Additionally, haploid cells can reproduce by budding to form haploid colonies.

circumstances, haploid petite cells can be obtained and mated with normals. On the basis of the results of these matings, three different petite varieties can be distinguished (Fig. 12–5).

In order to account for these bizarre results, Ephrussi (1953) postulated the existence of a cytoplasmic hereditary factor necessary for the synthesis of respiratory enzymes. This cytoplasmic factor could interact with a nuclear gene in different ways to yield either petites or normals. On this basis, the different petites would be categorized as follows.

Type I petite	Nucleus normal. Cytoplasmic factor mutant or absent.
Type II petite	Nucleus normal. Cytoplasmic factor mutant and capable of suppressing expression of the normal factor. Cytoplasmic factor can be lost in successive vegetative generation.
Type III petite	Mutant nuclear gene that is recessive. Cytoplasm normal.

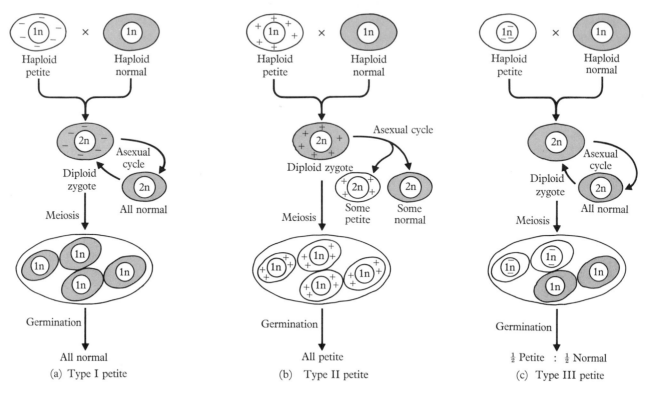

Fig. 12–5.   Crosses of different kinds of petites with normal yeast. (See text for explanation.)

On this basis, a mating of Type I with Type III ought to be possible, and it should produce some normal offspring because Type I would contribute a normal nucleus and Type III a normal cytoplasm. This does occur and is represented in Fig. 12–6.

The heterokaryon test (Chapter 4) is useful for identifying extranuclear genetic material in certain organisms, such as yeasts and fungi. When yeast heterokaryons are produced for normals and Type I petites, growth is normal. This would be expected because the normal's nuclei and cytoplasm can supply whatever it is that is lacking in the petites. However, when homokaryons are recon-

stituted, growth of both varieties is again normal (Fig. 12–7). If Type I petites are so because of a nuclear event, isolation of petite nuclei should result in petite growth. Only if they lack something cytoplasmic can these results be explained, because the reconstituted petite nuclei do not have their original cytoplasm, but rather a share of the heterokaryon cytoplasmic pool. If the "new" trait (normal growth in this case) can be correlated with the cytoplasm rather than with the nucleus and is maintained in successive generations, the factor must be cytoplasmic and renewable (Wright and Lederberg, 1957).

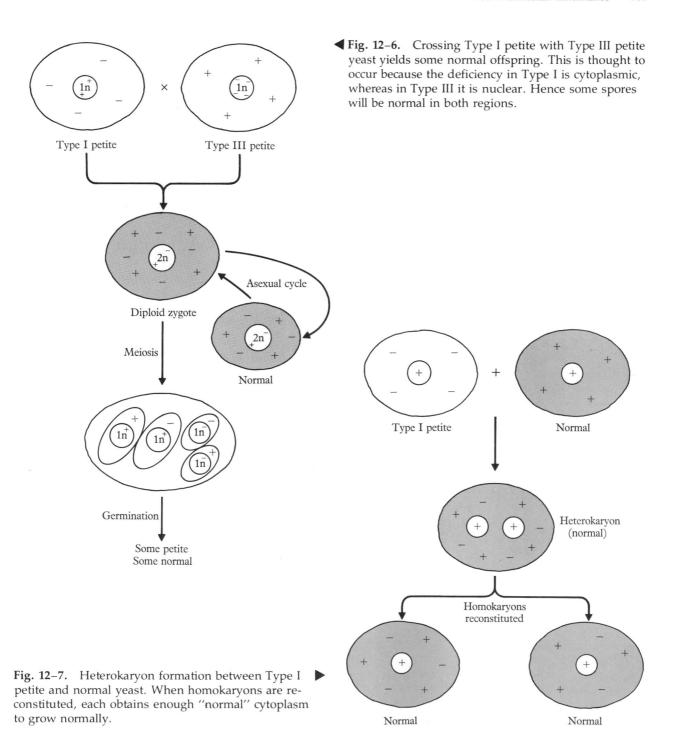

Fig. 12–6. Crossing Type I petite with Type III petite yeast yields some normal offspring. This is thought to occur because the deficiency in Type I is cytoplasmic, whereas in Type III it is nuclear. Hence some spores will be normal in both regions.

Fig. 12–7. Heterokaryon formation between Type I petite and normal yeast. When homokaryons are reconstituted, each obtains enough "normal" cytoplasm to grow normally.

In summary, the petite characteristics are thought to be partially due to cytoplasmic heredity because (1) non-Mendelian segregation occurs, (2) homokaryons, reconstituted from heterokaryons, show cytoplasmic association and perpetuation of the trait, and (3) nonrandom mutation occurs with certain agents.

A similar respiratory deficiency called "poky" is found in *Neurospora*. This example is a bit more clear because, unlike haploid yeast cells that merely fuse to form a diploid zygote, *Neurospora* produces gametes of unequal size and cytoplasmic contribution. Therefore, in addition to all the evidence from yeast, maternal inheritance can be seen as well.

As with chloroplasts, if one wishes to identify the source of the trait, a plausible cytoplasmic home must be located. Respiratory deficiencies have been correlated with faults in the mitochondria of the cells involved. Mitochondria contain enough double-stranded DNA to correspond to about twenty genes, RNA, ribosomes, and the enzymes necessary for protein synthesis. Interestingly, in some organisms the DNA appears to be linear, whereas in other organisms it appears to be circular (Fig. 12–8). Although the life cycle of mitochondria is less well understood than that of chloroplasts, evidence exists for reproductive autonomy here as well. In 1966, Mounolou and his co-workers found not only that a variety of petite yeast had observable mitochondrial defects, but that mitochondrial DNA was abnormal. More recently, Nagley and Linnane (1970) found other petite mutants that contained little or no mitochondrial DNA at all. As with chloroplasts and nuclear chromosomes, actinomycin-D, DNase, and RNase inhibited the appropriate reactions. Figure 12–9 represents a tentative plan of control in mitochondria, with some structural parts under the direction of the nucleus and others, notably the synthe-

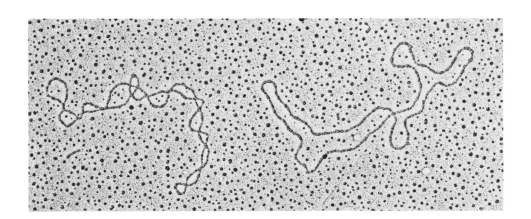

Fig. 12–8. Two mitochondrial DNA molecules isolated from mouse L cells grown in tissue culture. Each is approximately five microns in length. [Courtesy of Dr. Jerome Vinograd and Dr. Harumi Kasamatsu, California Institute of Technology.]

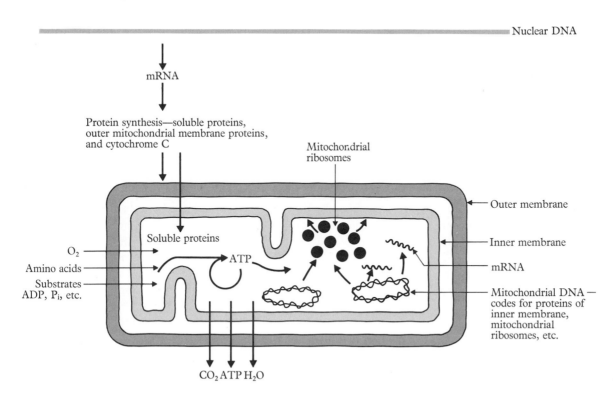

Nuclear DNA

mRNA

Protein synthesis—soluble proteins,
outer mitochondrial membrane proteins,
and cytochrome C

Mitochondrial
ribosomes

Outer membrane

Soluble proteins

O_2

Amino acids

Substrates
ADP, P_i, etc.

ATP

Inner membrane

mRNA

Mitochondrial DNA —
codes for proteins of
inner membrane,
mitochondrial
ribosomes, etc.

CO_2 ATP H_2O

Fig. 12–9. Proposed interactions between nuclear DNA and mitochondrial
DNA. The various components of this scheme are not drawn to scale. [Adapted
from D. B. Roodyn and D. Wilkie (1969), *The Biogenesis of Mitochondria.* London: Methuen and Company.]

sis of mitochondrial ribosomes and protein, being
directed by mitochondrial DNA. Note that more
than one circlet of DNA is present in the
mitochondria. The circlets are apparently duplicates of one another, such redundancy being
characteristic of mitochondrial DNA.

Ancestry of Eukaryotic Cells. Most eukaryotic
cells contain many mitochondria and most green
plant cells have many chloroplasts as well. It has

been extremely difficult to analyze chloroplasts
and mitochondria by the usual genetic techniques—if a cell has 1000 mitochondria and one
gene in one of them mutates, the chances are that
the mutation will be undetected. Luckily, certain agents such as acridine dyes, ultraviolet light,
and certain drugs appear to cause nonrandom, or
at least multiple, changes in these organelles, thus
permitting some kinds of study. Additionally, the
small size, uniparental inheritance, and intracellu-

lar location of these structures present difficulties. As a result, most of the evidence for partial genetic autonomy is indirect, yet some direct evidence does exist (Wilkie, 1970; Nagley and Linnane, 1970). The mountain of indirect evidence for the presence of semiconservative replication of double-stranded DNA in plasmids and for the presence of the necessary apparatus for protein synthesis in plasmids has been persuasive in establishing the genetic role of these organelles in the minds of most geneticists. Along with their genetic role has come a more firmly based revival of an old theory, namely, that chloroplasts and mitochondria were once free-living cells and now exist as obligatory, intracellular symbionts. The array of evidence that has been marshalled in support of this theory is extremely impressive (Margulis, 1970).

Plasmids of Extracellular Origin: Infective Heredity

This category of nonnuclear inheritance differs from that of chloroplast or mitochondrial in three ways. In the first place, the cause of the traits concerned is identified with particulate entities that are not normal or indispensable constituents of the cells involved. Secondly, the causative agents of these traits can be transferred via cell-free extracts of affected organisms to unaffected individuals, in which they become established and propagate. Finally, many of these causative agents are capable of moving into or out of the chromosomal DNA of the host, and are thus able to exist in either the nucleus or the cytoplasm.

Drosophila.   Two examples of infective heredity in *Drosophila* concern the traits of carbon dioxide sensitivity and the sex ratio of offspring. Most *Drosophila* can tolerate anaesthetic concentrations of CO_2, but certain strains are killed by these concentrations and are termed CO_2-sensitives. Carbon dioxide sensitivity is a heritable trait, but transmission is almost always through the maternal line. Introduction of extracts of sensitive flies

into normal flies is followed by heritable sensitivity in the latter after an initial incubation period. Sensitive flies can be "cured" by heat treatment. L'Heritier (1958) proposed that the cause of sensitivity is a virus-like particle, which was detected and designated *sigma*. Although the molecular consititution of sigma is not known, its X-ray inactivation pattern and growth cycle suggest that it is a virus.

Sigma exhibits other characteristics that it shares with some other infective traits. The behavior or sigma after transfer between strains and species of *Drosophila* suggests that an appropriate nuclear gene must be present if the successful life cycle of the particle is to continue. Sigma is not infective in other insect genera, which also may indicate its dependance upon appropriate genes. Secondly, in some ways sigma behaves episomally, that is, as a genetic element that can exist either as a free particle within the cytoplasm or integrated within the genome of the cell. The nature of its integration is not clear, because linkage between sigma and other genes has not been found. The number of sigma particles that can be found in females that transmit sensitivity to all their offspring is lower than in those females that produce some wild-type progeny. On this basis, it has been postulated that when sigma becomes integrated in the nucleus of the cell, all progeny must receive sigma, but it becomes correspondingly more difficult to recover the particle. If sigma is in fact an episome, an explanation of its incorporation only in female nuclei must be sought.

In a similar manner, the maternal inheritance of the "sex-ratio" trait in many *Drosophila* species results from an infective agent, which was shown later to be a protozoan spirochete (Paulson and Sakaguchi, 1961). Affected females produce predominantly female offspring which also show the trait. The few males that appear among the progeny do not pass the sex-ratio characteristic to their offspring. The skewed sex ratio results from differential survival of eggs and pupae. Most of those destined to be males die; the future females live.

Here again, heat treatment (keeping the eggs and pupae at 26°C) "cures" the condition and results in a more normal distribution of males to females. Although the sex-ratio trait was first discovered in *Drosophila bifasciata*, no evidence of spirochetes could be found in this species. Moreover, cytoplasmic extracts of these sex-ratio females are not infective, although in every other way the trait manifests itself identically to that in *D. willistoni*, *D. melanogaster*, or *D. pseudoobscura*. It is believed that sex-ratio originated in *D. bifasciata* by cytoplasmic infection, and subsequently the infective DNA became integrated into the genome of this species, perhaps as a result of the activity of a specific nuclear gene, whereas in the other *Drosophila* species, the infective agent remained cytoplasmic (Poulson and Sakaguchi, 1961). Ephrussi (1953) has postulated that infectious particles such as sex-ratio may represent a general mechanism for the acquisition of new genetic information and the subsequent expansion of the genome.

Infective heredity has also been demonstrated in mammals. The females of certain strains of mice are particularly susceptible to mammary cancer, whereas other strains appear to show resistance to the disease. Virus-like particles have been isolated from the milk of susceptible females. There is evidence that this extrachromosomal particle, acting with an appropriate nuclear gene found in the highly inbred susceptible strains, predisposes young females to mammary cancer when they mature.

Paramecium. One of the most fascinating and confusing stories in genetics deals with the pioneering work of T. M. Sonneborn (1937, 1965) and his colleagues with *Paramecium*. The complexity of *Paramecium* life cycles may make the story difficult, but once that is mastered, these investigations epitomize genetic "progress," because each new discovery led not to a solution in the final sense of the word, but rather to new mysteries and ingenious concepts that illustrate what can happen in the course of species adaptation.

An outline of certain components of the complicated life cycle of Paramecium including asexual reproduction, sexual reproduction, and autogamy (cf. Fig. 12–10).

1. *Asexual reproduction.* Paramecia can reproduce asexually by mitosis. Daughter cells receive a full diploid complement of DNA and about half of the parent cell's cytoplasmic constituents. In the discussion below, what is meant by generation 15 or 16 is asexual division 15 or 16 times, starting with an original parent cell.

2. *Sexual reproduction.* Paramecia of different mating types can exchange genetic material by conjugation. The details of the process are presented in Chapter 5, but two aspects are of concern here. First, the nature of conjugation is such that the two exconjugants are identical to each other with respect to nuclear genes, although genetic recombination during meiosis may render them unlike the two cells that originally met. The exconjugants may be homozygous or heterozygous for certain genes. Second, the cytoplasmic bridge that joins the conjugating pair may last for varying amounts of time. If the bridge disappears immediately after nuclear transfer, little or no cytoplasmic mixing will occur; the exconjugants will have identical nuclei, but they may differ cytoplasmically. If the bridge persists for a time, cytoplasmic materials pass from one cell to the other.

3. *Autogamy.* This process is analogous to self-fertilization, and is important to our discussion because it results in completely homozygous organisms. The two diploid micronuclei undergo meiosis to form eight haploid nuclei. Seven disintegrate and the eighth divides mitotically to form two identical haploid nuclei. These then fuse and form a diploid product that divides mitotically to form two new diploid micronuclei. If a population of heterozygotes undergo autogamy, by chance about half will become homozygous for one allele and the remainder will be homozygous for the other allele.

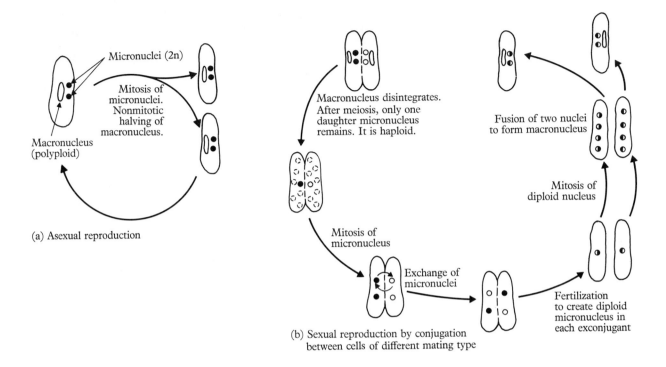

(a) Asexual reproduction

(b) Sexual reproduction by conjugation
between cells of different mating type

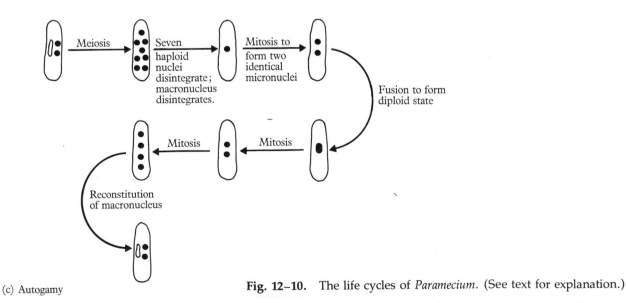

(c) Autogamy

Fig. 12–10. The life cycles of *Paramecium*. (See text for explanation.)

Sonneborn established many different stocks of *Paramecium aurelia*, each stock being derived from the asexual reproduction of one organism. He noticed that when individuals of certain stocks were combined with individuals of certain other stocks, one stock would live (the killers) and the other stock would die (the sensitives). Different kinds of killer stocks were discovered, each of which killed at a characteristic time or in a recognizable manner. Two kinds of killers will be of concern here: *hump-killers* and *mate-killers*.

a) *Hump-Killers.*   Hump-killers liberate a toxic substance into the medium and cause a peculiar aboral hump to develop in the sensitives just before they die. Hump-killers can be mated to sensitives, however, because the sensitives are not susceptible to hump-killing action during conjugation.

The first clue to the nature of hump-killing came from comparing two sets of conjugations (Fig. 12–11). Killers from stock 51 were mixed with sensitives of stock 52. The conjugal bridge lasted only long enough for nuclear exchange. Two kinds of exconjugants were formed even though their nuclei were identical. Those that were cytoplasmically derived from the sensitives remained sensitive. When stock 51 killers were mated with stock 47 sensitives, the bridge between them persisted for a time after nuclear exchange and all exconjugants were killers. It was concluded that hump-killing is a cytoplasmically localized trait, and a killer factor called *kappa* was postulated to be the causative agent.

As with sex-ratio and CO_2-sensitivity, a hump-killer can be "cured" by elevated temperatures or by certain drugs. Subsequently, the hypothetical kappa factors were identified as minute virus-containing particles that measure about 0.2μ in diameter, are membrane-bound, and contain DNA and RNA. They are not present in sensitive individuals. During asexual division, about half the hump-kappa particles are distributed to each daughter cell, whereupon the hump-kappas reproduce to reconstitute the full complement of

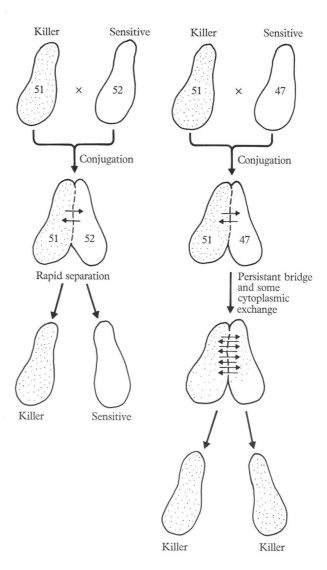

Fig. 12–11.   Conjugation of hump-killers (stock 51) with sensitives of stocks 52 and 47. The cytoplasmic bridge formed between 51 and 47 persists long enough for some cytoplasmic exchange to occur, whereas the bridge formed between 51 and 52 is very short-lived. Phenotypic difference in offspring of these two crosses is explained by transfer or lack of transfer of cytoplasmic particles called *kappa,* represented here by stippling.

more than 1600 particles per cell. Occasionally one or more of these particles escape. If just one is ingested by a sensitive *Paramecium*, the *Paramecium* will die. Hump-killers are resistant to the hump-killing action of other *Paramecia*.

Up to this point, the picture seemed to be clearly a case of infective inheritance—hump-kappa was thought of as a cytoplasmic particle capable of multiplication, having the ability to kill sensitive organisms while conferring upon its host a resistance to hump-kappa effects, and it can be infective during conjugation. Complications, however, soon arose. First of all, only certain species of *Paramecia* were sensitive to hump-killers, suggesting a nuclear component. Secondly, certain mutations of the hump-killer trait were discovered. These mutants were resistant to hump-killers, but they themselves did not kill. Finally, when exconjugants of stock 51 (killers) and stock 47 (sensitives) underwent autogamy, half of the progeny permanently remained killers but the other half did not (Fig. 12–12). As mentioned, mating stocks 51 and 47 results in cytoplasmic exchange. Hump-kappas are received by stock 47 and all exconjugants are killers. It was thus postulated that stock 51 was homozygous for a nuclear gene *K*, that stock 47 was homozygous for its recessive allele *k*, and that at least one *K* gene was necessary for preservation of hump-kappas. The killer exconjugants were of genotype *Kk*—they could kill and maintain hump-kappas. However, autogamy would result in some *KK* cells and some

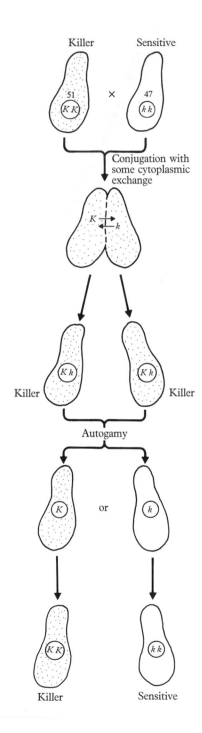

Fig. 12–12. Conjugation of killer stock 51 with sensitive stock 47, followed by autogamy of the exconjugants. The postulated genotypes are shown. *Paramecia* of genotypes *KK* or *Kk* are killers as long as they have hump-kappa particles. Those of genotype *kk* cannot maintain kappa particles and thus become sensitive to the action of hump-kappas. During autogamy, a group of heterozygotes (*Kk*) reverts to a state of homozygosis, about half becoming *KK* and half becoming *kk*.

kk ones. The former would remain killers, but the latter were unable to maintain their hump-kappas and became sensitive. Individuals of genotype *KK* or *Kk* can maintain hump-kappas. The *K* gene is for maintenance; the kappa particle is for killing. *Paramecia* can be of genotype *KK* or *Kk* and not contain any hump-kappas. They are just as sensitive as *kk* individuals until they acquire hump-kappa by conjugation with a killer strain. The relationships are summarized in the following table:

Genotype	Hump-kappa present	Phenotype
KK or *Kk*	Yes	Killers and resistant
KK or *Kk*	No	Nonkillers and sensitive
kk	No	Nonkillers and sensitive

b) *Mate-Killers.* A similar relationship emerged from studies of mate-killing. These killers contain cytoplasmic entities called *mu* particles. During conjugation between mate-killers and sensitives, mus or mu secretions are evidently transferred across the cytoplasmic bridge. The exconjugants derived from the sensitive organisms subsequently die without dividing further, or if the sensitive exconjugant divides, as occasionally happens, its daughter cells die. Possession of mus renders an organism resistant to the effects of mu. Hence, conjugation between mate-killers can be completed successfully and neither exconjugant dies. However, there are different kinds of mate-killers containing different mus. If two such dissimilar killers conjugate, one may kill the other.

Mu particles are maintained by the unlinked nuclear genes M_1 and/or M_2. *Paramecia* with the doubly recessive genotype $m_1m_1 \ m_2m_2$ cannot maintain mu particles, and are therefore sensitive to the effects of mu. As with kappa and the gene *K*, an individual containing one or more dominant alleles is also sensitive if it does not have cytoplasmic mu (Fig. 12–13).

The pattern of nuclear genes and cytoplasmic particles became confusing when results of autogamy were pursued. As mentioned above, mating hump-killer stock 51 with sensitive stock 47 results in presumptive heterozygotes for gene *K*. In a similar way, heterozygotes for gene *M* can be obtained, although the ones descended from the sensitive parent die. In either case, when the heterozygotes undergo autogamy, half are killers and half become sensitives, but the "sensitives" do not become so immediately. Kappas or mus persist for seven or eight cell generations in the homozygous recessives. In the next and subsequent generations, cells totally lacking in symbionts appear until virtually all are sensitives.

These bizarre results were investigated by Gibson and Beale (1961) in mate-killer strain 540. They postulated that the cells that maintain mu have either gene M_1 or M_2 and, moreover, that maintenance is accomplished by another cytoplasmic particle or substance. The name *metagon* was given to this hypothetical substance. Presumably, different metagons are required for different gene-particle combinations, and each M_1 or M_2

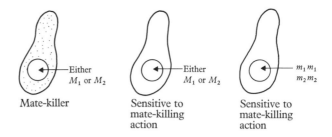

Fig. 12–13. The interaction of nuclear genes and *mu* particles in the trait of mate-killing. Doubly recessive *Paramecia* ($m_1m_1 \ m_2m_2$) are sensitives. They cannot support a population of mu particles. Possession of at least one dominant allele out of the four permits the existence of mu. *Paramecia* must have the appropriate genotype and the particles in order to be mate-killers.

cell maintains about 1000 metagon units in its cytoplasm. After autogamy, cells that become homozygous recessive cease making metagon but retain whatever metagon is present. If any metagon is present, irrespective of genotype, the mus are not destroyed and the cell will be a mate-killer. The amount of metagon present in the recessive cells diminishes with each successive division to a minimum. After this, only one daughter cell from each division receives metagon. The metagonless daughters can no longer maintain their mus and they become sensitive (Fig. 12–14).

The Physical Nature of Metagon.   Metagon was assumed to be a direct product of gene M_1 or M_2, since it was eventually lost if neither of these genes was present (Sonneborn, 1965). It was also found that metagon activity was destroyed by RNase.

To test the inference that metagon is RNA, recessive *Paramecia* that presumably had only a few particles of metagon left were supplied with RNA extracts of killer M_1M_2 cells. The extracts contained no visible evidence of mu particles. Theoretically, this should supply the recessive cells with more metagon, enabling them to produce killers for a longer time. This did occur. Secondly, M_1 or M_2 killers were fed RNase. Before they divided, the RNase was inactivated and these treated cells were supplied with fresh RNA from other killers. Theoretically, the fresh RNA should rescue the mus. It did.

Further testing established that metagon is indeed RNA, and that RNA with metagon activity will hybridize with DNA from killer organisms but not with DNA from recessive sensitives. It was concluded, therefore, that metagon is a very stable form of mRNA that leads possibly to the production of a protective protein substance for the maintenance of mu.

Summing up to this point, investigators have observed cytoplasmic inheritance of infective particles that are protected by long-lived mRNA, presumably produced by a nuclear gene. Excitement was generated because a link between nuclear and nonnuclear inheritance had been found, and also

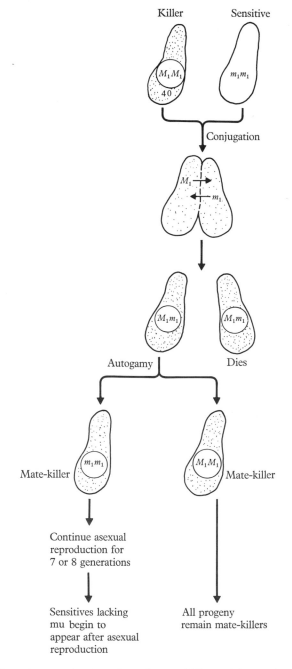

Fig. 12–14.   Conjugation of mate-killers with sensitives, followed by autogamy of the viable exconjugants. (See text for further description.)

because long-lived mRNA in eukaryotic organisms was something of a rarity at that time.

And then a new avenue of investigation opened, almost by accident. Sonneborn was working with two other ciliates, *Didinium* and *Dileptus*, predators that ingest *Paramecium* whole and alive. He found that certain strains of these ciliates, when fed killer *Paramecium*, liberated kappas or mus and killed sensitive *Paramecia* on contact. Evidently *Didinium* and *Dileptus* did not destroy the infective particles, and so the critical question was whether they merely eliminate these particles when they are part of their food or whether some other active process takes place. Various stocks of these two ciliates were fed killer *Paramecia* once, then were changed to a nonkiller diet. Stock 1 or 4 of *Didinium* maintained large populations of hump-kappa after the supply of kappa and gene *K* (from ingested *Paramecia*) was cut off. Stocks 1 and 3 treated mus in a similar manner. Stock 1 or 2 of *Dileptus* could also maintain hump-kappa and certain mu particles.

If *Didinium* and *Dileptus* are never fed killer *Paramecia*, they do not become *Paramecium* killers, but once certain strains receive either kappa or mu, they acquire killing abilities. Furthermore, after the ingested *Paramecia* have been fully consumed, the infective particles flourish and reproduce in successive generations of the foreign host. RNA extracts of *Didinium* do not normally show metagon activity, and metagon RNA does not bind to *Didinium* RNA.

These results, even more bizarre than those encountered within *Paramecium* alone, raise substantive questions about the existence of metagon as postulated. Byrne (1969) disputed the interpretations that were made and suggested that alternative explanations for these curious phenomena be sought.

Bacterial Symbionts with Episomal Activity. Infectivity and the ability of cells to get along with or without the infection were characterized as features of infective heredity. In order to identify such a genetic agent as an episome, one further requirement must be met. After a cell acquires the element by infection or conjugation, that element must be able to reproduce itself in either of two alternative states—as an autonomous particle or as an integral part of the genome. While certain traits such as CO_2-sensitivity in higher organisms behave in some ways as episomes, and while some forms of cancer may be caused by episomes, the only proved episomes known to date occur in bacteria. It may be added parenthetically that cells containing integrated episomes behave differently in some important ways from cells containing autonomous episomes or from cells free from episomal contamination. Also, behavior may change during the transition from one state to another. For instance, the disease diphtheria is caused by a bacterial toxin. Production of this toxin is limited to diphtheria bacilli that contain integrated episomes in the process of becoming autonomous. Pure bacilli or bacilli with stable integrated episomes cannot produce the toxin (Campbell, 1969).

The concept of episomal activity of infective particles brings together two areas of investigation that share many similar features. One concerns sexuality and conjugation in bacteria (see Chapters 5 and 10). The other concerns the study of temperate bacteriophage and lysogeny (see Chapter 7). Later, other kinds of infective heredity were shown to exhibit episomal behavior as well (Campbell, 1969). From the ever-growing list of episomal agents, just three are discussed here—sexuality and lysogeny, because they are historically important in this area of genetics, and infectious drug resistance, because of its medical implications.

Fertility Factors. As discussed in Chapter 10, Lederberg and Tatum (1946) found that donor cells of *E. coli* K12 contained a small, transmissible element that was called the *F-factor* (for fertility). Conjugation between donor cells (F^+) and recipients (F^-) resulted in two kinds of changes: (1) The F^- cells were converted to F^+ and the original F^+ cells remained F^+, indicating that the F-factor was capable of reproduction and transmission; (2) in a very low percentage of F⁻ cells, some donor-cell genes were apparently substituted in the recipient genome.

It was later discovered that some F⁺ populations changed in some way that facilitated recombination. Such changed cells were termed Hfr and, as noted in Chapter 10, the Hfr condition was associated with integration of the F-factor into the donor chromosome. It was subsequently postulated that the original discoveries of genetic recombination during conjugation were probably due to the presence of a few Hfr cells or to F′ cells (see below). Different F-factors enter the chromosome at different locations, but the location is constant for any one Hfr strain (cf. Fig. 10–14).

The F-factor is here reconsidered from the standpoint of episomal inheritance. Like most other genetic systems, the F-factor is composed of DNA. It consists of about 250,000 nucleotide pairs and is, therefore, long enough to contain many genes. It can exist autonomously as F⁺ or it can reversibly convert to the integrated state. Because it is about $1/40$th as long as the bacterial chromosome, the chromosome is that much longer when it is integrated. A cell containing either state of F develops two specific characteristics. First, the cell becomes resistant to entrance by additional F-factors and, in the case of F⁺, replication of the factor is strictly controlled so that only one F⁺ is generally present. Secondly, F⁺ and Hfr cells present a changed surface to the world, and as part of the change, these cells can grow F-fimbriae through which genetic transfer takes place. When conjugation of F⁺ with F⁻ occurs, the F-factor replicates and the replica is transferred, thus converting the F⁻ to F⁺. During conjugation of Hfr with F⁻, one copy of the genome of the Hfr strain begins to migrate through the cytoplasmic bridge into the recipient cell. How much is transferred depends upon the length of time during which the two cells are in contact. If complete transferral does occur, a rare event, the F-factor is the last part of the genome to be transferred. The chance of any gene in the transferred portion of the chromosome recombining with its homologous segment in the F⁻ chromosome is 50 percent (Campbell, 1969).

Hfr and F⁺ cells are apparently interconvertible. They can be "cured" of their F-factor or, on occasion, F⁺ cells can spontaneously become F⁻ cells. When Hfr cells convert to F⁺, the F-factor may take with itself a bit of the host DNA. Such an F-factor, containing a few added genes, is designated F′, and during conjugation with an F⁻ cell, these genes will be transferred along with the F-factor. Recombination of these genes with those in the F⁻ cell may then occur. Different F′ strains transfer different genes, depending on the Hfr strain from which they were derived, and therefore in accordance with the point of their original integration into the host cell's DNA.

The salient features of the F-factor are summarized below, and comparison of these features with those of lysogeny follows.

1. Fertility is infectious—F⁺ × F⁻ yields all F⁺ cells.

2. In the autonomous state, fertility is curable by acridine dyes, X-rays, and other treatment. Different treatment causes the interconversion of autonomous to integrated F-factor.

3. Fertility is an hereditary trait—fission of F⁺ cells yields F⁺ progeny. F⁻ and Hfr cells also "breed true."

4. The fertility factor can enter the host DNA (Hfr) or be physically independent of the chromosome (F⁺) and autonomous, but because the cells become "immune" to further F-factors and because replication of F is controlled, a cell may be F⁺ or Hfr, but not both simultaneously.

5. The F-factor is composed of DNA. Variant F-factors are known that are mutant in their ability to replicate or to initiate transfer.

6. A cell containing F changes its behavior by becoming immune to more F and by growing F-fimbriae. The presence of F-fimbriae makes the cell more susceptible to some viral infections because these viruses specifically attach to the fimbriae.

7. When an F-factor leaves the chromosome, it can transfer some bacterial genes.

8. The F-factor obviously has two kinds of active genes: genes for replication and genes for transfer. In the case of Hfr, these include the mobilization of the host's DNA. Other genes are responsible for the new characteristics of an infected cell (immunity to superinfection and changed surface characteristics) that may not be directly related to F-factor survival.

Lysogeny.   In Chapter 7, two kinds of viral life cycles were discussed. A virus may infect a cell and enter the vegetative state, whereby many new virus particles are produced. On the other hand, a non-pathogenic life cycle may ensue. There are some theoretical reasons that point to this mode of viral activity as the main road of evolution, whereas the virulent state may be a "mistake" or "accident," possibly due to wrong combinations of host cell and virus. We are beginning to realize that many cells isolated in the laboratory are already lysogenic. This is particularly true of bacteria. The nonpathogenic virus is said to be in a temperate state and is called *provirus* (*prophage* for bacterial viruses, the topic of this discussion). The host cell is termed *lysogenic*, that is, capable of being lysed at some time in the future but also capable of transmitting the virus to its progeny cells without external reinfection by that virus.

Lysogeny is specific. When a temperate virus converts or is converted experimentally to the virulent state, the virus that is liberated is one of the same type as the original infecting virus, even though the temperate state may have existed for many years and encompassed thousands of host-cell divisions. For this reason, it is recognized that the provirus is genetically identical to the original invader, and therefore contains information on how to be virulent. This information is obviously repressed by the lysogenic host cell and is derepressed when the virus leaves the host chromosome and becomes virulent again. For some temperate bacteriophage, the prophage acts as a bacterial gene, that is, a particular phage enters the bacterial chromosome at a particular place.

An integrated phage is replicated along with the bacterial genome and is passed along to the daughter cells in the same way as any other part of the chromosome. Under some appropriate stimulus such as ultraviolet irradiation, some bacteria burst and liberate many virus particles. This is the basis of our knowledge that viruses were present—it is also part of the definition of lysogeny. More to the point, however, is the fact that the viral genomes were obviously multiplying at a rate faster than that of the bacterium. The viral DNA not only maintained its autonomy but also its ability to reverse the lysogenic control process so that now it controls the host.

Although lysogeny differs from fertility in the manner of control of replication when the element is autonomous, the two phenomena share many other features. A lysogenic cell is "immune" to superinfection by the same phage. Lysogenic cells can be cured by acridine dyes or other treatment, and spontaneous curing is also known to occur. The virus in a lysogenic cell can move into and out of the chromosome, thus going from the virulent state to the temperate, and vice versa. When it does leave the chromosome, it can take with itself genes of the host. In fact, host genes may be packaged in a viral envelope as a substitute for viral genes. When these "viruses" infect new cells, recombination of the new and the old host genes may be effected by transduction.

This virus itself contains the genes that determine immunity to superinfection and the choice of which state it shall enter. Clearly, genes to make new virus particles must be represented in an integrated virus, whereas genes that produce enzymes permitting the viral DNA to enter the host DNA must be repressed in the virulent state. However, the control mechanism to determine which sequence of genes, virulent or temperate, shall act is unknown, although treatment with small doses of X-ray, for example, can help out the control mechanism.

The same features apply to the list of F-factor characteristics as well—with two exceptions. In the autonomous state, the replication of F is kept under host control, whereas that of the virulent virus is not; and the F-factor is associated with F-fimbriae, whereas the virus is not. These two features may be related if the F-factor is considered to be a virus that is deficient in its ability to synthesize a protective and stable protein coat. The fimbriae may, in effect, be a temporary viral tail for the injection of DNA, and the inability of the F-factor to surround itself with protein may result in its replicative control by the host. In any event, fertility factors and temperate viruses are clearly episomal.

Although this discussion of lysogeny applies to bacteria and bacteriophage, it is becoming increasingly clear that mammalian cells can also acquire genetic information by infection and can perpetuate this information intracellularly. In addition, much current speculation and research on genetic engineering has been devoted to the possibility of using lasers to excise defective genes and "friendly" viruses as the mode of gene replacement. Much recent research on some kinds of cancer points to a phenomenon similar to or identical to lysogeny as the causal agent of the disease.

Infectious Drug Resistance.   One of the more medically significant episomal systems deals with a phenomenon called *infectious drug resistance.* Before 1955, it was thought that the resistance of a bacterial pathogen to an antibiotic was simply a matter of the antibiotic acting as a single selective agent in exposing previously mutant bacteria. In 1955, however, a far more ominous story began to

unfold (Watanabe, 1967). A Japanese woman returning from Hong Kong developed a stubborn case of dysentery. The pathogen was identified as a typical dysentery bacillus of the genus *Shigella.* The bacillus, however, was resistent to four drugs: sulfanilamide, streptomycin, chloramphenicol, and tetracycline. Further such cases subsequently came to light in Japan and elsewhere, and epidemics of dysentery increased.

Multiple mutation of any one bacillus was ruled out for a number of reasons. First of all, the rise in multiple-drug resistance increased too quickly to be accounted for by spontaneous mutation. Secondly, patients were found to contain both sensitive and resistent bacilli. Furthermore, if one antibiotic was administered to a patient harboring only sensitive bacilli, these patients soon acquired bacilli resistant to all four drugs.

It was then found that many people who carried resistant *Shigella* also carried *E. coli* that were similarly drug-resistant. It appeared that resistance to all four drugs was being transferred *en masse* from *E. coli* (normal digestive tract inhabitants) to sensitive, infecting *Shigella*, and that the transfer was taking place within the digestive tract. The possibility of such a transfer was confirmed *in vitro.* Transfer by bacterial transformation and transduction were ruled out on grounds of frequency, the obviously larger size of the DNA required for all four characteristics, and the fact that *in vitro* transformation and transduction conditions did not seem to work. It was noted that multiple resistance transfer could be correlated with bacterial pairing, but that after pairing and conjugation were completed, drug resistance developed more rapidly in the exconjugants than could be accounted for by chromosomal replication and recombination. Therefore, it was postulated that infectious drug resistance was caused by an extra-chromosomal element. It was subsequently identified and named *R-factor* (for resistance).

R multiplies faster than the host's DNA. As we have seen, it is infectious and can be eliminated by acridine dyes. The R-factor has been known to transduce host-cell genes and, therefore, it must be capable of integration and episomal activity. Finally, R seems to promote conjugation by causing R$^+$ cells to form projections that are similar or identical to F-fimbriae. Evidently the R-factor is transferred in the same manner as the F-factor, although the cell seems less and less able to make fimbriae as R-factor accumulates. This is not so for a cell that is newly infected by R. Hence, rapid proliferation of R can be indicative of new infections.

Although F and R are transferred in much the same way, they are different episomal elements. The double-stranded DNA of the R-factor has been mapped by transduction studies. It is circular and contains a segment called *delta* that controls replication and transferability (Meynell, 1973). Other segments of the resistance factor determine resistance to a variable number of drugs. Some R-factors may enlarge by incorporation of new drug-resistant genes that develop by mutation of chromosomal genes, whereas others enlarge by forming an enlarged aggregate of nonchromosomal elements. R-plasmids evidently work by synthesizing enzymes to inactivate the drugs involved, whereas chromosomal genes for drug resistance generally work by altering the ribosomes in such a manner that the antibiotic cannot affect their ability to synthesize protein. R-factors vary, often in a way consistent with the drugs being used in a particular area. Doubtless they were detected because of wholesale use of "preventive" antibiotic treatment, leading to many single mutations that later became associated as an "aggregate-R" factor. In addition, R-factors seem to be acquiring resistant genes for an increasing number of drugs, and they keep up with the "progress" in new drugs to an almost uncanny degree. Finally, R-factors can be transferred to many different bacterial species such as *Salmonella, Vibrio cholerae, Pasteurella pastis*, as well as to over 90 percent of the agents of urinary-tract infections.

It is evident that "progress" in antibiotics is paralleled by "progress" in R-factors, and wholesale treatment with ranges of antibiotics (such as occurs in meat preservation and livestock feeding) would seem to bode more harm than good. As a final ominous note, one of the discoverers and key workers in this field has suggested that R be used in bacteriological warfare research (Watanabe, 1967). By this means, bacterial infections resulting from germ warfare may be rendered medically untreatable.

A CONCLUDING FABLE

If a strict, classical geneticist were transported through time and launched into the midst of non-Mendelian heredity, he might conclude that either genetics had gone mad or that geneticists had. He might concede to delayed Mendelian heredity because no one ever really argued the point of genetic interaction with the environment, even in a cellular context. As for organelles that might be infectious, sex that is not only "curable" but probably viral, interspecific transfer of viable genetic material, multiple cellular gene pools, and the possibility that viral diseases are really lysogenic accidents—to swallow all that is really asking a lot.

Perhaps what is needed is a new point of view. This has been outlined by Richmond (1970) for bacteria and, with modification, might be extendable to eukaryotic cells. If we look at the chromosome as the inviolable genetic material of the bacterial cell, we are then faced with an additional array of genetic entities that are not only bewildering but apparently meaningless. These entities place a bacterium in a state of constant genic flux, and in some cases they negate the notion of discrete bacterial species. On the other hand, if all the genetic material of a bacterial cell, including the chromosome, R- and F-factors, lysogenic viruses, and other plasmids that we have not discussed, were thought of as part of a whole, we are in a meaningful evolutionary ballpark. All these bits of genetic material are replicons—that is, they are all composed of genes that autonomously replicate and can be passed to the progeny. The major distinctions between the chromosomes and the plasmids then become matters of size and the cell's ability to survive the loss of one or the other. In addition to considering all the replicons as part of a whole (even if it is a labile whole), geneticists need to revise the definition of a bacterial species as an ecological unit that derives its identity not only from its chromosomal genes but also from its environment. When the environment changes, the range of variation exhibited by any species similarly changes. It may, in fact, overlap the characteristics of a supposedly different species.

When a chromosomal gene mutates, the bacterium usually suffers a blow. The characteristic controlled by that gene is either lost or becomes changed to a form that is probably less desirable for the cell. If mutation is a constant process, then the bacterium is constantly on the debit side of the genetic ledger. But when it gains a plasmid, it gains six to ten new genes all at once. One important question immediately presents itself: Could plasmid genes totally replace chromosomal genes? The answer is almost surely negative. In all probability, only the chromosome carries the all-important information for cell division. However, even if replacement of genes is only fractional, there must still be a relationship between the origin of the chromosome and the origin of the plasmid. This is especially important in the case of episomal plasmids, because entrance into the chromosome is a very intimate process in cellular existence. What facts and deductions point to a relationship?

1. In order for plasmids to be accurately distributed to daughter cells, a point of attachment between the plasmid and the cell becomes vital. Such an interaction implies two complementary genetic parts: plasmid DNA to attach, and chromosomal DNA to code for a cellular attachment site. Two mutants are known in both *E. coli* and *Staphylococcus aureus* that affect plasmid distribution and stability. In each case, one of the mutations behaves chromosomally, whereas the other appears to be nonchromosomal.

2. All plasmids contain genes governing their own replication. In addition, they carry other genes that confer new and often desirable metabolic properties on the host. The existence of resistance factors that may permit a cell to live when it would otherwise die is a prime example. The F-plasmid contains genes that are active in arginine biosynthesis, and other plasmids are known that specify certain bacterial antigens.

3. In order to isolate plasmid DNA, it usually must be transferred by infection or transduction to a new host species. Isolation from the normal host species is usually unsuccessful, because its DNA and the DNA of its native plasmid are too similar in base ratios, and therefore in density, to be separable.

4. Plasmids can fragment and reform in new plasmid combinations. There is experimental evidence to show that both this process and integration of episomes depend on the presence of homologous genes on both chromosome and plasmid or between plasmids. Additionally, episomes have the property of chromosomal gene "pickup." Therefore, any gene in the bacterium may find itself on the chromosome at one time and on any one of a number of extrachromosomal elements at another time. This genetic flux may be the basis of gene homology, and it certainly suggests an intimate evolutionary relationship between the replicons of any cell (Fig. 12–15).

5. There are clear-cut genetic mechanisms to ensure the transmission of plasmids between neighboring cells. Whether by conjugation (transmissible plasmids) or by transduction (nontransmissible plasmids), interpenetration of gene pools can occur between cells that by other standards are unrelated.

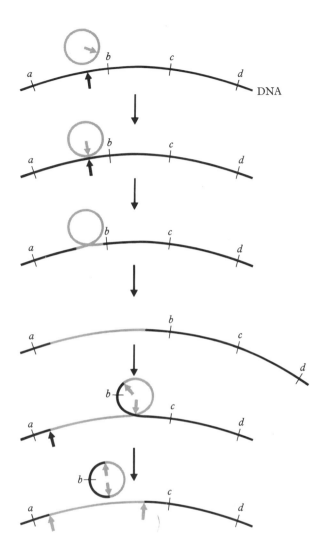

Fig. 12–15. A schematic view of the mechanism that may underlie the ability for continuous gene flow between replicons. The arrows indicate complementary genetic regions that permit attachment of one replicon to another. [After M. H. Richmond (1970), "Plasmids and chromosomes in prokaryotic cells," *Organization and Control in Prokaryotic Cells.* Cambridge: Cambridge University Press, pp. 249–277.]

On the basis of these remarks, it has been postulated that the genetic complement of bacteria is organized into separate and separable replicons. The chromosome would be the "master" replicon because (1) it controls the rate of DNA replication throughout the cell, (2) it contains the gene for cell division, and (3) it has the genetic information for cell survival under "normal" conditions. The other replicons emerge as discrete packages of "ecological" DNA to permit the colonization of specific and unusual ecological niches. All the information pertaining to one specific kind of niche may thus be brought together for ease of transmission to neighboring cells. These extra genes can be lost as a package and, if they are no longer essential for cell survival, the cell will not die. If the extra genes are continuously needed, they may become relatively permanent residents of the master replicon. Finally, because it is known that the ecological replicons can be transferred between certain species (for example, *E. coli* to *Shigella*), the genetic difference between these species is not absolute and interpenetration between their gene pools exists.

How might this theory be extended to eukaryotic cells? They too have plasmids, but the interaction between the replicons is not as fluid and the cells are not in as much flux. On the other hand, most eukaryotic cells are found in multicellular organisms. These creatures are more highly protected from environmental shifts, not only because of larger size (and, hence, less exposed surface), but also because they have developed mechanisms to ensure stability of their internal environments. The ability to protect seems to be a leitmotif of eukaryotic existence. DNA is protected in complex chromosomes. Plasmids have complicated and protective membranes. Chromosomes and plasmids are protected from each other. The organism is protected by specific multicellular devices. Species are protected by chromosomal limitations of sex which is now no longer "curable." Bacteria are busy enough just living. The luxury of multicellularity has given rise to "leisure" time and energy. Perhaps the key to the prokaryotic-eukaryotic transition is to be found in the evolution of protection at all levels.

QUESTIONS

1. By means of diagrams, show why an F_2 ratio of three long pollen grains to one short pollen grain in sweet peas is considered abnormal.

2. Why is delayed Mendelian inheritance often considered to be maternal inheritance?

3. How would you determine if a particular trait was caused by a cytoplasmic factor or was an example of delayed Mendelian inheritance?

4. By means of diagrams, carry the following cross to the F_2 generation. Explain each proposed genotype and phenotype.

$$Dd\;♀ \qquad × \qquad dd\;♂$$
$$\text{(dextral)} \qquad\qquad \text{(sinistral)}$$

5. In the water flea, the following results have been reported concerning eye color. Explain.

   a) P   Dark-eyed females × Light-eyed males
      F_1 All dark-eyed throughout life

   b) P   Light-eyed females × Dark-eyed males
      F_1 All light-eyed upon hatching,
            becoming darker at adulthood

6. What is the functional relationship between the *iojap* gene and chloroplasts?

7. Explain how autogamy can produce complete homozygosity in *Paramecium*.

8. Speculate on the evolutionary development and inclusion of kappa particles in *Paramecium*.

9. Why might metagon be considered a link between eukaryotic mRNA and RNA viruses?

10. What new evidence would be required in order to reclassify the virus-like milk factor as an episome?

11. What are the similarities and differences between F' and transducing viruses? between R-factors and transducing viruses?

13

Population Genetics: Measuring Genetic Stability

It has often been said that the evolutionary "action" in biology takes place at the level of the population. This is not to deny the fundamental, underlying importance of molecular, chromosomal, cellular, or organismal events, but rather to stress that the central theme of biology, evolution, operates on groups of organisms, not on individuals or their component parts. If all organisms on earth arose by descent with modification of their genes and phenotypes, such individual modifications can only become meaningful in the context of a group within which there is variation and exchange of genetic material via the usual methods of meiosis and fertilization, or via other methods such as transduction and transformation.

If population genetics is to concern itself with groups, then the study of an individual gene would be futile. In studies of populations, the individual provides a transitory "package" that is important only in the relative ease with which that package can be opened and the genetic gift passed on to a new generation. What does become important is the *gene pool* of the population. Let us first define a *population* as a group of organisms (of the same species) that exhibits a free flow of genetic material, that is, within the limits of sexuality there is random interbreeding. The gene pool consists of all the genes, in all their allelic forms, in the reproductive gametes of a population. The gene pool is really the genetic endowment of one generation to the next, and because of this, only the gametes that accomplish something are important.

If the gene pool of a population is constant over many generations, the population itself is stable and considered to be not evolving. Changes in the gene pool will reflect changes in the population. The more rapid or extreme the gene pool shifts, the faster the population undergoes change. Obviously one cannot study entire gene pools. The amount of work would be prohibitive, even if one possessed the tools to discriminate between genes and identify all of them. What is usually done is to select a gene that is phenotypically measurable or discernable—namely, a gene marker.

Once one grasps the concept of gene pools, one is ready for the concept of *gene frequency* (really allelic frequency). In any diploid individual, the frequency of a given allele is either 100 percent, 50 percent, or zero. But in a population, the frequency of a given allele is a function of the percent of individuals in the population who have that gene, and so its frequency may have any value from zero to 100 percent. A stable gene pool is one in which gene frequencies, whatever their value, remain constant from one generation to another.

Study and analysis of gene frequencies and changes in them have both theoretical and practical value. They provide insight into mechanisms of evolution and, as a matter of fact, they may often reveal evolutionary changes in progress. They can reveal both where a population came from in terms of taxonomic relationships, and where it is probably going in terms of adaptive change. Observable changes in the frequencies of genes that entail physical handicaps of one kind or another have proved useful to national agencies that are responsible for anticipating future needs of society in terms of health care, educational programs, and the like.

A well-documented example of the theoretical usefulness of studies of gene frequencies is that concerning the A–B–O blood grouping in man. With some notable exceptions, such as the Basques in the high Pyrenees between France and Spain, the data that these studies provide has proved to be a valuable adjunct to the information gained from anthropological study and comparative linguistics in unraveling the mysteries of human migrations in prehistoric times.

In sharp contrast to this are studies that will enable specialists to know how much of our resources must be set aside in order to treat victims of such genetic diseases as phenylketonuria, Tay-Sachs syndrome, or diabetes mellitus, to name a few.

In a very different way, the tools and analyses of population geneticists have become important weapons in ideological wars against those who would limit some kinds of human freedom. One can say, for instance, that there is no such entity as a genetically pure race of men and, therefore, any political philosophy that is based on the pure-race concept stands precariously on the edge of a pit of biological quicksand. One can also say that common political programs of eugenics must fail because the majority of deleterious genes are recessive and, in equilibrium, they will be carried and masked in heterozygotes for more generations than any known political regime has been able to last.

Finally, population genetics is important because it puts in clear focus an important issue that most people do not often think about: the individual versus the group (society). Should symptomatic treatment for hemophilia be automatically given? To do so is humane to the individual but contrary to the long-term good of society, because genes that normally would be weeded out or kept under control by natural selection are permitted to remain and perhaps increase in frequency. Certainly this aspect of genetics does not produce answers for such moral issues—it merely points out that the issues exist and that there are forms of action that can be taken. In these overpopulated times, decisions on such matters may profoundly influence the future of the human species.

MODEL SYSTEM FOR POPULATION STABILITY

In order to obtain short-term practical information and/or long-term evolutionary data about interactions within the gene pool of a population, one must be able to describe a model of genetic stability.

Shortly after the rediscovery of Mendel's laws in 1900, most geneticists assumed that a population would reach stability when dominant phenotypes were three times as numerous as their recessive counterparts. This seemed like a logical assumption based on the 3:1 monohybrid ratio obtained by Mendel. By 1908, however, this idea was abandoned by the independent analyses of G. Hardy and W. Weinberg. These men demonstrated that in a stable population, the frequency of dominant and recessive alleles remains constant at whatever value applied to the previous generation, that is, present and future gene frequencies are identical to past frequencies. In stable situations, gene frequencies are not influenced by dominance or recessiveness, nor does reproduction *per se* alter the variability in a population.

It is important to differentiate between gene (or allele) frequency and genotype frequency. Assume an artificial population made up of 100 individuals, of which 20 are homozygous for the *a* allele, 20 are heterozygous, and 60 are homozygous for the *A* allele. The following frequencies are listed:

Number of individuals	Genotype	Percent	Genotype frequency
60	A/A	60	0.6
20	A/a	20	0.2
20	a/a	20	0.2
100		100	1.0

From these data, the gene frequencies are calculated:

Genotype	Number of individuals	Numbers of each kind of allele in the genotype	Percent of total number of genes	Gene frequency
A/A	60 ———— 120A			
A/a	20 < ⟨ 20A ⟩	$\frac{140}{200}$ = 70		A = 0.7
	20a			
a/a	20 ———— 40a ⟩	$\frac{60}{200}$ = 30		a = 0.3
	100	200	100	1.0

The Hardy-Weinberg principle predicts that the gene frequencies of 0.7 for *A* and 0.3 for *a* will be maintained in future generations if the population is stable. Moreover, the genotype frequencies will change if the population begins mating at random, until these frequencies reach an equilibrium that will remain invariable as long as the population continues to be stable with random mating. The equilibrium values can be predicted by the expansion of the binomial $(p + q)^2 = p^2 + 2pq + q^2$, where p is the frequency of *A* and q is the frequency of *a*.

Genotype	Present genotype frequency	Genotype frequency after one generation of random mating	General formula
A/A	0.6	0.49	p^2
A/a	0.2	0.42	$2pq$
a/a	0.2	0.09	q^2
	1.0	1.00	

There are two further statements to be made about the Hardy-Weinberg equilibrium. If *A* and *a* are the only two alleles for this gene, then

$$p + q = 1,$$

or 100 percent of the alleles have been accounted for in the population. Secondly, if p^2, $2pq$, and q^2 are the frequencies of all the possible genotypes (with respect to this gene) in the population, then their total is 100 percent of the genotypes. Therefore, there are two conditions which must be met if the Hardy-Weinberg equilibrium is to apply:

1. $p + q = 1$
2. $(p + q)^2 = p^2 + 2pq + q^2 = 1$

If these conditions are not met, as in the case of multiple alleles, then modifications must be made in the model in order to study the genes in question.

The following points summarize the theoretical data of Hardy and Weinberg.

1.   Gene frequencies are stable in stable populations.

2.   After one generation of random mating, and continuing for as long as such matings last, genotype frequencies are predicted by the general statement:

p^2	$+$	$2pq$	$+$	q^2
(frequency of one homozygote)		(frequency of heterozygotes)		(frequency of the other homozygote)

where p is the frequency of one allele (the dominant one if dominance exists), and q is the frequency of the other allele (the recessive one).

How are these generalizations obtained? To illustrate, out of a hypothetical colony of 1000 people, 700 individuals are of blood group M, 200 are of blood group MN, and 100 are of blood group N.

Pheno-type	Geno-type	Number of individuals	Genotype frequency	Gene frequency L^M	Gene frequency L^N
M	L^ML^M	700	0.7	$\frac{1400}{2000} = 0.7$	—
MN	L^ML^N	200	0.2	$\frac{200}{2000} = 0.1$	$\frac{200}{2000} = 0.1$
N	L^NL^N	100	$\frac{0.1}{1.0}$	—	$\frac{200}{2000} = 0.1$
		1000		p = 0.8	q = 0.2

If the frequency of L^M is 0.8, then 80 percent of all gametes produced by this population will carry the L^M allele. Similarly, 20 percent of all gametes will carry the L^N allele. Now, suppose that among the 1000 colonizers, mating is random with respect to this characteristic, and that no benefit or detriment ensues from having one or the other of these alleles. In such a situation, the chance that a gamete carrying a particular allele will participate in zygote formation will simply be a function of the frequency of that allele. Put another way, the genotype frequencies of the zygotes will depend on the gene frequencies of the gametes.

Sperm	Eggs	
	$L^M = 0.8$	$L^N = 0.2$
$L^M = 0.8$	$L^ML^M = (0.8)^2$ $= 0.64$	$L^ML^N = (0.8)(0.2)$ $= 0.16$
$L^N = 0.2$	$L^ML^N = (0.8)(0.2)$ $= 0.16$	$L^NL^N = (0.2)^2$ $= 0.04$

Therefore in the next generation, the genotype frequencies are as follows:

$$L^ML^M = 0.64$$
$$L^ML^N = 0.16 + 0.16 = 0.32$$
$$L^NL^N = 0.04$$

It is obvious that after one generation of random mating, the genotype frequencies have changed.

Genotype	Frequency in original population	Frequency after random mating
L^ML^M	0.7	0.64
L^ML^N	0.2	0.32
L^NL^N	0.1	0.04

Have the gene frequencies changed? Suppose that the new generation, the generation produced by random mating, consists of 1500 individuals.

Genotype	Frequency × 1500	Actual number of individuals
L^ML^M	0.64 × 1500	= 960
L^ML^N	0.32 × 1500	= 480
L^NL^N	0.04 × 1500	= 60

$$p = \text{frequency of } L^M = \frac{2(960) + 480}{3000} = \frac{2400}{3000} = 0.8$$

$$q = \text{frequency of } L^N = \frac{2(60) + 480}{3000} = \frac{600}{3000} = 0.2$$

What will happen in the next generation if random mating continues? According to the Hardy-Weinberg equilibrium, the gene frequencies and the genotype frequencies should remain constant. Whether this prediction is accurate can be seen by an analysis of what random mating really means. There are nine kinds of mating that can occur based on the genotypes of males and females:

Mating type	Females	×	Males
a	L^ML^M		L^ML^M
b	L^ML^M		L^ML^N
c	L^ML^M		L^NL^N
d	L^ML^N		L^ML^M
e	L^ML^N		L^ML^N
f	L^ML^N		L^NL^N
g	L^NL^N		L^ML^M
h	L^NL^N		L^ML^N
i	L^NL^N		L^NL^N

Because this trait is not sex-linked, six mating permutations are obtained by grouping together some of these categories. The frequencies of these permutations are based on the frequencies of the parental genotypes.

Mating type	Genotype of parents	Frequency of genotypes	Frequency of mating
a	$L^ML^M \times L^ML^M$	$(p^2)(p^2)$ = (0.64)(0.64)	0.4096
b d	$L^ML^M \times L^ML^N$	$2(p^2)(pq)$ = 2(0.64)(0.32)	0.4096
c g	$L^ML^M \times L^NL^N$	$2(p^2)(q^2)$ = 2(0.64)(0.04)	0.0512
e	$L^ML^N \times L^ML^N$	$(pq)(pq)$ = (0.32)(0.32)	0.1024
f h	$L^ML^N \times L^NL^N$	$2(pq)(q^2)$ = 2(0.32)(0.04)	0.0256
i	$L^NL^N \times L^NL^N$	$(q^2)(q^2)$ = (0.04)(0.04)	0.0016
			1.0000

Each of these mating permutations will produce offspring in Mendelian proportions of the genotypes L^ML^M, L^ML^N, and L^NL^N (that is, in the $L^NL^N \times L^ML^N$ mating, one-half of the offspring will have the L^NL^N genotype and the other half will carry the

L^ML^N genotype). Therefore, it is possible to partition the mating frequencies among the kinds of offspring expected for each kind of cross.

Genotypes of parents	Frequency of mating type	Ratio of offspring
$L^ML^M \times L^ML^M$	0.4096	All L^ML^M
$L^ML^M \times L^ML^N$	0.4096	$1/2 L^ML^M : 1/2 L^ML^N$
$L^ML^M \times L^NL^N$	0.0512	All L^ML^N
$L^ML^N \times L^ML^N$	0.1024	$1/4 L^ML^M : 1/2 L^ML^N : 1/4 L^NL^N$
$L^ML^N \times L^NL^N$	0.0256	$1/2 L^ML^N : 1/2 L^NL^N$
$L^NL^N \times L^NL^N$	0.0016	All L^NL^N

Genotypes of parents	Frequency of offspring		
	L^ML^M	L^ML^N	L^NL^N
$L^ML^M \times L^ML^M$	0.4096		
$L^ML^M \times L^ML^N$	0.2048	0.2048	
$L^ML^M \times L^NL^N$		0.0512	
$L^ML^N \times L^ML^N$	0.0256	0.0512	0.0256
$L^ML^N \times L^NL^N$		0.0128	0.0128
$L^NL^N \times L^NL^N$			0.0016
	0.6400	0.3200	0.0400

Therefore, after the second generation of random mating, the genotypic frequencies remain $L^ML^M = 0.64$, $L^ML^N = 0.32$ and $L^NL^N = 0.04$. Similarly, the gene frequencies are stable at p = 0.8 and q = 0.2.

Alternatively, the same result could have been obtained by noting that the gene frequencies of generation 1 (p = 0.8; q = 0.2) would be accurately reflected as gamete frequencies in both males and females.

Sperm	Eggs	
	$L^M = p = 0.8$	$L^N = q = 0.2$
$L^M = p = 0.8$	$L^ML^M = p^2 = 0.64$	$L^ML^N = pq = 0.16$
$L^N = q = 0.2$	$L^ML^N = pq = 0.16$	$L^NL^N = q^2 = 0.04$

Obviously, the total number of offspring equals 100 percent, or a frequency of 1. Therefore:

$$p^2 + 2pq + q^2 = 1.$$

The preceding was a genetic example that was simplified by codominance. All three genotypic frequencies can be determined directly, and since all possible information was available, the calculated gene frequencies are sure to be accurate. In situations where two alleles show complete dominance and recessiveness, it is impossible to observe directly all genotypic frequencies. The only certain genotypic frequency is q^2, the frequency of the recessive homozygotes in the population. From q^2, q can be estimated, and because $p + q = 1$, p can be derived, and from that p^2 and $2pq$ can be calculated.

Now consider the "taster" gene to see how this type of problem can be worked out. This is a gene that determines whether or not an individual can taste the chemical substance phenylthiocarbamide (PTC). The allele for tasting (T) is dominant to that for the inability to taste PTC (t). In a population consisting of 2200 individuals, 1606 were tasters while 594 were nontasters. According to the Hardy-Weinberg equilibrium $p^2 + 2pq + q^2$, where q^2 is the frequency of nontasters (tt), the following calculations can be made:

$$q^2 = \frac{594}{2200} = 0.27$$
$$q = \sqrt{0.27} = 0.52$$
$$p + q = 1$$
$$p = 1 - q = 1 - 0.52 = 0.48$$
$$p^2 = (0.48)^2 = 0.23$$
$$2(pq) = 2(0.48)(0.52) = 0.50$$

Therefore, of 1606 tasters, $0.23\,(2200) = 506$ were of genotype TT, and $0.50\,(2200) = 1100$ were of genotype Tt.

In considering frequencies of X-linked alleles, A and a for example, five genotypes occur:

AA	Aa	aa
AY	aY	

It is relatively easy to estimate p and q in the hemizygous sex, because genotype frequency equals gene frequency. If the frequency of gene A is p, the genotypic frequency of AY males is also p because no allelic associations can occur.

Genotype	Genotype frequency
AA	p^2
Aa	$2pq$ $\quad p^2 + 2pq + q^2 = 1$
aa	q^2
AY	p
aY	q $\quad p + q = 1$

In an equilibrium population, then, the frequency of homozygous recessive females should be the square of the frequency of recessive males—if 0.06 is the frequency of red-green color-blind males in the population ($q = 0.06$), then 0.0036 should be the frequency of red-green color-blind females ($q^2 = 0.0036$). This statistical relationship illustrates the familiar consequence of X-linked traits caused by recessive alleles, namely, many more males are affected than females.

The same type of mathematical logic can be applied to a situation in which there are more than two forms of a gene in the population. Much work has been done on the A–B–O blood group system in various human populations. Because three alleles (I^A, I^B, and i) contribute to this system, a trinomial rather than a binomial system is necessary.

$$p = \text{frequency of } I^A$$
$$q = \text{frequency of } I^B$$
$$r = \text{frequency of } i$$

The same kind of statements can be made about this allelic system.

1. $p + q + r = 1$ 2. $(p + q + r)^2 = 1$

$$\begin{array}{r} p + q + r \\ \times\ p + q + r \end{array}$$

$(p + q + r)r = \quad\quad\quad pr\quad\ + \ qr + r^2$
$(p + q + r)q = \quad\quad pq\quad + q^2 + \ qr$
$(p + q + r)p = p^2 + \ pq + \ pr$

Frequency of genotypes $\{\ p^2 + 2pq + 2pr + q^2 + 2qr + r^2$
$\quad I^AI^A\ \ I^AI^B\ \ I^Ai\ \ I^BI^B\ \ I^Bi\ \ ii$

Frequency of phenotypes $\{\ p^2 + 2pr\quad 2pq\quad q^2 + 2qr\quad r^2$
$\quad\quad A\quad\quad\quad AB\quad\quad B\quad\quad O$

Assume that a hypothetical population of 100 individuals exhibits the following distribution:

Phenotype	Number	Phenotype frequency
A	45	0.45
B	13	0.13
AB	6	0.06
O	36	0.36
	100	1.00

These data can be analyzed thus:

1. Blood group O = genotype ii = frequency of r^2

$$r^2 = 0.36$$
$$r = \sqrt{0.36} = 0.6$$

2. Directly, p or q cannot be calculated, but they can be calculated indirectly by noting that $p + q + r = 1$. Therefore $p = 1 - (q + r)$, and $q = 1 - (p + r)$.

There is sufficient information to calculate the combined frequencies of $(q + r)$ and $(p + r)$. First solve for p, and having done so, subtract $(p + r)$ from 1 to get q:

Solving for p:

$$p = 1 - (q + r)$$
$$q + r = \sqrt{(q + r)^2} = \sqrt{q^2 + 2qr + r^2}.$$

But $q^2 + 2qr$ = frequency of phenotype B, and
r^2 = frequency of phenotype O.

Therefore,

$$q + r = \sqrt{\text{frequency of B} + \text{frequency of O}}, \text{ or}$$
$$q + r = \sqrt{0.13 + 0.36} = \sqrt{0.49} = 0.7.$$

Therefore, $p = 1 - 0.7 = 0.3$.

Solving for q:

$$q = 1 - (p + r)$$
$$p + r = \sqrt{(p + r)^2} = \sqrt{p^2 + 2pr + r^2}.$$

But $p^2 + 2pr$ = frequency of phenotype A, and
r^2 = frequency of phenotype O.

Therefore,

$$p + r = \sqrt{\text{frequency of A} + \text{frequency of O}}, \text{ or}$$
$$p + r = \sqrt{0.45 + 0.36} = \sqrt{0.81} = 0.9.$$

Therefore, $q = 1 - 0.9 = 0.1$.

Finally, $p = 0.3$, $q = 0.1$, and $r = 0.6$.

Pheno-type	Geno-type	Genotype frequency	Number of people	
			Calculated	Observed
A	$I^A I^A$ $I^A i$	$p^2 = 0.09$ $2pr = 0.36$	$0.45(100) = 45$	45
B	$I^B I^B$ $I^B i$	$q^2 = 0.01$ $2qr = 0.12$	$0.13(100) = 13$	13
AB	$I^A I^B$	$2pq = 0.06$	$0.06(100) = 6$	6
O	ii	$r^2 = 0.36$	$0.36(100) = 36$	36

Without a doubt, the calculations fit the observed data. It is well that they did, because the problem was mathematically rigged to eliminate the morass of significant figures and interminable square roots in order to illustrate the principles involved. In all fairness, the same method should be tested in a factual situation. Blood grouping of 1000 Englishmen provided the following distribution:

Phenotype	Number	Phenotype frequency
A	417	0.417
B	86	0.086
AB	30	0.030
O	467	0.467
	1000	1.000

To estimate r:

$$r^2 = \text{frequency of O} = 0.467$$
$$r = \sqrt{0.467} = 0.683.$$

To estimate p:

$$p = 1 - (q + r)$$
$$p = 1 - \sqrt{(q + r)^2} = 1 - \sqrt{q^2 + 2qr + r^2}$$
$$p = 1 - \sqrt{\text{frequency of B} + \text{frequency of O}}$$
$$p = 1 - \sqrt{0.086 + 0.467}$$
$$p = 1 - \sqrt{0.553}$$
$$p = 1 - 0.744$$
$$p = 0.256.$$

To obtain q, it would be possible to simply add together p and r and subtract from 1. However, because errors have been introduced by estimating square roots, to obtain q in this manner would only compound the error. Hence, q should be calculated as p was:

To estimate q:

$$q = 1 - (p + r)$$
$$q = 1 - \sqrt{(p + r)^2} = 1 - \sqrt{p^2 + 2pr + r^2}$$
$$q = 1 - \sqrt{\text{frequency of A} + \text{frequency of O}}$$
$$q = 1 - \sqrt{0.417 + 0.467}$$
$$q = 1 - \sqrt{0.884}$$
$$q = 1 - 0.940$$
$$q = 0.060$$

Finally,

$$p = 0.256$$
$$q = 0.060$$
$$r = 0.683$$
$$\overline{0.999}$$

It was stated that $p + q + r = 1$, and yet they do not. Why? These values of p, q, and r are estimates, close to be sure, but not exact because the square roots were not exact. A correction formula (Bernstein, 1930) that has proved accurate in the past can be used to give more exact values. The difference (d) between 1 and 0.999 is 0.001. This difference is apportioned between p, q and r in the following manner:

Esti-mates	Correction factor	Rounded off to significant values
p 0.256	$p(1 + \frac{1}{2} d) = 0.2561280$	0.256
q 0.060	$q(1 + \frac{1}{2} d) = 0.060030$	0.060
r 0.683	$(r + \frac{1}{2} d)(1 + \frac{1}{2} d) = 0.683842$	0.684
		1.000

At first, it seems odd to note that in this case the correction affects only r. After all, r was obtained in the easiest possible way—from the frequency of blood group O. However, this estimate of r was probably more inexact than that for p and q, because many i genes were concealed in two kinds of heterozygotes ($I^A i$ and $I^B i$), whereas the I^A and I^B genes (p and q, respectively) are "concealed" in only one class of heterozygote each.

CHECKING ASSUMPTIONS

How do these calculations compare with the observed facts (see table below)?

Referring to Table 3–1 (p. 45), one finds that with one degree of freedom, a value of $x^2 = 0.0235$ corresponds to a probability between 0.90 and 0.95, which means that one can accept the estimates of gene frequencies for p, q, and r in the population studied.

Two points should be considered about the chi-square test. Why only one degree of freedom, and why a x^2 value at all? In testing goodness of fit for gene frequency calculations, usually only one degree of freedom is allowed, because once the frequency for one allele in a set is established and the total number of individuals specified, all other data can be calculated from this. This, in fact, was done. The general rule for this use of the x^2 is: Number of phenotypes minus number of alleles equals degrees of freedom. In this example, four blood groups were constructed from three alleles.

	A	B	AB	O	Total
Observed (o)	417	86	30	467	1000
Expected (e)	$(p^2 + 2pr)1000$	$(q^2 + 2pr)1000$	$(2pq)1000$	$(r^2)1000$	1000
	= 415.74	= 85.68	= 30.72	= 467.86	
Difference (d)	+1.26	+0.32	−0.72	−0.86	0
d^2	1.5876	0.1024	0.5184	0.7396	
d^2/e	0.0038	0.0012	0.0169	0.0016	

$x^2 = \Sigma\, d^2/e = 0.0038 + 0.0012 + 0.0169 + 0.0016 = 0.0235$

The use of the x^2 seems at first glance to be the epitome of an exercise in circular reasoning. After all, phenotypic frequencies were measured, gene frequencies were calculated from them, and then these calculations were used to arrive back at phenotypic frequencies again. Why aren't the numbers the same at the beginning and at the end? First of all, as indicated, mathematical errors are introduced when rounding off square roots and other calculations. These errors can increase exponentially. Except when each genotype is phenotypically distinct (as in the M–N blood group), these calculations are, at best, estimates. Therefore, the x^2 can be used to see if these estimates are acceptable. Second, by using the Hardy-Weinberg law, one assumes some facts. If the x^2 test of the data is acceptable, the original assumptions are valid. If the data are unacceptable, it is a clue that one or more of the assumptions is or are unjustified. For example, no sampling errors were assumed, and the sample was used as if gene frequencies were uniform throughout the population. It was also assumed that random mating is taking place with respect to the alleles in the population being studied. As seen, nonrandom mating will not affect gene frequencies, but will have a decided influence on gene *associations* —that is, genotype and phenotype frequencies. It was further assumed that mutation from one allele to another is not occurring, that natural selection is not taking place, and

that there is no significant immigration or emigration. In short, it was assumed that the population is a stable, ideal one. Use of the x^2 test helps to indicate if these assumptions are warranted.

Just as the assumptions of stability and random mating were tested when dealing with multiple alleles, so too must these assumptions be tested in other instances.

Testing assumptions with the M–N blood group, discussed earlier, is easy. If the observed p^2, $2pq$, and q^2 frequencies fit a x^2 test for expected values, that would be check enough. But what about cases of true dominance such as the taster gene? In this situation, q is only estimated—it cannot be pinpointed because the $2pq$ class is not measurable. When q is obtained only from data on homozygous recessive individuals, one can only infer p^2 and $2pq$. The inference *assumes* that stability and random mating exist. In such a situation, how can this assumption be tested? To do a x^2 test on observed and expected frequencies of p^2, $2pq$, and q^2 is not possible, because one cannot observe two of the three classes. To do a x^2 test on the measurable classes ($p^2 + 2pq$) and (q^2) would not prove that random mating is occurring, because one still could not obtain the distribution between p^2 and $2pq$.

Snyder (1934) suggested the use of two parameters or equilibrium ratios, neither of which depends on recognizing heterozygotes. Assuming

Frequencies for ♂ parent	Frequencies for ♀ parent		
	AA p^2	Aa $2pq$	aa q^2
AA p^2	1. $AA \times AA$ $p^2 \times p^2 = p^4$	2. $Aa \times AA$ $2pq \times p^2 = 2p^3q$	3. $aa \times AA$ $q^2 \times p^2 = p^2q^2$
Aa $2pq$	4. $AA \times Aa$ $p^2 \times 2pq = 2p^3q$	5. $Aa \times Aa$ $2pq \times 2pq = 4p^2q^2$	6. $aa \times Aa$ $q^2 \times 2pq = 2pq^3$
aa q^2	7. $AA \times aa$ $p^2 \times q^2 = p^2q^2$	8. $Aa \times aa$ $2pq \times q^2 = 2pq^3$	9. $aa \times aa$ $q^2 \times q^2 = q^4$

random mating and stability exist, expected values can be calculated, and these can be compared to observations. The first of these ratios concerns recessive offspring produced by two dominant parents. Obviously, both parents must be heterozygous in such a case, but in the observations, one need note only dominant-to-dominant matings and the recessive offspring therefrom. The second ratio concerns recessive offspring of dominant-to-recessive matings. One can arrive at expected values by consulting the table on page 327 which enumerates parental and zygotic frequencies in all equilibrium combinations.

The first of the theoretical equilibrium ratios to obtain is the frequency of recessive offspring produced by two dominant parents compared to all offspring of dominant matings. It is obvious that recessives can be produced only if both parents are heterozygous.

Frequency of heterozygote $\times$ heterozygote matings = $4p^2q^2$ (Box 5). Of these, one-quarter will be recessive = $^1/_4(4p^2q^2) = p^2q^2$.

Frequency of all offspring of two dominant parents = p^4 (Box 1) + $2p^3q$ (Box 2) + $2p^3q$ (Box 4) + $4p^2q^2$ (Box 5) = $p^4 + 4p^3q + 4p^2q^2$.

$$\frac{p^2q^2}{p^4 + 4p^3q + 4p^2q^2} = \frac{p^2q^2}{p^2(p^2 + 4pq + 4q^2)} =$$

$$\frac{q^2}{p^2 + 4pq + 4q^2} = \frac{q^2}{(p + 2q)^2} = \frac{q^2}{(p + q + q)^2} =$$

$$\frac{q^2}{[(p + q) + q]^2} = \frac{q^2}{(1 + q)^2}$$

The second equilibrium ratio compares the frequency of recessive offspring produced by one dominant and one recessive parent compared to all offspring of dominant-to-recessive matings. Again, to produce a recessive, the dominant parent must be a heterozygote.

Frequency of heterozygote-to-recessive matings = $2pq^3$ (Box 8) + $2pq^3$ (Box 6) = $4pq^3$. Of these, half will be recessive = $^1/_2(4pq^3) = 2pq^3$.

Frequency of all offspring of dominant-to-recessive matings = p^2q^2 (Box 7) + $2pq^3$ (Box 8) + p^2q^2 (Box 3) + $2pq^3$ (Box 6) = $2p^2q^2 + 4pq^3$.

$$\frac{2pq^3}{2p^2q^2 + 4pq^3} = \frac{pq^3}{p^2q^2 + 2pq^3} = \frac{q^3}{pq^2 + 2q^3} =$$

$$\frac{q}{p + 2q} = \frac{q}{p + q + q} = \frac{q}{(p + q) + q} = \frac{q}{1 + q}$$

Now there are two theoretical ratios,

$$\frac{q^2}{(1 + q)^2} \quad \text{and} \quad \frac{q}{1 + q},$$

neither of which depends on recognition of heterozygotes, that can be compared with actual measurements. Snyder (1934) did just this with a population of 800 married couples to see if this group was at equilibrium with respect to the taster gene. The number of nontasters was 461 (q^2), which represents a frequency of 0.288. The value of q was calculated from q^2:

Number of homozygous recessives = 461

Size of sample = 1600

$$q^2 = \frac{461}{1600} = 0.2881$$

$$q = \sqrt{0.2881} = 0.5368$$

Parental matings	Number of couples	Offspring		
		Taster	Nontaster	Total
Taster $\times$ Taster	425	929	130	1059
Taster $\times$ Nontaster	289	483	278	761
Nontaster $\times$ Nontaster	86	Information not needed		
	800			

Offspring from taster-to-taster matings

	Taster	Nontaster
Observed	929	130
Expected	$1059 - \dfrac{q^2}{(1+q)^2} = 930$	$\dfrac{q^2}{(1+q)^2} = 129$
Difference	-1	$+1$
d^2	1	1
$\dfrac{d^2}{e}$	$\dfrac{1}{930} = 0.001$	$\dfrac{1}{129} = 0.008$

$$\sum \frac{d^2}{e} = 0.009$$

Offspring from taster-to-nontaster matings

	Taster	Nontaster
Observed	483	278
Expected	$761 - \dfrac{q}{1+q} = 495$	$\dfrac{q}{1+q} = 266$
Difference	-12	$+12$
d^2	144	144
$\dfrac{d^2}{e}$	$\dfrac{144}{495} = 0.290$	$\dfrac{144}{266} = 0.541$

$$\sum \frac{d^2}{e} = 0.831$$

At one degree of freedom, even without the Yates correction (cf. Chapter 3), each of these values is acceptable alone and also acceptable when summed. This population can be considered to be at equilibrium with random mating taking place. The estimates of q (0.5368) and p (0.4632) can be taken as valid as well as the assumption that the genotypes distribute as follows:

$q = 0.5368$	$q^2 = 0.2881 = 461$ people; homozygous tt
$p = 0.4632$	$2pq = 0.4973 = 796$ people; heterozygous Tt
	$p^2 = 0.2146 = 343$ people: homozygous TT

The situation with X-linked genes can be more complicated in some circumstances. If the values of p and q can be separately and accurately obtained for males and females, and if these agree, then a chi-square value on the distribution of genotypes will suffice. If the values of p and q differ between the sexes, one must determine if the difference is significant. These statistical methods are not discussed here—interested students can consult a volume on statistical genetics. The point remains, however, that a natural population cannot justifiably be assumed to be at equilibrium if observed data can yield only estimates of gene and genotypic frequencies.

QUESTIONS

1. Why don't dominant alleles increase in frequency to 100 percent in a stable population?

2. Once the frequency of alleles in a population changes, what events must then occur in order to regain an equilibrium state? Will the new equilibrium be identical to the old one?

3. In a small population, 30 percent is of blood group M, 40 percent is of blood group MN, and 30 percent is of blood group N. What are the gene frequencies and the genotype frequencies?

4. In a group of 812 individuals, 91 percent are tasters. What are the gene frequencies and genotype frequencies in this population? If all the tasters married nontasters, what proportion of them would have some nontaster offspring?

5. Among a group of people, the following distribution of blood types was noted:

A	212
B	60
AB	24
O	236

Calculate the gene frequencies of the alleles involved.

6. In a small population, the frequencies of the blood-group alleles were $I^A = 12$ percent; $I^B = 2$ percent;

i = 86 percent. Calculate the expected distribution (in percent) of each of the A–B–O phenotypes.

7. Assume only for the purposes of this problem that the Rh blood types are controlled by a single pair of alleles. Further, assume that Rh-negative individuals (genotype rr) comprise 15 percent of a stable population numbering 1000 individuals. Calculate the frequency of the alleles R and r, and the expected distribution of genotypes.

8. Among North American whites, about 8 percent of the males are red-green color-blind. What proportion of the females would be similarly afflicted?

9. In a large, randomly mating population, 80 percent of the individuals have round heads and 20 percent have egg-shaped heads. Assuming that head shape is controlled by one pair of alleles, is the round-head allele necessarily dominant? Explain.

10. With respect to genotype frequency of the taster trait, is the following population in equilibrium? Explain.

Matings	Number of couples	Offspring Taster	Offspring Nontaster
Taster × Taster	400	694	106
Taster × Nontaster	250	260	240
Nontaster × Nontaster	50	3	93

11. In a population of 360 individuals, 304 are of blood type M, 52 of blood type MN, and 4 of blood type N. Calculate the expected genotypic frequencies. Is this population at equilibrium?

12. Are the following groups in genotypic equilibrium? Explain.

Group	AA	Aa	aa
a	0.72	0.20	0.08
b	0.12	0.80	0.08
c	0.08	0.01	0.91
d	0.25	0.25	0.25

14

Changes in Populations

> The expression of the ideas of Darwin in precise language is a taxing domain of applied mathematics whose intrinsic difficulty is the inherent complexity of the phenomena it must describe.*

The Hardy-Weinberg equilibrium is neither a theoretical nor a pragmatic end in itself. It is nonmagnetic in the sense that once gene frequencies have been disturbed, there is no implicit tendency for them to return to their former equilibrium values. The Hardy-Weinberg principle really offers a means of testing assumptions about genetic stability in order to detect and measure genetic change. A skeptical student once remarked that the calculations amount to a simple statement of the obvious: When no change occurs, no change is evident! True as the sentiment underlying this remark may be, the point remains that evolution is the *sine qua non* of biology, and if by studying genetic stability more can be learned about genetic change, the better will be our understanding of the mechanisms of evolution.

Because modern, theoretical population genetics is based in large measure upon highly complex mathematical models that may or may not be entirely valid, only the barest elements of population change will be considered in this chapter. Whenever feasible, examples will be with humans, although much of the important work has been done with *Drosophila*. Students interested in delving more deeply into the subject are referred to the works on population genetics cited in the bibliography.

Genetic variability is grist for the mill of genetics. If all people had brown eyes, eye color would be difficult to analyze as an inherited trait. In fact,

---

*P. Morrison (1970), "Population genetics, particularly the human kind, and other matters," *Sci. Amer.* **223:** 126–128.

every example of inheritance cited in this book illustrates the central role of mutation. Noting the ways in which DNA may change or the mode of inheritance of newly formed alleles does not directly contribute much to understanding evolution, however. Rather, mutations must be studied in the context of the populations that carry them. The fate of these mutations in time depends upon how the population treats the phenotypes that correspond to them.

There are two basic forces to which new genes are subjected. The first is the force of selection; the second is the force of genetic drift. Both of these phenomena involve differential reproduction of alternative genotypes, but the causal factors for selection and drift differ. Selection is a consequence of reproductive fitness, that is, the capacity to reproduce more offspring, for whatever reason, than another member within the population. A new allele may be selected because it contributes directly to the fitness of the individual within a particular environment; or it may be selected indirectly as a part of a genome that has other selectively advantageous genes. "Free rides" of this kind are limited by the shuffling processes of sexual reproduction and recombination, however, so that at some point the opportunity will arise to apply differential selection to each of the components of the gene pool. Because of the requirement of fitness, selection imposes a temporary direction upon evolution. It is only temporary because the selective factors themselves are subject to change.

Genetic drift is random in direction from the very start. It has been compared to a bottleneck through which only a part of the whole can pass. The causes of the bottleneck are not necessarily genetic and, as implied, involve either small populations or populations undergoing a radical reduction in members. For example, as revealed by the studies of Shine and Gold (1970), the isolated islanders of St. Helena have a much greater proportion of harmful recessive genes than their ancestral populations in England, Africa, and China. In-

breeding alone cannot account for the frequencies of stillbirths, albinism, or deaf-mutism. The explanation for this phenomenon is genetic drift. The founders of St. Helena must have carried with them recessive genes that were rare in populous England, China, or regions of Africa. By chance the founders were not representative of their homelands and, because the population numbered in the few hundreds, whatever mutant genes they had were no longer rare. Their presence was promptly revealed as a consequence of a greatly increased probability of homozygosis. Genetic drift here, as elsewhere, resulted from atypical "sampling" introduced into a new population. Moreover, the effective reproductive population is often small in such situations, even though one may not be aware of it. Limited mobility, discontinuous environmental ranges, seasonal fluctuations in number, and social organization are all examples of factors that can so fragment a population as to bring its effective size down to the limits required for genetic drift.

All of these forces interact in nature. Having been changed by genetic drift, a new, small population must offer its phenotypes to winnowing by natural selection and will, in time, present new mutations for selection as well.

MUTATION

A mutation must be recurrent in order to have any effect upon a population. Single mutations, even if they are selectively advantageous, are generally lost by a process that can be called a multiplication of "ifs." *If* a mutation occurs, *if* it is in the germ cell, *if* it does not affect the viability of the germ cell, and *if* that germ cell is used for fertilization—these are some of the "ifs." The resulting zygote will be heterozygous for the mutation, because its other parent most probably did not have it. *If* the zygote matures, it likely will mate with an individual lacking the mutation. *If* any offspring result, there is a 50 percent chance that each of them will not inherit the new allele, although the chance of conserving the mutation increases with increasing numbers of offspring.

The situation with recurrent mutations is quite different. It is analogous to a bag of red marbles, from which one-by-one a marble is removed and replaced by a green one. Given enough time, the bag will be full of green marbles. The population bag for a particular gene will complete the changeover from 100 percent of one allele to 100 percent of its alternative at a rate consistent with the mutation rate of that allele (Table 14–1).

Table 14–1. Recurrent, one-way mutation of A to a.

Let p_0 = initial frequency of A
   u = mutation rate of A to a.

After one generation, up_0 of the A alleles will have mutated. Hence, the new frequency (p_1) of A will be:

$$p_1 = p_0 - up_0 = p_0(1 - u).$$

After a second generation, the frequency (p_2) of A will fall to:

$$p_2 = p_1 - up_0,$$

but since $p_1 = p_0(1 - u)$, then

$$\begin{aligned} p_2 &= p_0(1 - u) - up_0(1 - u) \\ &= p_0 - up_0 - up_0 + u^2p_0 \\ &= p_0 - 2up_0 + u^2p_0 \\ &= p_0(1 - 2u + u^2) \\ &= p_0(1 - u)^2 \end{aligned}$$

After n generations, the frequency of A will fall to:

$$p_n = p_0(1 - u)^n$$

and, in time, to zero.

Mutation is not a one-way street—as A mutates to a, back mutation of a to A ($A \rightleftharpoons a$) will also occur, although the back mutation rate may not be equivalent to the "forward" mutation rate ($A \rightleftharpoons a$). The actual *number* of genes mutating depends upon both the rate of mutation and the frequency with which the allele is encountered in the population. Each time one A changes to a, the frequency of A falls and that of a rises. Given constant mutation rates, fewer A's will be available for mutation

and the subsequent total of $A \rightarrow a$ changes will fall. Conversely, as more a's are formed, the total number of back mutations will increase. Eventually an equilibrium will be established at which the number of forward mutations is equal to the number of back mutations (Table 14–2). It is not necessary that the frequencies of the two alleles be identical; it is the product of frequency times mutation rate that matters. For example, if A mutates to a three times as frequently as a mutates to A, then one can see on Table 14–2 that equilibrium will be reached when the frequency of A is 0.25 and the frequency of a is 0.75. The tendency to reach mutational equilibrium is sometimes spoken of as *mutation pressure* because, unlike the Hardy-Weinberg equilibrium, mutational equilibrium is magnetic and exerts a constant pull on gene frequencies.

Table 14–2. Mutational equilibrium.

Let p_0 = initial frequency of A
   p = equilibrium frequency of A
   q_0 = initial frequency of a
   q = equilibrium frequency of a
   u = mutation rate of A to a
   v = mutation rate of a to A.

After one generation, the frequency of A will be:

$$p_1 = p_0 + (vq_0 - up_0).$$

After n generations, the frequency of A would be:

$$p_n = p_{n-1} + (vq_{n-1} - up_{n-1}).$$

Equilibrium will be established when:

$vq - up = 0,$

that is, when $vq = up$.

But since $q = 1 - p$, then

$v(1 - p) = up$
   $v - vp = up$
      $v = up + vp$
      $v = p(u + v)$

$$p = \frac{v}{u + v}$$

EXAMPLE:
If $u = 3v$, then
$p = \dfrac{v}{3v + v}$
$p = \dfrac{v}{4v}$
$p = \dfrac{1}{4} = 0.25$
Since $q = 1 - p$, then
$q = 0.75.$

A population can be subjected to a process that is analogous to recurrent mutation in its effects. Migration of individuals from a different population or emigration of a disproportionate distribution of genes can radically alter the gene pool. If migration is one-way, the genetic composition of the accepting group can be changed. If two groups exchange migrants, divergencies between the groups will be damped and a greater degree of homogeneity will result. Remember, however, that migration is only a second-order change, because it was genetic mutation that caused the populations to diverge in the first place. How closely migration mimics the effects of ordinary mutation in a population will depend upon the genetic dissimilarity between the two groups, the number of migrants, and the size of the accepting population.

Calculation of equilibrium frequencies of alleles reveals the following significant points. Forward mutation rates are usually far greater than reverse rates. Under some circumstances, if the difference is great enough, what can be observed in nature is tantamount to recurrent one-way change. Adaptations that involve loss or gain of organs may be related to mutations of this genre. Loss of eyes in cave-dwelling fish and perhaps even the gill-to-lung transition may have been due to the back mutation rate being swamped by the forward rate. Conversely, not as many mutant *Drosophila* are seen in laboratory populations as would be expected from calculations of mutational equilibria alone. There should be a slow but steady decline in the frequency of wild-type flies. Obviously, mutation pressure toward equilibrium values is not the whole story.

It has been calculated that the frequency of mutation is of the order of one change per 5×10^5 DNA doublings. In many cases, frequencies are even lower. For a newly introduced allele to become established and reach any sort of equilibrium would take thousands of generations at the very least (more than recorded human history). If during that time the environment changes radically, the pressure of selection can also undergo

upheaval. New mutant forms of the gene will undoubtedly arise. There may be widespread immigration or emigration and genetic drift may occur. Because of all these factors, not all genes are believed to have reached their balance point. Mutational equilibria become important, not as predictors of precise future frequencies, but rather as a directional force that helps to maintain the existence of alternative alleles so that variation in the gene pool is continued.

There has been much speculation that with increasing use of radiation and chemicals that are now known to be mutagenic, mutation rates in human beings may be increased. This kind of thinking usually leads to an analysis of the costs of mutation to a population. Cost includes not only monetary outlay that may be needed for medically treating genetic infirmities, but also the social cost of increased variation as it interacts with social stability and the evolutionary implications involved. Presently about 6 percent of all newborn infants have defects that are partly or wholly due to genetic origin, and half of these may be due to recurrent mutation (Neel, 1970).

Crow (1959) has argued that the effects from technological sources of radiation may differ between long-lived species and short-lived species. Because human beings are representatives of a long-lived group, his distinctions become relevant. For most organisms, the other natural causes of mutation far outweigh the effects of radiation. For example, it has been estimated that in mice only 0.1 percent of the mutation rate is ascribable to natural radiation. To raise this figure significantly, the level of radiation would have to be increased more than would seemingly be possible in a peace-time world. In addition, selection appears to play a role in the mutation rate itself. If the rate per unit time for lethal mutations were the same for a man as for a fruit fly, each man would carry about twenty new lethal alleles and the species would shortly be extinct. The mutation rate *per generation* is more nearly alike between most species; this implies adjustments in rate that recognize reproductive life span. However, what adjustments may be made in mutation rates accelerated by radiation are not known. Hence, organisms that are exposed to increased radiation and other mutagenic agents for a long generational period may indeed suffer the genetic effects of such technology.

It has been observed that variability via mutation is the basis of evolution. If the mutation rate in man is raised, does it necessarily follow that the rate of progressive evolution will also be increased? The answer is probably no for a number of reasons. Most mutations are not beneficial in the same environment, nor would they be beneficial in a foreseeably changed one. Haldane (1957) estimated that the numerical value of the mutation rate is equivalent to the reduction in fitness of a population. If the fitness of a group declines relative to other groups, something is bound to happen. History has shown us that the "something" is usually extinction. Of present-day vertebrate families, 98 percent are derived from eight Mesozoic species. Only twenty-four (out of tens of thousands) of Mesozoic vertebrate species have descendants that are alive today. Clearly, extinc-

tion is a sure way to have one's ecological niche won by a better adapted population.

One might ask, could there not be controlled selection to weed out the potential increase of harmful genetic variables? In point of fact, most of the selection already practiced by a population is directed toward preservation of the status quo by adjusting to minor and temporary environmental changes and by constant weeding out of recurring harmful mutations. Lowered infant mortality has raised the effective reproductive rate in many human populations, which has placed these populations in a difficult position even to try to maintain the status quo. To add a large dollop of selectivity to the problem would probably not be feasible.

Sewall Wright (cf. Lerner, 1968) created a model of the human population that could help in finding answers to questions of social cost. As more becomes known about the influence of genes on behavior, more definite insights could be gained from this model. Wright divided the population into 13 major groups according to their contributions to and their drain on society (Fig. 14–1).

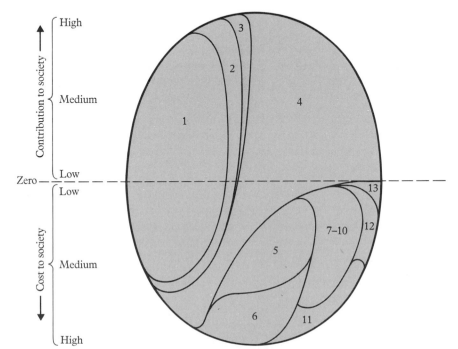

Fig. 14–1. The relative contributions and costs of various human phenotypic classes. (See text for explanation.)

Groups 1 through 4 appear as a unit because the individuals in these groups give back to their society at least as much as they take from it. The bulk of the population appears here specifically as members of Group 1 (low cost balanced by low contribution) and Group 2 (higher cost, higher contribution). Group 3 includes those few individuals who contribute greatly to society but who ask little from it in terms of health care or education. In Group 4, the costs are higher than those for Group 3, but these are outweighed by the members' service to society. Wright maintains that an increased mutation rate in this vast sector might be personally unfortunate but could easily be borne by a medically and technologically efficient society.

At the other end of the scale are Groups 11, 12, and 13. The individuals included here give minimal contributions to society for a variety of natural reasons. Individuals in Group 11 have a normal life span but suffer from complete mental and/or physical incapacity. Group 12 includes completely incapacitated individuals whose life spans are also foreshortened. Fetal and embryonic deaths are placed in Group 13. Costs, especially for Group 11, are high for this sector of society. Because of many of the causes of disease for these individuals are genetic, an increase in the mutation rate would further raise the cost and increase the number of members in these groups.

Most of the unknowns lie in an analysis of Groups 5 through 10. The members of Group 5 have the capacity for greater contribution, but for personal reasons they do not live up to their capacity. According to Wright, the genetic component of variation in this group is probably low, the environment playing a more influential role. Because of this, an increase in mutation rates would not severely affect this group.

Group 6 includes the bulk of the antisocial elements of a population. Little is known of the genetics of criminality and so any educated guesses about the effect of an increased mutation rate cannot be made. Furthermore, the problem may be compounded and distorted by value judgments and preconditioned thinking about what antisocial behavior involves. If a high correlation is found between genes and antisocial behavior, then the costs to society would greatly increase with an increase in mutation.

Group 7 individuals are subnormal in physical health. Group 8 includes individuals of low mental ability but who are sufficiently developed to care for themselves. Individuals who suffer physical accidents early in life or who have degenerative diseases fall into Group 9. Group 10 includes those who have had postadolescent mental breakdowns. In all of these categories, there is a net loss to society for a variety of reasons. This is not to deny the individual contributions of a Beethoven (Group 9), or a Darwin (Group 7), or even all those famous men reputed to have been syphilitic (Group 9). Rather it is to concentrate on populations of human beings living in an overpopulated, social-security-numbered world. For these groups of individuals, an analysis of the effects of increasing mutation hinges on the kinds of genetic mechanisms that underlie their characteristics. If the mutations that play a role in these conditions are presently balanced by selection and are simply acting as dominant or recessive alleles, then an increased mutation rate will surely tax society. If, on the other hand, the genetic components of these problems are minimal, or if the alleles involved express themselves in more complicated ways (such as giving a selective advantage to heterozygotes), then an increased mutation rate may not be felt as strongly. The importance of Wright's model lies in the assumption that the social aspects of society cannot be divorced from the purely genetic aspects. The interaction of genetic inheritance and social inheritance will become even more critical as society faces the issues of finite resources.

In contrast to Wright, who concedes that some of the cost of harmful mutations may be balanced by beneficial ones, Muller (1955) takes a far more pessimistic view of increases in mutation rates that may be caused by technological radiation and chemicals. Part of Muller's pessimism arises from things normally thought of as blessings. In the "Utopia of Inferiority" to which Muller had society heading, medicine, improved sanitation, and artificial, technological aids are keeping alive and permitting to reproduce persons who in more primitive times would have died quite early in life. What we are doing, in effect, is relaxing selection. Muller calculated that the effect of technologically saving genetically defective persons will result, in ten generations, in the whole of human society being in the same mutational state as the local survivors of the Hiroshima blast. He envisions a future in which all leisure hours are spent in medical treatment, much of the working time of a population being taken up with producing aids to keeping people alive. "After 1,000 years the population in all likelihood would be as heavily loaded with mutant genes as though it were descended from the survivors of the hydrogen-uranium bomb fallouts, and the passage of 2,000 years would continue the story until the system fell of its own weight or changed." The basis of Muller's argument constitutes what has been called *genetic load*. Muller points out that all mutations are equivalent in their effect, namely, that each causes one genetic death (either death literally, or the inability or reduced capacity to reproduce). This is easy to understand for a lethal, dominant gene, but even if the new mutant is sublethal and causes only partial reproductive failure, it is only a matter of time before its cumulative effect equals one genetic death. The less harmful the mutations, the more they will tend to accumulate in the population, and the greater the number of generations will ensue before the "death point" is reached. In populational terms, it matters little if one individual leaves no descendants or if ten individuals each have a one-tenth reduction in offspring. As mutation increases, so will the genetic load of a population—this process, according to Muller, cannot be reversed until we not only employ medicine and technology to ease suffering and save lives, but also harness those agencies in our society whose business it is to persuade and use their talents to "sell" selective reproduction of the genetically undamaged. For those who have faith that a miracle will come along, Muller and many other populationists offer bitter food for thought.

SELECTION

In many respects, the real test of mutation is natural selection. Three basic kinds of selective effects are recognized, all of which result from the cumulative forces of the environment acting to determine the fate of an allele.

Stabilizing Selection

Stabilizing selection tends toward the elimination of phenotypic extremes. The phenotypes that cluster about the mean are preserved in this manner by populations that are environmentally well adapted after a long period of trial with recurrent mutations. In a study of 6693 female births in a London hospital (Karn and Penrose, 1951), it was found that the chances of survival for one month were optimum for those babies whose birth weight was eight pounds. The average mortality within the first month (294 deaths versus 6693 births) was 4.1 percent, but for the eight-pounders it was 1.2 percent. Deviations above or below eight pounds were being selected against because, for a variety of reasons, a newborn female weighing eight pounds is the most fit.

Directional Selection

Under *directional selection*, one of the extremes in a phenotypic range becomes the most fit. Directional selection in animal husbandry is often achieved by mating for the expression of a character such as yield in eggs or milk, or weight. Natural directional selection is believed to come into play when the environment in which a well-adapted population lives changes. With environmental shifts, new mutations that were once disadvantageous may now become desirable, and the optimum "average" that was previously most fit must now give way to a better-adapted phenotype. The evolution of man from his simian ancestors may have been due to directional selection imposed by severe climatic changes.

A classic case of directional selection took place during Darwin's life, though it is not known if he ever observed this demonstration of his theories. During his lifetime, inhabitants of England began to suffer the fallout of soot that accompanied the industrial revolution. The extent of the sooty deposits can be imagined from the fact that around Sheffield, 50 tons of smoke particles per square mile drift downward to cover the cityscape and landscape per month. Lichen-covered rocks, light-colored tree trunks and boughs, and "earthen-hued" ground became almost black in the process, and one of the most rapid cases of evolution in multicellular organisms was witnessed, because with the changing color of the world came the changing camouflage color of many moths that use the rocks and tree trunks as daily resting places. In former times these moths matched the surroundings; in present times they do as well. The change has been termed *industrial melanism* (Kettlewell, 1956). In 1848, the first dark form of the normally light, peppered moth was recorded in England. Today, 90 percent of the moths caught in Britain's industrial areas are dark, and the reason for the increased fitness of the dark forms becomes quite clear in films made by Nilo Tinbergen which show that not only are the light forms more conspicuous to human eyes, they are to birds as well. Light moths are simply selectively eaten, whereas the dark forms are protected by the dark, polluted surroundings.

Disruptive Selection

Finally, selection is *disruptive* when both extremes in a phenotypic range are selected for, thus preserving differences in the gene pool of a population. Naturally occurring disruptive selection may occur when a variety of new ecological niches becomes available to a population. If the phenotypic extremes become isolated from each other so that the gene flow of interbreeding ceases, then subspeciation can occur and can, in time, lead to the emergence of two new species (Thoday and Gibson, 1970).

Selection is not only a long-term process that, fed by mutation, can lead to evolutionary change. It may be short-term as well. If an environment fluctuates in a cyclic manner, selection for the most fit phenotype can also be cyclical around a mean value. If environmental change ceased altogether, selection would still continue processing new mutations in order to keep the adaptivity of a population at optimum levels. Finally, because selection works on phenotypes and not individual genes, if mutation in man were to cease abruptly, the mixing forces of sexual reproduction would continue to provide new material for selection for hundreds, if not thousands, of generations (Crow, 1959).

FITNESS

The relationship between genotype, phenotype, and environment lies in the term "fitness," which means the reproductive success of one phenotype as opposed to alternative phenotypes. The basis of fitness may be any one of the obvious aspects of the environment, such as food supply, temperature, humidity, or elevation, or any of a host of subtle, inconspicuous features of the particular part of the world that a population inhabits. As complex as the causes of relative fitness may be, its effects are measured solely by the numbers of progeny produced by contrasting phenotypes. If carriers of gene A each have 100 progeny while those carrying gene a have only 90, the fitness of A is 1.00 while that of a is 0.9. *Selection pressure* can be tersely defined as the force acting upon a phenotype to reduce its fitness or adaptive value. In mathematical terms, $s = 1 - W$, where s is the selection coefficient and W is the fitness or adaptive value. In our previous example, if the fitness of A is 1.00, then the selection coefficient is 0; if the fitness of a is 0.9, then the selection coefficient is 0.1.

The fitness of many human genotypes has been measured in comparison to the normal homozygote (fitness 1.0). Genes that are genetically lethal, such as dominant retinoblastoma and recessive infantile amauritic idiocy, obviously have a fitness of zero. Some genes that are harmful alone but in combination with others that are beneficial have a fitness greater than 1.0. For instance, in the malarial regions of east Africa, heterozygotes for sickle-cell anemia have a relative fitness of 1.26. Other genes lie between these extremes—for example, dominant achondroplasia is at 0.196, and X-linked hemophilia in males is at 0.29 (Schull, 1963).

Selection can occur in gametes, in which case, as in haploid organisms, it depends only upon relative fitness because all genes, dominant or recessive, are exposed. On the other hand, stabilizing selection runs different time courses for alleles in diploid organisms that are dominant, recessive, incompletely dominant, or sex-linked. Some examples are given in Table 14–3.

Note that with low initial frequencies, the number of generations required to accomplish even minor changes becomes large, because recessive alleles can be masked in heterozygotes. Depending upon the mutation rate for various alleles,

any change via selection may be counterbalanced or even overridden by mutation at low gene frequencies or low selective coefficients.

Table 14–3 can be used in reverse if the recessive gene is being selected for, that is, if selection pressure is against a dominant gene. If there is a selection pressure of 0.20 for the recessive, then it will take 4512 generations to increase its frequency from 0.001 to 0.01.

Selection is more rapid if alleles are codominant, because they cannot be hidden in heterozygotes. Table 14–4 illustrates the acceleration of change.

SELECTION AND VARIABILITY

Implied in the preceding tables is the idea that under conditions of stabilizing selection, one allele will tend toward extinction while its opposite becomes universally represented. The outcome of this would be a reduction in the variability of the gene pool. It is known that this does not occur. There is enormous variability even in populations of domesticated animals undergoing rigorous artificial selection. If a farmer or livestock breeder changes his ideas on what is a desirable phenotypic extreme, he usually finds that there is

Table 14–3. Number of generations necessary for a given change in frequency (q) of a deleterious recessive gene under different selection coefficients (s = 1 − W).

Change in gene frequency		Change in frequency of homozygotes		Number of generations for different selection values						
From q_0	To q_n	From q_0^2	To q_n^2	s = 1 (lethal)	s = 0.80	s = 0.50	s = 0.20	s = 0.10	s = 0.01	s = 0.001
0.99	0.75	0.980	0.562	1	5	8	21	38	382	3,820
0.75	0.50	0.562	0.250	1	2	3	9	18	176	1,765
0.50	0.25	0.250	0.062	2	4	6	15	31	310	3,099
0.25	0.10	0.062	0.010	6	9	14	35	71	710	7,099
0.10	0.01	0.010	0.0001	90	115	185	462	924	9,240	92,398
0.01	0.001	0.0001	0.000001	900	1,128	1,805	4,512	9,023	90,231	902,314
0.001	0.0001	0.000001	0.00000001	9,000	11,515	18,005	45,001	90,023	900,230	9,002,304

Table 14–4. Approximate number of generations necessary for gene frequency changes under different selection coefficients when dominance is absent.

Frequency change				Number of generations						
For deleterious allele		For favored allele		$s = 1$	$s = 0.80$	$s = 0.50$	$s = 0.20$	$s = 0.10$	$s = 0.01$	$s = 0.001$
From q_0	To q_n	From q_0	To q_n	(lethal)						
0.99	0.78	0.01	0.25	3	4	7	17	35	350	3496
0.75	0.50	0.25	0.50	1	1	2	5	11	110	1099
0.50	0.25	0.50	0.75	1	1	2	5	11	110	1099
0.25	0.10	0.75	0.90	1	1	2	5	11	110	1099
0.10	0.01	0.90	0.99	2	3	5	12	24	240	2395
0.01	0.001	0.990	0.999	2	3	5	12	23	231	2314
0.001	0.0001	0.9990	0.9999	2	3	5	12	23	230	2304

a large amount of genetic variability that has survived his previous selective techniques. Not only may new mutations occur, but many different genotypes may produce similar phenotypes, and genetic differences in polygenic systems may be masked, thus preserving genetic variability. According to Neel (1970), new biochemical techniques have revealed such a large degree of previously hidden variability as to challenge the fundamental models of present-day population genetics.

Stabilizing selection may itself act to preserve variability by working toward balanced, adaptive combinations of genes both in populations and in individuals. It has been shown that at low frequencies, even harmful recessive genes are difficult to eliminate and, at the level of the cell or individual, epistasis or position effects may further serve to maintain genes which by themselves would be selected against. Selection at the level of the population may be frequency-dependent. It is easy to envision that some genes may benefit a population if they are rare, but may be harmful to the group if their frequency is high. Finally, selection may favor *balanced polymorphism*, a situation defined as one in which two or more discontinuous

forms or phases of a population exist in the same habitat in such proportions that even the least frequent cannot be accounted for simply by recurrent mutation. Excluded from this definition are continuous polygenic traits such as skin color or height in man, traits that are affected differently by seasonal or other environmental variation, or genes in which the balance of new mutation versus selection results in the maintenance of the mutant genes (for example, phenylketonuria). Polymorphisms are genetically based, intermediate forms being absent. Balanced polymorphism can result from disruptive selection (Thoday, 1960) or when the heterozygote enjoys a degree of fitness that surpasses each of the homozygotes, that is, when an allele is sometimes advantageous. Even alleles that are lethal in homozygous combination may be preserved if the heterozygote is in some way more reproductively fit than the homozygote. An extreme case of polymorphism was described by Bennett (1964) at the T locus in mice. There are three dominant alleles and a number of recessive alternatives at this locus. The homozygotes T_1T_1 and tt are embryonically lethal combinations, whereas the heterozygote T_1t is tailless but viable. In man, balanced polymorphism is observed with regard to a number of characteristics (Clarke, 1969).

The classic balanced polymorphic situation in man involves the allele Hb^s, which is responsible for the abnormal hemoglobin of sickle-cell anemia. In some regions of east Africa, this allele has a frequency of at least 20 percent, even though homozygous individuals have a survival value of less than 21 percent as compared to other hemoglobin genotypes. But the heterozygotes have a 25 percent survival advantage because they are less likely to contract falciparum malaria or hookworm. Furthermore, if a heterozygote does get malaria, only rarely will lethal complications such as cerebral malaria or blackwater fever result. That the environment is the driving force serving to maintain the polymorphic state is illustrated by the fact that in nonmalarial regions, the frequency of the Hb^s allele is selectively kept at the low levels characteristic of simple deleterious alleles.

For a similar reason, the frequency of the sex-linked allele for glucose-6-phosphate dehydrogenase deficiency (G6PD) is maintained at about 10 percent in malarial populations. This is in spite of the allele's effects—anemia induced by certain drugs, and neonatal and aspirin-induced jaundice.

Another example of balanced polymorphism is seen in the maintenance of the A, B, AB, and O blood groups at different frequencies in different populations. It has been well documented that zygotic selection operates in cases of maternal-foetal incompatibility, especially when the mother is of blood group O and the foetus is of blood group A, B, or AB. Prezygotic selection is believed to favor the production of O-type sperm, especially since an A- or B-type sperm may be eliminated by contact with the anti-A or anti-B antibodies in the cervical secretions of O-type mothers. What then prevents total selection for blood group O? There is some evidence that heterozygotes (I^Ai, I^Bi, I^AI^B) are selected for, although the reasons for such differential survival remain obscure. The incidence of duodenal and gastric ulcer is higher in type O individuals than would be statistically expected. There has been some suggestion that the frequencies of the A–B–O alleles may have been influenced by epidemics of the past. In areas where plague was rampant, there is a shortage of people of type O. Similarly, there is a dearth of type A individuals in areas beset in the past by smallpox epidemics. There is also the possibility that different blood types are associated with differing capacities to synthesize certain enzymes, and that the possession of a particular A or B antigen may also influence production of antibodies to various universally present bacterial agents (for example, *E. coli*). Finally, the A–B–O system can interact with other serological systems, including the Rh genes which themselves may be in a condition of polymorphism (Bresler, 1961). All of these factors, and no doubt others as well, interact to maintain each of the A–B–O genes at a level greater than that of recurrent mutation.

SELECTION AND MATING SYSTEMS

The rate and efficiency of selection is closely related to mate selection. Until this point, mating has been assumed to be random. However, mating of like with like (*assortive*) or unlike (*disassortive*) can be demonstrated both in man and in fruit flies. This should not be surprising because for practical reasons, most organisms avoid the broadening effects of travel and tend to mate with the most closely available partner. Many social forces can alter human mating patterns. According to Neel (1970), these patterns depart so far from what they have been assumed to be that totally new models of human population genetics may be necessary. We can envision many small subgene pools arising within a population. As a consequence, and for other social reasons as well, inbreeding can occur. Although mating between relatives by itself need not necessarily change gene frequencies, it can allow selection to proceed at a rate that would be more rapid than otherwise. The reason is simple. As noted, selection against harmful recessive genes can be a herculean task, often reaching the point at which new mutations balance out selective forces because the bulk of the recessive alleles lie hidden and protected in heterozygotes. Under systems of inbreeding, these recessives come to the surface more and more often due to increased incidence of homozygosis and are therefore exposed to selective forces. Inbreeding can especially change the degree of homozygosity of quantitative characteristics, resulting in a phenomenon called *inbreeding depression.* As more and more harmful recessive genes are brought to the homozygous state, survival and reproductive fitness will be progressively lowered. An example of this phenomenon in rats is shown in Table 14–5.

Table 14–5.   Effect of approximately thirty generations of inbreeding (parent-to-offspring, brother-to-sister) on fertility and mortality in rats.

Year	Percent nonproductive matings	Average litter size	Percent mortality from birth to 4 weeks of age
1887	0	7.50	3.9
1888	2.6	7.14	4.4
1889	5.6	7.71	5.0
1890	17.4	6.58	8.7
1891	50.0	4.58	36.4
1892	41.2	3.20	45.5

[From Ritzema Bos (1894), *Biol. Centralbl.* **14:**75.]

Selection against dominant genes is more rapid than against recessives and without particular regard for homozygosity. Therefore, inbreeding will not change the rate of dominant-gene selection as much. Recessive mutants that are beneficial will also enjoy greater chance for homozygosity in an inbred group, but because few recessive mutants are beneficial, the net effect will not be great.

Offsetting inbreeding depression, and thus slowing down the rate of natural selection, is disassortive mating. Self-sterility genes in some plants ensure that disassortive mating will occur, and in many organisms patterns of sexual maturity or behavior lead toward mating patterns that favor heterozygosity. When different strains are purposely crossed, one often observes that the hybrid is more vigorous than the parent. This phenomenon has been called *heterosis* and has been economically important, for example in corn and chickens. Heterosis may be due to simple masking of harmful recessive genes by dominants, or it may be that it is another case of heterozygote superiority to either of the homozygotes. However, whatever the cause, hybrid corn seed, for example, can produce high-yield plants under a broad range of environmental conditions. The parent inbred strains would be affected much more by environmental fluctuations.

RACE AND SPECIES FORMATION

The appearance of new, unfilled ecological niches combined with the raw material of mutation and the shaping forces of selection and drift can lead to increasing divergence within populations. When two groups within a species differ sufficiently within gene pools as to be recognizably different, but do not differ to an extent that precludes successful interbreeding, race formation has occurred. Races are defined as populations of organisms that differ in the relative frequency of some genes. Race formation is a reversible process; sufficient gene flow between the races results in their assimilation and disappearance into the whole. Races, then, are tentative evolutionary steps caused by flimsy reproductive barriers and partial isolation. If the isolating mechanisms become firm and gene flow between the groups ends, the races may diverge to the point at which reproduction between them is no longer possible. The process of change can then become irreversible, assimilation cannot occur, and speciation will have taken place.

In addition to variation around an optimum phenotype (birth weight, for example) or polymorphic variation within a population, one can often find geographic variation within the range of a species. This should not be surprising on the basis of the relationship between selection and the environment. Oftentimes there is a natural environmental gradient in altitude, temperature, or moisture that runs in a particular direction through the range of a species. In such cases, matching gradients (clines) of particular characteristics can be found. These may not be simple—an east-west moisture gradient may interact with a north-south temperature curve, and both may show traces of altitudinal variation. In any case, if no real geographical barriers exist, reproduction along the cline will be possible and, although races may be recognized at the extremes, the species remains unitary. However, if something occurs that isolates a part of the cline, further events may take place. Included in the list of "somethings" are ecological changes that act as barriers, or geological changes such as glaciation, flooding or flood-recession, or island formation. Once a part of the species is isolated, it will find new kinds of adjustments necessary and selection will begin to test the gene pool in a different way than before. Because the climate, or the parasites, or the predators may now be different, alleles or combinations of alleles that were not previously useful may now become beneficial. Eventually, the isolate will become increasingly different from the parent group.

If permanent geographical barriers are introduced, chance alone (plus time) may result in the acquisition and incorporation of mutations that result in sexual isolation. The essence of a new species lies in its inability to interbreed with the parent group. Hence, geographical separatism alone will not suffice. In cases where geographical barriers are not permanent, there appears to be evidence that natural selection for sexual isolating mechanisms occurs in the area of overlap. Therefore, once an initial geographical separation has permitted an accelerated rate of divergence, speciation can take place by chance *allopatrically* (in separated groups) or *sympatrically* (within an area of overlap).

Numerous examples of allopatric speciation can be found among the fauna on each of the rims of the Grand Canyon. Indefinite separatism has resulted in sexual isolation and speciation. An example of speciation that is both allopatric and sympatric has been proposed by Mengel (1964) for several species of northern wood warblers. In the

Pliocene about 14 million years ago, the parental population of warblers inhabited southeastern North America. This original range was maintained until the withdrawal of the first glaciation, at which time the parental stock expanded northward and westward across Canada. The second glaciation isolated a western group and also forced the parental group back to its original range. With the withdrawal of the glaciers, expansion again took place with possible contact between the first isolate and the parental group. Two subsequent glaciations again temporarily isolated portions of the warbler expansion, resulting in three western species and the proposed original eastern group. Other species of warblers exist today as well, but their history is more problematical.

Sexual Isolating Mechanisms

A variety of mechanisms that ensure sexual isolation of sympatric species exist and are often combined in various complex ways. These are divided basically into premating and postmating mechanisms.

Premating mechanisms include ecological, seasonal, or temporal isolation. Two groups may occupy the same region but have reproductive cycles that do not match in time or place. Furthermore, the two groups may be mechanically isolated in that the act of mating is physically impossible. A premating method of isolation that is now under broad study is species-specific behavior. Many species of animals have finely tuned, innate courtship rituals. If an act of a male is not followed by the appropriate response of the female or vice versa, mating will not occur. Many such species-specific behavioral devices are known. The call-site and timbre of frogs, the length of time between firefly flashes, and courtship dances in many birds are only a few examples.

Postmating sexual isolation is represented by mortality of the gametes, of the zygote, or of the embryo, as well as hybrid sterility. Isolation of this kind has been shown in some cases to be a direct result of gene incompatibility or nuclear-cytoplasmic irregularity.

Rapid Speciation

Speciation can be a fairly rapid process if the chromosomes undergo direct change. The appearance of inversions or translocations will render the individual partly sterile, and the beginnings of sexual isolation will have occurred. If for other reasons the rare individuals possessing these aberrations are at a strong selective advantage, this kind of change can be perpetuated and lead to the emergence of a new species.

A relatively common mode of speciation in plants, though rare in animals, has been the mechanism of polyploidy. Autotetraploids, usually arising from somatic doubling or the formation of unreduced gametes, are the most common variety, although higher chromosome multiples are possible. Again, reproductive isolation is immediate because an autotetraploid crossed with a normal diploid yields triploids, which are usually sterile. New species, natural and man-made, have also arisen by the formation of amphidiploids. In this case, unreduced gametes from different species will result in an organism phenotypically different and reproductively isolated from either parent.

The extent and simplicity of rapid speciation in nature is not clear. There is much documentation on the chromosomal differences between closely related species of *Drosophila*. Certainly it is theoretically possible by the means outlined above, and natural species do occur that appear to be polyploids of other species. However, polyploidy and chromosomal aberrations tend to be relatively unstable states (tetraploid plants often produce haploid gametes or reproduce vegetatively), and it may be that in the wild, strong selective advantage or geographical isolation is as necessary for them as it is for the slower modes of species formation.

Biochemical Approaches to Speciation

Whether or not the causes of sexual isolation can be shown to be genetic, speciation is, in fact, the compilation of a substantially different gene pool that is reflected in the protein "fingerprint" of the species. Based on this logic and the not necessarily valid assumption that genetic changes occur at a uniform rate, there has been an attempt to trace out not only degrees of relationship between vertebrate species, but also the temporal pattern of speciation.

One avenue of approach has been DNA hybridization studies. The more closely related two species are, the greater is the number of pairing sites that will occur if single strands of two species' DNA are mixed. A second type of analysis has been concerned with common metabolic pathways, and in large measure these types of studies have formed the basis of modern phylogenies. Finally, analyses have been made to show species-specific amino acid differences in similar proteins native to different species. Of interest here have been myoglobin and the hemoglobins, partly because their occurrence is so widespread and also because these proteins have been finely analyzed (Zuckerkandl and Pauling, 1965).

The biochemical approach to race formation and speciation has also been used as a tool for studying human races. Not surprisingly, genetic differences are apparent in man just as they are in other species of organisms. That this fact is used to shed more heat than light on a socially explosive situation is a sad commentary on the presumed intelligence of man.

Human Races

About a thousand generations of modern man have lived since the time of known paleontological history. Large-scale natural evolutionary change within these time limitations is of a low degree of probability. Nevertheless, certain variances in the human gene pool have occurred that are no doubt related to the interaction of environmental variation and chance.

The role of chance in human variation is revealed in the phenomenon of genetic drift. Nowadays with our large populations and easy com-

Table 14–6. Serological differences with known genetic causes.

Blood group antigens	European	American Indians	Asians	Pacific group	African group
A–B–O	Variable but generally high A; low B	High O; little or no B except in some South American tribes	Generally high B	High A; low B	Variable
MN complex	M and N about 0.5 each	M high (0.8 to 0.9)	Similar to Europeans	High N except Micro-nesians	Unique S^μ variant present
Rh complex	r as high as 0.5; R' relatively high (0.4)	r low to zero; R^2 between 0.5 and 0.6	r low to zero	R low to zero	r low to zero
Lutheran system	Present	____	Absent	Absent	Present
P group	78% positive	____	41% positive	____	95% positive
Kidd	70% positive	Between 70–90% positive	About 50% positive	____	About 95% positive
Duffy (Fy^a, Fy^b)	Fy^a present	Fy^a present	Fy^a present	Fy^a present	Fy^a low to zero; Fy^b present
Diego system	Absent	Present in certain tribes	Present	Absent	Absent

[Data from Race and Sanger (1962).]

munication, drift probably plays a minor role if any. However, in the past human beings lived in small, isolated bands that were firmly tied to the exigencies of nature. In all probability, people today are not uniformly descended from those of 6000 years ago. In the past, whole tribes vanished because of warfare, drought, or disease. Similarly, an advantageous invention or two may have led to the multiplication and expansion of another tribe. Obviously, populations isolated by chance alone need not be representative of the whole. Therefore, if human ancestry is uneven, it should not be surprising that any indications of genetic drift should have become magnified, although different studies offer different and often conflicting interpretations of the data.

Probably of greater importance than genetic drift is the phenomenon of selection. Skin pigmentation, stature, and degrees of hairiness can be correlated with differing amounts of available sunlight and temperature. Recently there have been suggestions that overexposure to sunlight can induce a dangerously high level of vitamin D synthesis in light-complexioned people, resulting in calcium deposits of a pathological nature.

Biological, though not necessarily anthropological, distinctions of race are usually made on the basis of physically observable differences in specific physical features and serological proteins. Five main races have been proposed by Boyd (1964) and others, though these groups have been shuffled around and subdivided into so many smaller, more homogeneous groups as to suggest that race is merely a linguistic concept. A typical classification is the following:

1. European group—including Lapps, Southern Europeans, and North Africans.
2. African group.
3. Asian group—including the Indian subcontinent.
4. American group—including all original inhabitants.
5. Pacific group—Melanesians, Polynesians, and Australians.

Some of the gene frequencies that have been tested are listed in Table 14–6. To these genotypic analyses must be added the purely phenotypic observations of characteristics such as pigmentation, stature, cranial shape, etc. It is crucial to note the "nonabsoluteness" of these data. Races are not absolutes. The phrase "pure race" is a contradiction in terms because all parts of a species interact within the gene pool. Furthermore, each of the entries in Table 14–6 is an average or a trend. In some cases, variability within the group is so high that no trend is discernible, and in other cases, two trends are at variance.

Human history has produced groups of human beings that differ in the frequency of certain genes. The frequency curves overlap time and time again. As people continue to move and to create new patterns of reproduction, different frequencies and different areas of overlap will occur. Perhaps the future will exhibit very different distributions of gene frequencies and quite different phenotypic races.

QUESTIONS

1. What is meant by "fitness"?

2. Why is genetic drift unlikely in large populations?

3. What causes mutation pressure to reach an equilibrium?

4. How does genetic death differ from medical death?

5. The mutation rate per generation is relatively uniform for many mammalian species. How could this have come about?

6. How can directional selection lead to speciation?

7. Under what circumstances would disruptive selection exist?

8. Are there any parallels between mutation pressure and selection pressure? Explain.

9. Why is the A–B–O blood group considered to be a stable, balanced polymorphic system?

10. Why are gene frequencies not changed by inbreeding?

11. What is the probable evolutionary outcome of a situation in which heterozygote superiority exists?

12. Why is simple reproductive isolation insufficient for evolutionary change?

13. Which is likely to be more rapid, allopatric or sympatric speciation? Why?

14. Why do you think the human species has not produced subspecies?

Bibliography

Aaronson, S. A., and G. J. Tordaro (1969), "Human diploid cell transformation by DNA extracted from the tumor virus SV40," *Science* **166**:390–391.

Abdel-Hameed, F. (1972), "Hemoglobin concentration in normal diploid and intersex triploid chickens: genetic inactivation or canalization?" *Science* **178**:864–865.

Aboud, M., and M. Burger (1970), "The effect of catabolite repression and of cyclic 3'–5' adenosine monophosphate on translation of the lactose messenger RNA in *Escherichia coli*," *Biochem. Biophys. Res. Commun.* **38**:1023–1032.

Adams, J. M., and M. R. Capecchi (1966), "N-formyl-methionyl-sRNA as the initiator of protein synthesis," *Proc. Nat. Acad. Sci.* **55**:147–155.

Ames, B. N., and P. E. Hartman (1963), "The histidine operon," *Cold Spring Harbor Symp. Quant. Biol.* **28**:349–356.

Ashburner, M. (1970), "Function and structure of polytene chromosomes during insect development," *Adv. Insect Physiol.* **7**:1–95.

Astaurov, B. L. (1969), "Experimental polyploidy in animals," *Ann. Rev. Genetics* **3**:99–126.

Atlas, D., S. Levit, L. Schechter, and A. Berger (1970), "On the active site of elastase: partial mapping by means of specific peptide substrates," *FEBS Letters* **11**:281–283.

Auerbach, C., and J. M. Robson (1946), "Chemical production of mutations," *Nature* **157**:302.

Avery, A. G., S. Satina, and J. Rietsema (1959), *Blakeslee: The Genus Datura*. New York: The Ronald Press.

Avery, O. T., C. M. MacLeod, and M. McCarty (1944), "Studies on the chemical nature of the substance inducing transformation of pneumococcal types. Induction of transformation by a desoxyribonucleic acid fraction isolated from pneumococcus Type III," *J. Exptl. Med.* **79**:137–158.

Bacci, G. (1965), *Sex Determination*. Oxford: Pergamon.

Baglione, C. (1963), "Correlations between the genetics and chemistry of human hemoglobins," *Molecular Genetics, Part I*, J. H. Taylor (ed.). New York: Academic Press, pp. 405–475.

Baltimore, D. (1970), "RNA-dependent DNA polymerase in virions of RNA tumour viruses," *Nature* **226**:1209–1211.

Barlow, P., and C. G. Vosa (1970), "The Y chromosome in human spermatozoa," *Nature* **226**:961–962.

Barr, M. L., and E. G. Bertram (1949), "A morphological distinction between neurones of the male and female, and the behaviour of the nucleolar satellite during accelerated nucleoprotein synthesis," *Nature* **163**:676–677.

Barski, G., S. Sorieul, and F. Cornefert (1960), "Production dans des cultures in vitro de deux souches cellulaires en association, de cellules de caractère 'hybride'," *Comptes Rend. Acad. Sci.* (Paris) **251**:1825–1827.

Bateson, W. (1909), *Mendel's Principles of Heredity.* Cambridge: Cambridge University Press.

Bateson, W., and E. R. Saunders (1902), "Experimental studies in the physiology of heredity," *Rep. Evol. Comm. Roy. Soc.* **1**:1–160.

Bateson, W., E. R. Saunders, and R. C. Punnett (1905), "Experimental studies in the physiology of heredity," *Rep. Evol. Comm. Roy. Soc.* **2**:1–55 and 80–89.

Bauerle, R. H., and P. Margolin (1966), "The functional organization of the tryptophan gene cluster in *Salmonella typhimurium*," *Proc. Nat. Acad. Sci.* **56**:111–118.

Beadle, G. W. (1945), "Genetics and metabolism in *Neurospora*," *Phys. Rev.* **25**:643–663.

Beadle, G. W., and E. L. Tatum (1941), "Genetic control of biochemical reactions in *Neurospora*," *Proc. Nat. Acad. Sci.* **27**:499–506.

Beams, H. W. (1964), "Cellular membranes in oogenesis," *Cellular Membranes in Development*, M. Locke (ed.). New York: Academic Press, pp. 175–219.

Beaudet, A. L., and C. T. Caskey (1971), "Mammalian peptide chain termination, II Codon specificity and GTPase activity of release factor," *Proc. Nat. Acad. Sci.* **68**:619–624.

Beckwith, J., and D. Zipser (eds.) (1970), *The Lactose Operon.* New York: Cold Spring Harbor Laboratory.

Beermann, W. (1952), "Chromomerenkonstanz und spezifische Modifikationen der Chromosomenstructur in der Entwicklung und Organdifferenzierung von *Chironomus tentans*," *Chromosoma* (Berlin) **5**:139–198.

———— (1966), "Differentiation at the level of the chromosomes," *Cell Differentiation and Morphogenesis*, W. Beermann, et al. (eds.). Amsterdam: North Holland Publishing Company, pp. 24–54.

Beermann, W., and U. Clever (1964), "Chromosome puffs," *Sci. Amer.* **210**:50–58.

Belling, J. (1931), "Chromomeres in liliaceous plants," *Univ. Calif. (Berkeley) Publs. Bot.* **16**:153.

Bennett, D. (1964), "Embryological effects of lethal alleles in the t-region," *Science* **144**:263–267.

Benzer, S. (1955), "Fine structure of a genetic region in bacteriophage," *Proc. Nat. Acad. Sci.* **41**:344–354.

———— (1961), "On the topography of the genetic fine structure," *Proc. Nat. Acad. Sci.* **45**:1607–1620.

Blakeslee, A. F. (1934), "New Jimson weeds from old chromosomes," *J. Heredity* **25**:80–108.

Bloom, W. (1937), "Transformation of lymphocytes into granulocytes *in vitro*," *Anat. Rec.* **69**:99–116.

Boivin, A., R. Vendrely, and C. Vendrely (1948), "Biochimie de l'hérédité—L'acide desoxyribonucléique du noyau cellulaire, dépositaire des caractères héréditaires; arguments d'ordre analytique," *Comptes Rend.* **226**:106.

Bonner, J. T. (1969), "Hormones in social amoebae and mammals," *Sci. Amer.* **220**:78–91.

Boveri T. (1887), Über Differenzierung der Zellkerne während der Furchung des Eies von *Ascaris megalocephala*," *Anat. Anz.* **2**:688–693.

Boyd, W. C. (1964), "Modern ideas on race, in the light of our knowledge of blood groups and other characters with known mode of inheritance," *Taxonomic Biochemistry and Serology*, C. A. Leone (ed.). New York: The Ronald Press, pp. 119–169.

Brammer, W. J., H. Berger, and C. Yanofsky (1967), "Altered amino acid sequences produced by reversion of frameshift mutants of tryptophan synthetase *A* gene in *E. coli*," *Proc. Nat. Acad. Sci.* **58**:1499–1506.

Bregger, T. (1918), "Linkage in maize; the C aleurone factor and waxy endosperm," *Am. Nat.* **52**:57–61.

Brenner, S., G. Streisinger, R. W. Horne, S. P. Champe, L. Barnett, S. Benzer, and M. W. Rees (1959), "Structural components of bacteriophage," *J. Mol. Biol.* **1**:281–292.

Bresler, J. B. (1961), "Effect of ABO–Rh interaction on infant hemoglobin," *Hum. Biol.* **33**:11–24.

Breuer, M. E., and C. Pavan (1955), "Behavior of polytene chromosomes of *Rhynchosciara angelae* at different stages of larval development," *Chromosoma* **7**:371–386.

Brewin, N. (1972), "Catalytic role for RNA in DNA replication," *Nature New Biology* **236**:101.

Bridges, C. B. (1917), "Deficiency," *Genetics* **2**:445–465.

_____ (1925), "Sex in relation to chromosomes and genes," *Am. Nat.* **59:** 127–137.

_____ (1936), "The bar 'gene' a duplication," *Science* **83:** 210–211.

Brinkhous, M., and J. B. Graham (1950), "Hemophilia in the female dog," *Science* **111:** 723–724.

Britten, R. J., and E. H. Davidson (1969), "Gene regulation for higher cells: a theory," *Science* **165:** 349–357.

_____ (1971), "Repetitive and non-repetitive DNA sequences and a speculation on the origins of evolutionary novelty," *Quart. Rev. Biol.* **46:** 111–138.

Brown, D. D. (1967), "The genes for ribosomal RNA and their transcription during amphibian development," *Current Topics Develop. Biol.* **2:** 47–73.

Brown, D. D., and E. Littna (1964), "RNA synthesis during the development of *Xenopus laevis*, the South African clawed toad," *J. Mol. Biol.* **8:** 669–687.

Brown, D. D., P. C. Wensink, and E. Jordan (1972), "A comparison of the ribosomal DNA's of *Xenopus laevis* and *Xenopus mulleri:* the evolution of tandem genes," *J. Mol. Biol.* **63:** 57–73.

Brown, J. C., and A. E. Smith (1970), "Initiator codons in eukaryotes," *Nature* **226:** 610–612.

Burns, R. O., J. Calvo, R. Margolin, and H. E. Umbarger (1966), "Expression of the leucine operon," *J. Bacteriol.* **91:** 1570–1576.

Buznikov, G. A., A. N. Kost, N. F. Kucherova, A. L. Mndzhoyan, N. N. Suvorov, and L. V. Berdysheva (1970), "The role of neurohumours in early embryogenesis," *J. Embryol. Exptl. Morph.* **23:** 549–569.

Cairns, J. (1963), "The chromosome of *Escherichia coli*," *Cold Spring Harbor Symp. Quant. Biol.* **28:** 43–46.

Cairns, J., and C. I. Davern (1967), "Mechanism of DNA replication in bacteria," *J. Cell Phys.* **70** (Supp. I): 65–76.

Callan, H. G. (1963), "The nature of lampbrush chromosomes," *Int. Rev. Cytol.* **15:** 1–34.

Campbell, A. M. (1969), *Episomes.* New York: Harper and Row.

Carr, D. H. (1971), "Chromosome studies in selected spontaneous abortions: polyploidy in man," *J. Genetics* **8:** 164–174.

Case, M. E., and N. H. Giles (1960), "Comparative complementation and genetic maps of the *pan*–2 locus in *Neurospora crassa*," *Proc. Nat. Acad. Sci.* **46:** 649–676.

Cattanach, B. M., and C. E. Pollard (1969), "An XYY sex-chromosome constitution in the mouse," *Cytogenetics* **8:** 80–86.

Cerda-Olmedo, E., P. C. Hanawalt, and N. Guerola (1968), "Mutagenesis of the replication point by nitrosoguanidine: map and pattern of replication of the *Escherichia coli* chromosome," *J. Mol. Biol.* **33:** 705–719.

Chambers, D. A., and G. Zubay (1969), "The stimulatory effect of cyclic adenosine 3'–5' monophosphate on DNA directed synthesis of β-galactosidase in a cell-free system," *Proc. Nat. Acad. Sci.* **63:** 118–122.

Chargaff, E. (1947), "On nucleoproteins and nucleic acids of microorganisms," *Cold Spring Harbor Symp. Quant. Biol.* **12:** 28–34.

Chen, P. S. (1967), "Biochemistry of nucleo-cytoplasmic interactions in morphogenesis," *The Biochemistry of Animal Development*, Vol. 2, R. Weber (ed.). New York: Academic Press, pp. 115–191.

Clark, A. J., and A. D. Margulies (1965), "Isolation and characterization of recombination deficient mutants of *E. coli* K12," *Proc. Nat. Acad. Sci.* **53:** 451–459.

Clarke, C. A. (1969), "Polymorphism," *Selected Topics in Medical Genetics*, C. A. Clarke (ed.). London: Oxford, pp. 22–45.

Cleland, R. E. (1962), "The cytogenetics of *Oenothera*," *Adv. in Genet.* **11:** 147–237.

Clough, E., R. L. Pyle, W. C. D. Hare, D. F. Kelly, and D. F. Patterson (1970), "An XXY sex-chromosome constitution in a dog with testicular hypoplasia and congenital heart disease," *Cytogenetics* **9:** 71–77.

Cohn, Z. A. (1968), "The differentiation of macrophages," *Differentiation and Immunology*, K. B. Warren (ed.). New York: Academic Press.

Comings, D. E., and T. A. Okada (1970), "Whole mount electron microscopy of meiotic chromosomes and the synaptonemal complex," *Chromosoma* **30:** 269–286.

Conklin, E. G. (1903), "The cause of inverse symmetry," *Anat. Anz.* **23:** 577–588.

Cooper, D. W., J. L. Vander-Berg, G. B. Sharman, and W. E. Poole (1971), "Phosphoglycerate kinase polymorphism in kangaroos provides further evidence for paternal X inactivation," *Nature New Biology* **230:** 155–157.

Correns, C. (1909), "Vererbungs versuche mit blass (gelb) frunen und buntblattrigen Sippen bei *Mirabilis jalapa, Urtica pilulifera* und *Lanaria annua*," *Z. Vererb. Lehre.* **1:** 291–329.

Court-Brown, W. M. (1968), "Males with an XYY sex chromosome complement," *J. Med. Genet.* **5**:341–359.

Creighton, H. B., and B. McClintock (1931), "A correlation of cytological and genetical crossing-over in *Zea mays*," *Proc. Nat. Acad. Sci.* **17**:492–497.

Crew, F. A. E. (1923), "Complete sex transformation in the domestic fowl," *J. Heredity* **14**:360–362.

Crick, F. H. C. (1958), "On protein synthesis," *Symp. Soc. Exptl. Biol.* **12**:138–163.

—— (1966), "Codon-anticodon pairing: the wobble hypothesis," *J. Mol. Biol.* **19**:548–555.

Crick, F. H. C., L. Barnett, S. Brenner, and R. J. Watts-Tobin (1961), "General nature of the genetic code for proteins," *Nature* **192**:1227–1232.

Crow, J. F. (1959), "Ionizing radiation and evolution," *Sci. Amer.* **201**:138–160.

Curtis, A. S. G. (1962), "Morphogenetic interactions before gastrulation in the amphibian *Xenopus laevis*—the cortical field," *J. Embryol. Exptl. Morph.* **10**:410–422.

Darlington, C. D. (1931), "Meiosis," *Biol. Rev.* **6**:221–264.

Darlix, J. L., A. Sentenac, and P. Fromageot (1971), "Binding of termination factor rho to RNA polymerase and DNA," *FEBS Letters* **13**:165–168.

Darnell, J. E., W. R. Jelinek, and G. R. Molloy (1973), "Biogenesis of mRNA: genetic regulation in mammalian cells," *Science* **181**:1215–1221.

Davidson, E. H. (1968), *Gene Activity in Early Development*. New York: Academic Press.

Davidson, R. L. (1973), "Somatic cell hybridization: studies on genetics and development," Module in Biology No. 3. Reading, Mass.: Addison-Wesley Publishing Company.

Davis, B. D. (1971), "The role of subunits in the ribosome cycle," *Nature* **231**:153–157.

Davis R. W., and N. Davidson (1968), "Electron-microscopic visualization of deletion mutations," *Proc. Nat. Acad. Sci.* **60**:243–250.

Dawid, I. B. (1965), "Deoxyribonucleic acid in amphibian eggs," *J. Mol. Biol.* **12**:581–599.

—— (1966), "Evidence for the mitochondrial origin of frog egg cytoplasmic DNA," *Proc. Nat. Acad. Sci.* **56**:269–275.

Dawson, G. W. P. (1962), *An Introduction to the Cytogenetics of Polyploids*. Philadelphia: F. A. Davis.

de Crombrugghe, B., B. Chen, W. Anderson, P. Nissley, M. Gottesman, and I. Pastan (1971), "*Lac* DNA, RNA polymerase and cyclic AMP receptor protein, cyclic AMP, *lac* repressor and inducer are the essential elements for controlled *lac* transcription," *Nature New Biology* **231**:139–142.

de Harven, E. (1968), *The Nucleus*, A. J. Daltan and F. Hagenau (eds.). New York: Academic Press, pp. 197–227.

De Lucia, P., and J. Cairns (1969), "Isolation of an *E. coli* strain with a mutation affecting DNA polymerase," *Nature* **224**:1164–1166.

Denis, H. (1968), "Role of messenger ribonucleic acid in embryonic development," *Adv. Morph.* **7**:115–150.

De Terra, N. (1969), "Cytoplasmic control over the nuclear events of cell reproduction," *Int. Rev. Cytol.* **25**:1–29.

Deys, B. F., K. H. Grzeschick, A. Grzeschick, E. R. Jaffé, and M. Siniscalco (1972), "Human phosphoglycerate kinase and inactivation of the X-chromosome," *Science* **175**:1002–1003.

Dobzhansky, T. (1937), *Genetics and the Origin of Species*. New York: Columbia University Press.

Dobzhansky, T., and J. Schultz (1934), "The distribution of sex factors in the X-chromosome of *Drosophila melanogaster*," *J. Genetics* **28**:349–385.

Doonan, S., C. A. Vernan, and B. E. C. Banks (1970), "Mechanisms of enzyme action," *Prog. Biophys. Mol. Biol.* **20**:249–327.

Drake, J. W. (1969), "Mutagenic mechanisms," *Ann. Rev. Genetics* **3**:247–268.

—— (1970), *The Molecular Basis of Mutation*. San Francisco: Holden-Day.

Dressler, D. (1970), "The rolling circle for ϕX DNA replication, II. Synthesis of single-stranded circles," *Proc. Nat. Acad. Sci.* **67**:1934–1942.

Dube, S. K., P. S. Rudland, B. F. C. Clark, and K. A. Marcker (1969), "A structural requirement for codon-anti-codon interaction on the ribosome," *Cold Spring Harbor Symp. Quant. Biol.* **34**:161–166.

Dubnoff, J. S., and U. Maitra (1971), "Isolation and properties of polypeptide chain initiation factor FII from *Escherichia coli*: evidence for a dual function," *Proc. Nat. Acad. Sci.* **68**:318–323.

Duesberg, P. H., and C. Colby (1969), "On the biosynthesis and structure of double-stranded RNA in vaccinia virus-infected cells," *Proc. Nat. Acad. Sci.* **64**:396–403.

Ellgaard, E. G., and U. Clever (1971), "RNA metabolism during puff induction in *Drosophila melangaster*," *Chromosoma* (Berlin) **36**:60–78.

Elliot, A. M. (1959), "A quarter century exploring *Tetrahymena*," *J. Protozool.* **6**:1–7.

Ellis, C. H., Jr. (1966), "The genetic control of sea urchin development, a chromatographic study of protein synthesis in the *Arbacia punctulata* embryo," *J. Exptl. Zool.* **163**:1–21.

Elsasser, W. M. (1958), *The Physical Foundation of Biology.* New York: Macmillan (Pergamon).

Englesberg, E., C. Squires, and F. Meronk, Jr. (1969), "The L-arabinose operon in *Escherichia coli* B/r: a genetic demonstration of two functional states of the product of a regulator gene," *Proc. Nat. Acad. Sci.* **62**:1100–1107.

Ephrussi, B. (1953), *Nucleo-Cytoplasmic Relations in Microorganisms.* Oxford: Clarendon Press.

Eron, L., R. Arditti, G. Zubay, S. Connaway, and J. R. Beckwith (1971), "An adenosine 3'-5' cyclic monophosphate-binding protein that acts on the transcription process," *Proc. Nat. Acad. Sci.* **68**:215–218.

Falconer, D. S. (1963), "Quantitative inheritance," *Methodology in Mammalian Genetics*, W. J. Burdelle (ed.). San Francisco: Holden-Day, pp. 193–216.

Fantoni, A., A. de la Chapelle, R. A. Rifkind, and P. A. Marks (1968), "Erythroid cell development in fetal mice: synthetic capacity for different protein," *J. Mol. Biol.* **33**:79–81.

Ferguson-Smith, M. A. (1965), "Karyotype-phenotype correlations in gonadal dysgenesis and their bearing on the pathogenesis of malformations," *J. Med. Genet.* **2**:142–155.

Fincham, J. R. S., and P. R. Day (1965), *Fungal Genetics* (2nd edition). Oxford: Blackwell.

Fitzgerald, P. H., L. A. Brehant, F. T. Shannon, and H. B. Angus (1970), "Evidence of XX/XY sex chromosome mosaicism in a child with true hermaphroditism," *J. Med. Genet.* **7**:383–388.

Ford, C. E., P. E. Polani, J. H. Briggs, and P. M. F. Bishof (1959), "A presumptive XXY/XX mosaic," *Nature* **183**:1030–1032.

Fox, M. S., and M. K. Allen (1964), "On the mechanism of deoxyribonucleate integration in pneumococcal transformation," *Proc. Nat. Acad. Sci.* **52**:412–419.

Fraccaro, M., M. G. Bott, F. M. Salzano, R. W. Ross Russell, and W. I. Cranston (1962), "Triple chromosomal mosaic in a woman with clinical evidence of masculinisation," *Lancet* **i**:1379–1381.

Fraccaro, M., P. Davies, M. G. Bott, and W. Schutt (1962), "Mental deficiency and undescended testis in two males with XYY chromosomes," *Folia Hered. Path.* (Milano) **11**:211–220.

Fraenkel-Conrat, H. (1956), "The role of the nucleic acid in the reconstitution of active tobacco mosaic virus," *J. Am. Chem. Soc.* **78**:882–883.

Fraenkel-Conrat, H., and R. C. Williams (1955), "Reconstitution of tobacco mosaic virus from its inactive protein and nucleic acid components," *Proc. Nat. Acad. Sci.* **41**:690–698.

Freese, E. (1959), "The specific mutagenic effect of base analogues on phage T4," *J. Mol. Biol.* **1**:87–105.

Freese, E., E. Bautz, and E. Bautz-Freese (1961), "The chemical and mutagenic specificity of hydroxylamine," *Proc. Nat. Acad. Sci.* **47**:845–855.

Gall, J. G. (1954), "Lampbrush chromosomes from oocyte nuclei of the newt," *J. Morph.* **94**:283–353.

———— (1968), "Differential synthesis of the genes for ribosomal RNA during amphibian oogenesis," *Proc. Nat. Acad. Sci.* **60**:553–560.

Garrod, A. E. (1902), "The incidence of alkaptonuria: a study in chemical individuality," *Lancet* **ii**:1616–1620.

———— (1909), *Inborn Errors of Metabolism.* Reprinted with a supplement by Harris (1963). New York: Oxford University Press.

Georgiev, G. P. (1969), "Histones and the control of gene action," *Ann. Rev. Genetics* **3**:155–180.

Gibson, I., and G. H. Beale (1961), "Genic bases of the matekiller trait on *Paramecium aurelia* stock 540," *Genet. Res.* (Cambridge) **2**:82–91.

Gilbert, W., and D. Dressler (1968), "DNA replication: the Rolling Circle model," *Cold Spring Harbor Symp. Quant. Biol.* **33**:473–484.

Gilbert, W., and B. Müller-Hill (1967), "The *lac* operator in DNA," *Proc. Nat. Acad. Sci.* **58**:2415–2421.

Giudice, G., V. Mutolo, and G. Donatuti (1968), "Gene expression in sea urchin development," *Arch. Entwicklungsmech. organ.* **161**:118–128.

Godet, J., D. Schürch, J. P. Blanchet, and V. Nigon (1970), "Evolution des charactéristiques érythrocytaires au cours du développement postembryonnaire du poulet," *Exptl. Cell Res.* **60**:157–165.

Goldschmidt, R. (1942), "Sex determination in *Melandrium* and *Lymantria*," *Science* **95**:120–121.

Goodwin, B. C. (1966), "Histones and reliable control of protein synthesis," Ciba Foundation Study Group No. 24, *Histones, Their Role in the Transfer of Genetic Factors*. London: Churchill, pp. 68–76.

Gowen, J. W., and S.-T. C. Fung (1957), "Determination of sex through genes in a major sex locus in *Drosophila melanogaster*," *Heredity* **11**:397–402.

Granick, S., and A. Gibor (1967), "The DNA of chloroplasts, mitochondria and centrioles," *Prog. Nucleic Acid Res. Molec. Biol.* **6**:143–186.

Grant, P. (1965), "Informational molecules and embryonic development," *The Biochemistry of Animal Development*, Vol. I, R. Weber (ed.). New York: Academic Press, pp. 483–593.

Gregg, J. R., and S. Løvtrop (1960), "A reinvestigation of DNA synthesis in lethal amphibian hybrids," *Exptl. Cell Res.* **19**:621–623.

Griffith, F. (1928), "The significance of pneumococcal types," *J. Hyg.* (Cambridge) **27**:113–159.

Grobstein, C. (1955), "Inductive interaction in the development of the mouse metanephros," *J. Exptl. Zool.* **130**:319–340.

Gross, P. R. (1967), "The control of protein synthesis in embryonic development and differentiation," *Current Topics Develop. Biol.* **2**:1–46.

Gross, P. R., and G. H. Cousineau (1964), "Macromolecule synthesis and the influence of actinomycin on early development," *Exptl. Cell Res.* **33**:368–395.

Gurdon, J. B. (1962), "Adult frogs derived from the nuclei of single somatic cells," *Develop. Biol.* **4**:256–273.

———— (1967), "On the origin and persistence of a cytoplasmic state inducing nuclear DNA synthesis in frogs' eggs," *Proc. Nat. Acad. Sci.* **58**:545–552.

———— (1969), "Intracellular communication in early animal development," *Communication in Development*, A. Lang (ed.). New York: Academic Press, pp. 59–82.

Gurdon, J. B., and D. D. Brown (1965), "Cytoplasmic regulation of RNA synthesis and nucleolus formation in developing embryos of *Xenopus laevis*," *J. Mol. Biol.* **12**:27–35.

Gurdon, J. B., and R. A. Laskey (1970), "The transplantation of nuclei from single cultured cells into enucleate frogs' eggs," *J. Embryol. Exptl. Morph.* **24**:227–248.

Gurdon, J. B., and V. Uehlinger (1966), "'Fertile' intestine nuclei," *Nature* **210**:1240–1241.

Haggis, A. J. (1964), "Quantitative determination of deoxyribonucleic acid in embryos and unfertilized eggs of *Rana pipiens*," *Develop. Biol.* **10**:358–377.

Haldane, J. B. S. (1957), "The cost of natural selection," *J. Genetics* **55**:511–524.

Hall, B. D., and S. Spiegelman (1961), "Sequence complementarity of T2-DNA and T2-specific RNA," *Proc. Nat. Acad. Sci.* **47**:137–146.

Hampé, A. (1960), "Sur l'induction et la compétence dans les relations entre l'épiblaste et le mésenchyme de la patte de poulet," *J. Embryol. Exptl. Morph.* **8**:246–250.

Handler, P. (ed.) (1970), *Biology and the Future of Man.* New York: Oxford University Press.

Hardesty, B., W. Culp, and W. McKeehan (1969), "The sequence of reactions leading to the synthesis of a peptide bond on reticulocyte ribosomes," *Cold Spring Harbor Symp. Quant. Biol.* **34**:331–345.

Hardy, G. H. (1908), "Mendelian proportions in a mixed population," *Science* **28**:49–50.

Harris, H., and J. Watkins (1965), "Hybrid cells derived from mouse and man: artificial heterokaryons of mammalian cells from different species," *Nature* **205**:640–646.

Harris, T. M. (1964), "Pregastrular mechanisms in the morphogenesis of the salamander *Ambystoma maculatum*," *Develop. Biol.* **10**:247–268.

Hartmann, M. (1931), "Relative Sexualität und ihre Bedeutung für eine allgemeine Sexualitäts und Befruchtungstheorie," *Naturwiss.* **19**:230–232.

Haskell, E. H., and C. I. Davera (1969), "Pre-fork synthesis: a model for DNA replication," *Proc. Nat. Acad. Sci.* **64**:1065–1071.

Hayes, W. (1968), *The Genetics of Bacteria and Their Viruses* (2nd edition). New York: John Wiley and Sons.

Herbst, C. (1935), "Untersuchungen zur Bestimmung des Geschlechts. IV. Die Abhängigkeit des Geschlechts vom Kaliumgehalt des umgebenden Mediums bei *Bonellia virdis*," *Arch. Entwicklungsmech. organ.* **132**:576–599.

———— (1936). Untersuchungen zur Bestimmung des Geschlechts. V. Die Notwendigkeit des Magnesiums für die Lebenserhaltung und Weiterentwicklung des *Bonellia*—Larven und seine Bedeutung für die geschlechtliche Differenzierung derselben," *Arch. Entwicklungsmech. organ.* **134:** 313–330.

Hershey, A. D., and R. Rotman (1949), "Genetic recombination between host range and plaque-type mutants of bacteriophage in single bacterial cells," *Genetics* **34:**44–71.

Hershey, A. D., and M. Chase (1952), "Independent functions of viral protein and nucleic acid in growth of bacteriophage," *J. Gen. Physiol.* **36:**39–56.

Higurashi, M., I. Matsui, Y. Nakagome, and M. Naganuma (1969), "Down's syndrome: chromosome analysis in 321 cases in Japan," *J. Med. Genet.* **6:**401–404.

Holliday, R. (1970), "The organization of DNA in eukaryotic chromosomes,"*Organization and Control in Prokaryotic and Eukaryotic Cells.* Twentieth Symposium of the Society for General Microbiology, Cambridge University Press, pp. 359–380.

Holtzer, H., and S. R. Detweiler (1953), "An experimental analysis of the development of the spinal column. III. Induction of the skeletogenous cells," *J. Exptl. Zool.* **123:**335–370.

Hook, E. B. (1973), "Behavioral implications of the human XYY genotype," *Science* **179:**139–150.

Hotchkiss, R. D. (1954), "Cyclical behavior in pneumococcal growth and transformability occasioned by environmental changes," *Proc. Nat. Acad. Sci.* **40:**49–55.

Hotta, Y., and H. Stern (1971), "Analysis of DNA synthesis during meiotic prophase in *Lilium*," *J. Mol. Biol.* **55:**337–355.

Houlahan, M. B., G. W. Beadle, and H. G. Calhoun (1949), "Linkage studies with biochemical mutants of *Neurospora crassa*," *Genetics* **34:**493–507.

Humphrey, R. R. (1945), "Sex determination in ambystomid salamanders: a study of the progeny of females experimentally converted into males," *Amer. J. Anat.* **76:**33–65.

Humphreys, T. (1967), "The cell surface and specific cell aggregation," *The Specificity of Cell Surfaces*, B. D. Davis and L. Warren (eds.). Englewood Cliffs, N.J.: Prentice-Hall. pp. 195–210.

Ikeda, H., and J. Tomizawa (1965a), "Transducing fragments in generalized transduction by phage P1. I. Molecular origin of the fragments," *J. Mol. Biol.* **14:**85–109.

_____ (1965b), "Transducing fragments in generalized transduction by phage P1. II. Association of DNA and protein in the fragments," *J. Mol. Biol.* **14:**110–119.

Imamoto, F., N. Morikawa, and K. Sato (1965), "On the transcription of the tryptophan operon in *Escherichia coli*. III. Multicistronic messenger RNA and polarity for transcription," *J. Mol. Biol.* **13:**169–182.

Ingram, V. M. (1956), "A specific chemical difference between the globins of normal human and sickle cell anemia hemoglobins," *Nature* **178:**792–794.

Inui, N., and S. Takayama (1970), "Re-evaluation of the quantity of DNA in cell nucleus of two animal species by microspectrophotometry and autoradiography," *Cytologia* **35:**280–293.

Izawa, M., V. G. Alfrey, and A. E. Mirsky (1963), "The relationship between RNA synthesis and loop structure in lampbrush chromosomes," *Proc. Nat. Acad. Sci.* **49:**544–551.

Jacob, F., and J. Monod (1961), "Genetic regulatory mechanisms in the synthesis of proteins," *J. Mol. Biol.* **3:**318–356.

Jacob, F., and E. L. Wollman (1961), *The Sexuality and the Genetics of Bacteria.* New York: Academic Press.

Jaworska, H., and A. Lima-de-Faria (1969), "Multiple synaptinemal complexes at the region of gene amplification in *Acheta*," *Chromosoma* **28:**309–327.

Jenkins, M. T. (1924), "Heritable characters of maize. XX Iojap striping, a chloroplast defect," *J. Heredity* **15:**467–472.

Johnson, H. G., and M. K. Bach (1966), "The antimutagenic action of polyamines: suppression of the mutagenic action of an *E. coli* mutator gene and of 2-aminopurine," *Proc. Nat. Acad. Sci.* **55:**1453–1456.

Joseph, M. C., J. M. Anders, and A. I. Taylor (1964), "A boy with XXXXY sex chromosomes," *J. Med. Genet.* **1:**95–101.

Kaempfer, R. O. R., M. Meselson, and H. J. Raskas (1968), "Cyclic dissociation into stable subunits and re-formation of ribosomes during bacterial growth," *J. Mol. Biol.* **31:**277–289.

Kafiani, C. (1970), "Genome transcription in fish development," *Adv. Morph.* **8:**209–284.

Kalmus, H., and C. A. B. Smith (1960), "Evolutionary origin of sexual differentiation and the sex ratio," *Nature* (London) **186:**1004–1006.

Karn, M. N., and L. S. Penrose (1951), "Birth weight and gestation time in relation to maternal age, parity and infant survival," *Ann. Eugenics* **161**:147–164.

Karpechenko, G. D. (1927), "Polyploid hybrids of *Raphanus sativus* × *Brassica oleracea*," *Z. Indukt. Abst. Vererb.* **48**:1–85.

Kehl, H. G. (1965), "A demonstrable local and geometric increase in the chromosomal DNA of *Chironomus*," *Experientia* **21**:191–193.

Kettlewell, H. B. D. (1956), "Further selection experiments on industrial melanism in the Lepidoptera," *Heredity* **10**:287–301.

Kiger, J. A., and R. L. Sinsheimer (1971), "DNA of vegetative bacteriophage lambda. VI. Electron microscopic studies of replicating lambda DNA," *Proc. Nat. Acad. Sci.* **68**:112–115.

Kihlman, B. A. (1961), "Biochemical aspects of chromosome breakage," *Adv. in Genet.* **10**:1–59.

Kim, S. H., G. Quigley, F. L. Suddath, and A. Rich (1971), "High-resolution X-ray diffraction patterns of crystalline transfer RNA that show helical regions," *Proc. Nat. Acad. Sci.* **68**:841–845.

Knippers, R., J. M. Whalley, and R. L. Sinsheimer (1969), "The process of infection with bacteriophage øX174. XXX. Replication of double-stranded øX DNA," *Proc. Nat. Acad. Sci.* **64**:275–282.

Koch, R. E. (1971), "The influence of neighboring base pairs upon base-pair substitution mutation rates," *Proc. Nat. Acad. Sci.* **68**:773–776.

Kornberg, A. (1960), "Biological synthesis of deoxyribonucleic acid," *Science* **131**:1503–1508.

————— (1969), "Active center of DNA polymerase," *Science* **163**:1410–1418.

Kornberg, T., and M. L. Gefter (1972), "Deoxyribonucleic acid synthesis in cell-free extracts. IV. Purification and catalytic properties of deoxyribonucleic acid polymerase III," *J. Biol. Chem.* **247**:5369–5375.

Kroeger, H. (1968), "Gene activities during insect metamorphosis and their control by hormones," *Metamorphosis: A Problem in Developmental Biology*, W. Etkin and L. I. Gilbert (eds.). Amsterdam: North Holland Publishing Company, pp. 185–219.

Kühn, A. (1937), "Entwicklungsphysiologischgenetische Ergebnisse bei *Ephestia kuhniella*," *Z. Vererb. Lehre.* **73**:419–455.

Kurland, C. G. (1970), "Ribosome structure and function emergent," *Science* **169**:1171–1177.

Kushner, D. J. (1969), "Self-assembly of biological structures," *Bact. Rev.* **33**:302–345.

Lacks, S. (1962), "Molecular fate of DNA in genetic transformation of *Pneumococcus*," *J. Mol. Biol.* **5**:119–131.

Laird, C. D. (1971), "Chromatid structure: relationship between DNA content and nucleotide sequence diversity," *Chromosoma* **32**:378–406.

Lark, K. G. (1972), "Evidence for the direct involvement of RNA in the initiation of DNA replication in *Escherichia coli* 15T," *J. Mol. Biol.* **64**:47–60.

Lawley, P. D., and P. Brookes (1962), "Ionization of DNA bases or base analogues as a possible explanation of mutagenesis, with special reference to 5-bromodeoxyuridine," *J. Mol. Biol.* **4**:216–219.

Lederberg, J. (1955), "Recombination mechanisms in bacteria," *J. Cell. Comp. Physiol.* **45** (Supp. 2):75–107.

Lederberg, J., and E. Lederberg (1952), "Replica plating and indirect selection of bacterial mutants," *J. Bacteriol.* **63**:399–406.

Lederberg, J., and E. L. Tatum (1946), "Gene recombination in *Escherichia coli*," *Nature* **158**:558.

Lejeune, J. (1964), "The 21 trisomy—current stage of chromosomal research," *Progr. Med. Genet.* **3**:144–177.

Lengyel, P., and D. Söll (1969), "Mechanism of protein biosynthesis," *Bact. Rev.* **303**:264–301.

Lennox, E. (1955), "Transduction of linked characters of the host by bacteriophage P1," *Virology* **1**:190–206.

Lerner, I. M. (1968), *Heredity, Evolution and Society*. San Francisco: W. H. Freeman.

Lewin, S. (1970), "Possible basis for specificity in attachment of histones to DNA in relation to genetic control in mRNA synthesis," *J. Theor. Biol.* **29**:1–11.

L'Heritier, P. H. (1958), "The hereditary virus of *Drosophila*," *Adv. Virus Res.* **5**:195–245.

Lillie, F. R. (1917), "The freemartin: a study of the action of sex hormones in the foetal life of cattle," *J. Exptl. Zool.* **23**:371–452.

Lipmann, F. (1969), "Polypeptide chain elongation in protein biosynthesis," *Science* **164**:1024–1031.

Lutman, F. C., and J. V. Neel (1945), "Inherited cataract in the B geneology," *Arch. Ophthalmology* **33**:341–357.

Luyk:, P. (1970), "Cellular mechanisms of chromosome distribution," *International Review of Cytology*, G. H. Bourne and J. F. Dainelli (eds.). New York: Academic Press.

Lyon, M. F. (1961), "Gene action in the X-chromosome of the mouse (*Mus musculus L.*)," *Nature* (London) **190**:372–373.

MacGregor, H. D., and J. Kezer (1970), "Gene amplification in oocytes with 8 germinal vesicles from the tailed frog *Ascaphus truei* Stejneger." *Chromosoma* **29**:189–206.

Magasanik, B. (1970), "Glucose effects: inducer exclusion and repression," *The Lactose Operon*, J. Beckwith and D. Zipser (eds.). New York: Cold Spring Harbor Laboratory, pp. 189–219.

Maréchal, J. (1907), "Sur l'ovogénèse des sélaciens et de quelques autres chordates," *La Cellule* **24**:1–239.

Margulis, L. (1970), *Origin of Eukaryotic Cells*. New Haven: Yale University.

Martin, R. G. (1963), "The one operon-one messenger theory of transcription," *Cold Spring Harbor Symp. Quant. Biol.* **28**:357–361.

———— (1969), "Control of gene expression," *Ann. Rev. Genetics* **3**:181–216.

Marx, J. L. (1973), "Restriction enzymes: new tools for studying DNA," *Science* **180**:482–485.

Maugh, T. H., II (1973), "Molecular biology: a better artificial gene," *Science* **181**:1235.

McCarthy, B. J., and B. H. Hoyer (1964), "Identity of DNA and diversity of mRNA molecules in normal mouse," *Proc. Nat. Acad. Sci.* **52**:915–922.

McClintock, B. (1939), "The behavior in successive nuclear division of a chromosome broken at meiosis," *Proc. Nat. Acad. Sci.* **25**:405–416.

McClure, H. M., K. H. Belden, N. A. Pieper, and C. B. Jacobson (1969), "Autosomal trisomy in a chimpanzee: resemblance to Down's syndrome," *Science* **165**:1010–1011.

Mendel, G. (1866), "Versuche über Pflanzenhybriden," *Verh. Nuturf. Ver. Brünn*, **4**:3–44. Reprinted in translation in *The Origin of Genetics: A Mendel Source Book* (1966), C. Stern and E. R. Sherwood (eds.). San Francisco: W. H. Freeman and Company.

Mengel, T. (1964), "The probable history of species formation in some northern wood warblers (*Parulidae*)," *The Living Bird* **1964**:8–43.

Meselson, M., and F. W. Stahl (1958), "The replication of DNA in *Escherichia coli*," *Proc. Nat. Acad. Sci.* **44**:671–682.

Meselson, M., and J. J. Weigle (1961), "Chromosome breakage accompanying genetic recombination in bacteriophage," *Proc. Nat. Acad. Sci.* **47**:857–868.

Metafora, S., L. Felicetti, and R. Gambino (1971), "The mechanism of protein synthesis activation after fertilization of sea urchin eggs," *Proc. Nat. Acad. Sci.* **68**:600–604.

Meynell, G. G. (1973), *Drug-Resistance Factors and Other Bacterial Plasmids*. Cambridge, Mass: MIT Press.

Miescher, F. (1871), "Über die chemische Zusammensetzung der Eiterzellen," *Medicimisch-chemische Untersuchungen*, F. Hoppe-Seyler. Berlin: August Hirschwald, pp. 441–460.

Miller, I. L., and O. E. Landman (1966), "On the mode of entry of transforming DNA into *Bacillus subtilis*," *The Physiology of Gene and Mutation Expression*, M. Kohoutová and J. Hubáček (eds.). Prague: Academia.

Miller, O. L., and B. R. Beatty (1969), "Visualization of nucleolar genes," *Science* **164**:955–957.

Mirsky, A. E., and H. Ris (1949), "Variable and constant components of chromosomes," *Nature* **163**:666–667.

Moewus, F. (1936), "Faktorenaustausch, insbesondere der Realisateren bei *Chlamydomonas* Kreuzungen," *Ber. Deutsch. Bot. Ges.* **54**:48–57.

Monesi, V., and V. Salfi (1967), "Macromolecular synthesis during early development in the mouse embryo," *Exptl. Cell. Res.* **46**:632–635.

Moore, J. A. (1941), "Developmental rate of hybrid frogs," *J. Exptl. Zool.* **86**:405–422.

Morgan, T. H. (1910), "Sex limited inheritance in *Drosophila*," *Science* **32**:120–122.

———— (1911), "The application of the conception of pure lines to sex-limited inheritance and to sexual dimorphism," *Am. Nat.* **45**:65–78.

Moscona, A. A. (1962), "Analysis of cell recombinations in experimental synthesis of tissues *in vitro*," *J. Cell. Comp. Physiol.* **60** (Supp. 1):65–80.

———— (1968), "Cell aggregation: properties of specific cell-liquids and their role in the formation of multicellular systems," *Develop. Biol.* **18**:250–277.

Moses, M. J., and J. R. Coleman (1964), "Structural patterns and the functional organization of chromosomes," *The Role of Chromosomes in Development*, M. Locke (ed.). 23rd Symposium of the Society for the Study of Development and Growth. New York: Academic Press, pp. 11–49.

Moses, R. E., and C. C. Richardson (1970), "A new DNA polymerase activity of *Escherichia coli*. I. Purification and properties of the activity present in *E. coli* POLA1," *Biochem. Biophys. Res. Comm.* **41**:1557–1564.

Mounolou J. C., H. Jakob, and P. P. Slonimski (1966), "Mitochondrial DNA from yeast 'petite' mutants: specific changes of buoyant density corresponding to different cytoplasmic mutations," *Biochem. Biophys. Res. Comm.* **24**:218–224.

Mukai, T. (1964), "The genetic structure of natural populations of *Drosophila melanogaster*. I. Spontaneous mutation rate of polygenes controlling viability," *Genetics* **50**:1–19.

Muller, H. J. (1927), "Artificial transmutation of the gene," *Science* **66**:84–87.

———— (1928), "The measurement of gene mutation rate in *Drosophila*, its high variability, and its dependence upon temperature," *Genetics* **13**:279–357.

———— (1955), "Radiation and human mutation," *Sci. Amer.* **193**:58–68.

Nagley, P., and A. W. Linnane (1970), "Mitochondrial DNA deficient petite mutants of yeast," *Biochem. Biophys. Res. Comm.* **39**:989–996.

Neel, J. V. (1970), "Lessons from a 'primitive' people," *Science* **170**:815–822.

Newman, H. H. (1915), "Development and heredity in heterogenic teleost hybrids," *J. Exptl. Zool.* **18**:511–576.

Newton, A. A. (1970), "The requirements of a virus," *Organization and Control in Prokaryotic and Eukaryotic Cells.* 20th Symposium of the Society for General Microbiology. Cambridge: Cambridge University Press, pp. 323–358.

Nirenberg, M. S., and P. Leder (1964), "RNA codewords and protein synthesis: the effect of trinucleotides upon the binding of sRNA to ribosomes," *Science* **145**:1399–1407.

Nirenberg, M. W., and J. H. Matthaei (1961), "The dependence of cell-free protein synthesis in *E. coli* upon naturally occurring or synthetic polyribonucleotides," *Proc. Nat. Acad. Sci.* **47**:1588–1602.

Ochoa, S. (1968), "Translation of the genetic message," *Naturwiss.* **55**:505–514.

Ohno, S. (1967), *Sex Chromosomes and Sex-Linked Genes.* Berlin: Springer-Verlag.

Ohno, S., J. Poole, and I. Gustavsson (1965), "Sex-linkage of erythrocyte glucose-6-phosphate dehydrogenase in two species of wild hares," *Science* **150**:1737–1738.

Ohno, S., V. Wolf, and N. B. Atkin (1968), "Evolution from fish to mammals by gene duplication," *Hereditas* **59**:169–187.

Okada, Y., E. Terzaghi, G. Streisinger, J. Emrich, M. Inouye, and A. Tsugita (1966), "A frame-shift mutation involving the addition of two base pairs in the lysozyme gene of phage T4," *Proc. Nat. Acad. Sci.* **56**:1692–1698.

Okazaki, R., T. Okazaki, K. Sakak, K. Suzimoto, and A. Sugino (1968), "Mechanism of DNA chain growth. I. Possible discontinuity and unusual secondary structure of newly synthesized chains," *Proc. Nat. Acad. Sci.* **59**:598–605.

Parnell, F. R. (1921), "Note on the detection of segregation by examination of the pollen of rice," *J. Genetics* **11**:209–212.

Passmore, H. C., and D. C. Shreffler (1971), "A sex-limited serum protein variant in the mouse: hormonal control of phenotypic expression," *Biochem. Genetics* **5**:201–209.

Pastan, I., and R. L. Perlman (1969), "Stimulation of tryptophanase synthesis in *Escherichia coli* by cyclic 3,5-adenosine monophosphate," *J. Biol. Chem.* **244**:2226–2232.

Pateman, J. A., and D. J. Cove (1967), "Regulation of nitrate reduction in *Aspergillus nidulans*," *Nature* **215**:1234–1237.

Patton, J. L., and A. L. Gardner (1971), "Parallel evolution of multiple sex-chromosome systems in the phyllostomatid bats, *Carollia* and *Choeroniscus*," *Experientia* **27**:105–106.

Paul, J., and R. S. Gilmour (1966), "Template activity of DNA is restricted in chromatin," *J. Mol. Biol.* **16**:242–244.

———— (1968), "Organ-specific masking of DNA in differentiated cells," *Differentiation and Immunology*, Vol. I, K. B. Warren (ed.). Symposia of the International Society of Cell Biology, New York: Academic Press.

Pauling, L., H. A. Itano, S. J. Singer, and I. C. Wells (1949), "Sickle cell anemia, a molecular disease," *Science* **110**:543–548.

Penrose, L. S., and C. Stern (1958), "Reconsideration of the Lambert pedigree (*ichthyosis hystrix gravior*)," *Ann. Hum. Genetics* **22**:258–283.

Petrakis, N. L., M. Davis, and S. P. Lucia (1961), "The *in vivo* differentiation of human leukocytes into histiocytes, fibroblasts and fat cells in subcutaneous diffusion chambers," *Blood* **17**:109–118.

Philip, J., N. J. Brandt, B. Friis-Hansen, M. Mikkelsen, and I. Tygstrup (1970), "A deleted B chromosome in a mosaic mother and her cri du chat progeny," *J. Med. Genet.* **7**:33–36.

Piatigorsky, J., and A. Tyler (1967), "Radioactive labeling of RNA of sea urchin eggs during oogenesis," *Biol. Bull.* **133**:229–244.

Pontecorvo, G. (1953), "The genetics of *Aspergillus nidulans*," *Adv. in Genet.* **5**:142–238.

Poulson, D. F., and B. Salaguchi (1961), "Nature of sex-ratio agent in *Drosophila*," *Science* **133**:1489–1490.

Pritchard, R. H. (1955), "The linear arrangement of a series of alleles in *Aspergillus nidulans*," *Heredity* **9**:343–371.

Ptashne, M. (1967a), "Isolation of the λ phage repressor," *Proc. Nat. Acad. Sci.* **57**:306–313.

——— (1967b), "Specific binding of the λ phage repressor to λ DNA," *Nature* **214**:232–234.

Rabbitts, T. H., and T. S. Work (1971), "The mitochondrial ribosome and ribosomal RNA of the chick," *FEBS Letters* **14**:214–218.

Race, R. R., and R. Sanger (1962), *Blood Groups in Man* (4th edition). Philadelphia: F. A. Davis.

Ramakrishnan, T., and E. A. Adelberg (1965), "Regulatory mechanisms in the biosynthesis of isoleucine and valine," *J. Bacteriol.* **87**:566–573.

Rasmussen, H., and M. M. Pechet (1970), "Calcitonin," *Sci. Amer.* **223**:42–50.

Raven, C. P. (1961), *Oogenesis*. New York: Macmillan (Pergamon).

Revel, J. P., and S. Ito (1967), "The surface components of cells," *The Specificity of Cell Surfaces*, B. D. Davis and L. Warren (eds.). Englewood Cliffs, N.J.: Prentice-Hall, pp. 211–234.

Revell, S. H. (1959), "The accurate estimation of chromatid breakage, and its relevance to a new interpretation of chromatid aberrations induced by ionizing radiations," *Proc. Roy. Soc. London B.*, **150**:563–589.

Rhodes, A. J., and C. E. van Rooyen (1968), *Textbook of Virology* (5th edition). Baltimore: The William and Wilkins Company.

Rich, A., J. R. Warner, and H. M. Goodman (1963), "The structure and function of polyribosomes," *Cold Spring Harbor Symp. Quant. Biol.* **28**:269–285.

Richardson, B. J., A. B. Czuppen, and G. B. Sharman (1971), "Inheritance of glucose-6-phosphate dehydrogenase variation in kangaroos," *Nature New Biology* **230**:154–155.

Richardson, C. C. (1969), "Enzymes in DNA metabolism," *Ann. Rev. Biochem.* **38**:795–840.

Richardson, J. P. (1969), "RNA polymerase and the control of RNA synthesis," *Prog. Nucleic Acid Res. Mol. Biol.* **9**:75–116.

Richler, C., and D. Yaffe (1970), "The *in vitro* cultivation and differentiation capacities of myogenic cell lines," *Develop. Biol.* **23**:1–22.

Richmond. M. H. (1970), "Plasmids and chromosomes in prokaryotic cells," *Organization and Control in Prokaryotic and Eukaryotic Cells.* Cambridge: Cambridge University Press, pp. 249–277.

Riggs, A. D., and S. Bourgeois (1968), "On the assay, isolation and characterization of the *lac* repressor," *J. Mol. Biol.* **34**:361–364.

Ris, H. (1957), "Chromosome structure," *The Chemical Basis of Heredity,* W. D. McElroy and B. Glass (eds.). Baltimore: Johns Hopkins Press, pp. 23–69.

Roberts, J. (1969), "Termination factor for RNA synthesis," *Nature* **224**:1168–1174.

Robertson, W. R. B. (1916), "Chromosome studies. I. Taxonomic relationships shown in the chromosomes of *Tettigidae* and *Acrodidae*: V-shaped chromosomes and their significance in *Acrididae*, *Locustidae*, and *Grillidae*: chromosomes and variation," *J. Morph.* **27**:179–331.

Rosenberg, B. H., and L. F. Cavalieri (1968), "Shear sensitivity of the *E. coli* genome: multiple membrane attachment points of the *E. coli* DNA," *Cold Spring Harbor Symp. Quant. Biol.* **33**:65–72.

Rosenberg, B. H., F. M. Sirotnak, and L. F. Cavalieri (1959), "On the size of genetic determinants in pneumococcus and the nature of the variables involved in transformation," *Proc. Nat. Acad. Sci.* **45**:144–156.

Ross, J., E. M. Scolnick, G. J. Todaro, and S. A. Aaronson (1971), "Separation of murine cellular and murine leukemia virus DNA polymerases," *Nature New Biology* **231**:163–167.

Roth, S. (1968), "Studies on intercellular adhesive selectivity," *Develop. Biol.* **18**:602–631.

Rückert, J. (1892), "Zur Entwicklungsgeschichte des

Ovarialeies bei Selachiern," *Anat. Anz.* **7**:107–158.

Rudland, P. S., W. A. Whybrow, and B. F. C. Clark (1971), "Recognition of bacterial initiator tRNA by an initiation factor," *Nature New Biology* **231**:76–78.

Ryter, A. (1968), "Association of the nucleus and the membrane of bacteria: a morphological study," *Bacteriol. Rev.* **32**:39–54.

Salas, J., and H. Green (1970), "Proteins binding to DNA and their relation to growth in cultured mammalian cells," *Nature New Biology* **229**:165–169.

Sanger, F. (1955), "The chemistry of simple proteins," *Symp. Soc. Exptl. Biol.* **9**:10–31.

Sarabhai, A. S., A. O. W. Stretton, and S. Brenner (1964), "Co-linearity of the gene with the polypeptide chain," *Nature* **201**:13–17.

Saxén, L., et al. (1968), "Differentiation of kidney mesenchyme in an experimental model system," *Adv. Morph.* **7**:251–293.

Schaeffer, P. (1964), "Transformation," *The Bacteria*, Vol. V, I. C. Gunsalus and R. Y. Stanier (eds.). New York: Academic Press.

Schindler, A.-M., and K. Mikamo (1970), "Triploidy in man: report of a case and a discussion on etiology," *Cytogenetics* **9**:116–130.

Schneider, G., and C. W. Bardin (1970), "Defective testicular testosterone synthesis by the pseudo-hermaphrodite rat: an abnormality of 17β-hydroxy-steroid dehydrogenase," *Endocrinol.* **87**:864–873.

Schowing, J. (1961), "Influence inductrice de l'encéphale et de la chorde sur la morphogenèse du squelette cranien chez l'embryon de poulet," *J. Embryol. Exptl. Morph.* **9**:326–334.

Schreier, M. H., and H. Noll (1971), "Conformational changes in ribosomes during protein synthesis," *Proc. Nat. Acad. Sci.* **68**:805–809.

Scolnick, E. M., and C. T. Caskey (1969), "Peptide chain termination. V. The role of release factors in mRNA terminator codon recognition," *Proc. Nat. Acad. Sci.* **64**:1235–1241.

Scott, N. S., and R. M. Smillie (1967), "Evidence for the direction of chloroplast ribosomal RNA synthesis by chloroplast DNA," *Biochem. Biophys. Res. Comm.* **28**:598–603.

Sears, E. R. (1954), "The aneuploids of common wheat," University of Missouri Agricultural Experiment Station, *Res. Bull. No. 572*.

Shapiro, L. R., L. Y. F. Hsu, M. E. Calvin, and K. Hirschorn (1970), "XXXXY boy: a 15-month-old child with normal intellectual development," *Am. J. Dis. Children* **119**:79–81.

Shearer, R. W., and B. J. McCarthy (1967), "Evidence for ribonucleic acid molecules restricted to the cell nucleus," *Biochem.* **6**:283–289.

Shettles, L. B. (1964), "The great preponderance of human males conceived," *Amer. J. Obstet. Gynec.* **89**:130–133.

Shine, I., and R. Gold (1970), *Serendipity on St. Helena: A Genetical and Medical Study of an Isolated Community.* New York: Macmillan (Pergamon).

Short, R. V., J. Smith, T. Mann, E. P. Evans, J. Hallett, A. Fryer, and J. L. Hamerton (1969), "Cytogenetic and endocrine studies of a freemartin heifer and its bull co-twin," *Cytogenetics* **8**:369–388.

Schull, W. J. (1963), *Genetic Selection in Man.* Ann Arbor: University of Michigan Press.

Slizynska, H., and B. M. Slizynska (1947), "Genetical and cytological studies of lethals induced by chemical treatment in *Drosophila melanogaster*," *Proc. Roy. Soc. Edinb. B.* **62**:234–242.

Smillie, R. M., D. Graham, M. R. Dwyer, A. Grieve, and N. F. Tobin (1967), "Evidence for the synthesis *in vivo* of proteins of the Calvin cycle and of the photosynthetic electron-transfer pathway in chloroplast ribosomes," *Biochem. Biophys. Res. Comm.* **28**:604–610.

Smith, A. E., and K. A. Marcker (1970), "Cytoplasmic methionine transfer RNAs from eukaryotes," *Nature* **226**:607–610.

Smith, B. W. (1963), "The mechanism of sex determination in *Rumex hastatulus*," *Genetics* **48**:1265–1288.

Smith, J. M. (1971), "What use is sex?," *J. Theor. Biol.* **30**:319–335.

Smith, L. D., R. E. Ecker, and S. Subtelny (1966), "The initiation of protein synthesis in eggs of *Rana pipiens*," *Proc. Nat. Acad. Sci.* **56**:1724–1728.

Snyder, L. H. (1934), "Studies in human inheritance. X. A table to determine the proportion of recessives to be expected in various matings involving a unit character," *Genetics* **19**:1–17.

Sonneborn, T. M. (1937), "Sex, sex inheritance and sex determination in *Paramecium aurelia*," *Proc. Nat. Acad. Sci.* **23**:378–385.

———— (1965), "The metagon: RNA and cytoplasmic inheritance," *Am. Nat.* **99**:279–307.

Spencer, D. A., J. W. Eyles, and M. K. Mason (1969), "XYY syndrome, and XYY/XXYY mosaicism also showing features of Klinefelter's syndrome," *J. Med. Genet.* **6**:159–165.

Spirin, A. S. (1966), "On 'masked' forms of messenger RNA in early embryogenesis and in other differentiating systems," *Current Topics Develop. Biol.* **1**:1–38.

Srb, A. M., and N. H. Horowitz (1944), "The ornithine cycle in *Neurospora* and its genetic control," *J. Biol. Chem.* **154**:129–139.

Stadler, L. J. (1928), "Mutations in barley induced by X-ray and radium," *Science* **66**:186–187.

Steele, M. W., and W. R. Breg, Jr. (1966), "Chromosome analysis of human amniotic-fluid cells," *Lancet,* Feb. 19:383–385.

Steinberg, M. S. (1964), "The problem of adhesive selectivity in cellular interactions," *Cellular Membranes in Development*, M. Locke (ed.). New York: Academic Press, pp. 321–366.

Steitz, J. A. (1969), "Polypeptide chain initiation: nucleotide sequences of the three ribosomal binding sites in bacteriophage R17 RNA," *Nature* **224**:957–964.

Stern, C. (1931), "Zytologisch-genetische Untersuchungen als Beweise für die Morganische Theorie des Faktorenaustauschs," *Biol. Zbl.* **51**:547–587.

―――― (1936), "Somatic crossing over and segregation in *Drosophila melanogaster*," *Genetics* **21**:625–730.

Stone, L. S. (1950), "Neural retina degeneration followed by regeneration from surviving retinal pigment cells in grafted adult salamander eyes," *Anat. Rec.* **106**:89–109.

Streisinger, G., Y. Okada, J. Emrich, J. Newton, A. Tsugita, E. Terzaghi, and M. Inouye (1966), "Frameshift mutations and the genetic code," *Cold Spring Harbor Symp. Quant. Biol.* **31**:77–84.

Strickberger, M. (1968), *Genetics*. New York: Macmillan.

Strudel, G. (1955), "L'action morphogène du tube nerveux et de la corde sur las différentiation des vertèbres et des muscles vertébraux chez l'embryon de poulet," *Arch. Anat. Microscop. Morph. Exptl.* **44**:209–255.

Sturtevant, A. H. (1925), "The effects of unequal crossing over at the Bar locus in *Drosophila*," *Genetics* **10**:117–147.

―――― (1945), "A gene in *Drosophila melanogaster* that transforms females into males," *Genetics* **30**:297–299.

Stuy, J. H., and D. Stern (1964), "The kinetics of DNA uptake by *Haemophilus influenzae*," *J. Gen. Microbiol.* **35**:391–400.

Sugino, A., and R. Okazaki (1972), "Mechanism of DNA chain growth. VII. Direction and rate of growth of T4 nascent short DNA chains," *J. Mol. Biol.* **64**:61–85.

Sumner, A. T., J. A. Robinson, and H. J. Evans (1971), "Distinguishing between XY and YY-bearing human spermatozoa by fluorescence and DNA content," *Nature New Biology* **229**:231–233.

Sundaralingam, M., S. T. Rao, and J. Abola (1971), "Molecular conformation of dihydrouridine: puckered base nucleoside of transfer RNA," *Science* **172**:725–727.

Sunshine, G. H., D. J. Williams, and B. R. Rabin (1971), "Role for steroid hormones in the interaction of ribosomes with the endoplasmic membranes of rat liver," *Nature New Biology* **230**:133–136.

Sutton, W. S. (1903), "The chromosomes in heredity," *Biol. Bull.* **4**:231–248.

Swanson, D. W., and A. H. Stipes (1969), "Psychiatric aspects of Klinefelter's syndrome," *Am. J. Psychiat.* **126**:814–822.

Takeda, M., and R. E. Webster (1968), "Protein chain initiation and deformylation in *B. subtilis* homogenates," *Proc. Nat. Acad. Sci.* **60**:1487–1494.

Taylor, A. L., and M. S. Thoman (1964), "The genetic map of *Escherichia coli* K12," *Genetics* **50**:659–677.

Telfer, M. A., D. Baker, G. R. Clark, and C. E. Richardson (1967), "Incidence of gross chromosomal errors among tall criminal American males," *Science* **159**:1249–1250.

Temin, H. M. (1971), "The protovirus hypothesis: speculations on the significance of RNA-directed DNA synthesis for normal development and for carcinogenesis," *J. Nat. Cancer Inst.* **46**:iii–vii (editorial).

Temin, H. M., and S. Mizutani (1970), "RNA dependent DNA polymerase in virions of Rous sarcoma virus," *Nature* **226**:1211–1213.

Tennent, D. H. (1914), "The early influence of the spermatozoa upon the characters of echinoid larvae," *Carnegie Inst. Wash. Publ.* **182**:129.

Terman, S. A., and P. R. Gross (1965), "Translation level control of protein synthesis during early development," *Biochem. Biophys. Res. Comm.* **21**:595–600.

Terzaghi, E., Y. Okada, G. Streisinger, J. Emrich, M. Inouye, and A. Tsugita (1966), "Change of a sequence of amino acids in phage T4 lysozyme by acridine-induced mutations," *Proc. Nat. Acad. Sci.* **56**:500–507.

Tessman, I. (1965), "Genetic ultrafine structure in the T4rII region," *Genetics* **51**:63–75.

Thoday, J. M., and J. B. Gibson (1970), "The probability of isolation by disruptive selection," *Am. Nat.* **104**:219–230.

Thoday, J. M., and J. Read (1947), "Effect of oxygen on the frequency of chromosome aberrations produced by X-rays," *Nature* **160**:608.

Tiedemann, H. (1967), "Biochemical aspects of primary induction and determination," *The Biochemistry of Animal Development*, Vol. II, R. Weber (ed.). New York: Academic Press, pp. 3–55.

Timoféef-Ressovsky, N. W. (1934), "The experimental production of mutations," *Biol. Rev.* **9**:411–457.

Todd, N. (1970), "Karyotypic fissioning and canid phylogeny," *J. Theor. Biol.* **26**:445–480.

Toivonen, S., L. Saxén, and T. Vainio (1961), "Quantitative evidence for the two-gradient hypothesis in the primary induction," *Experientia* **17**:86–87.

Tomizawa, J., and N. Anraku (1964), "Molecular mechanisms of genetic recombination in bacteriophage. I. Effect of KCN on genetic recombination of phage T4," *J. Mol. Biol.* **8**:508–515.

Traub, P., and M. Nomura (1968), "Structure and function of *E. coli* ribosomes. V. Reconstitution of functionally active 30S ribosomal particles from RNA and proteins," *Proc. Nat. Acad. Sci* **59**:777–784.

Trujillo, J. M., B. Walden, P. O'Neil, and H. B. Anstall (1965), "Sex linkage of glucose-6-phosphate dehydrogenase in the horse and donkey," *Science* **148**:1603–1604.

Visconti, N., and M. Delbrück (1953), "The mechanism of genetic recombination in phage," *Genetics* **38**:5–33.

Vogel, H. (1964), "Genetics: a preliminary estimate of the number of human genes," *Nature* **201**:847.

Wallace, H., J. Morray, and W. H. R. Langridge (1971), "Alternative model for gene amplification," *Nature New Biology* **230**:201–203.

Watanabe, T. (1967), "Infectious drug resistance," *Sci. Amer.* **217**:19–27.

Watanabe, Y., and A. F. Graham (1968), "Structural units of reovirus ribonucleic acid and their possible functional significance," *J. Virol.* **1**:665.

Watson, J. D. (1970), *Molecular Biology of the Gene* (2nd edition). Menlo Park, Calif.: W. A. Benjamin.

Watson, J. D., and F. H. C. Crick (1953a), "Molecular structure of nucleic acids: a structure of doxyribose nucleic acid," *Nature* **171**:737–738.

———— (1953b), "The structure of DNA," *Cold Spring Harbor Symp. Quant. Biol.* **18**:123–131.

———— (1953c), "Genetical implications of the structure of deoxyribonucleic acid," *Nature* **171**:964–969.

Watterson, L. R., I. Fowler, and B. J. Fowler (1954), "The role of the neural tube and notochord in development of the axial skeleton of the chick," *Amer. J. Anat.* **95**:337–400.

Weinberg, W. (1908), "Über den Nachweis der Vererbung beim Menschen," *Papers on Human Genetics*, S. H. Boyer IV (ed.). Englewood Cliffs, N.J.: Prentice-Hall.

Weiss, P. (1950), "Perspective in the field of morphogenesis," *Quart. Rev. Biol.* **25**:177–198.

Weiss, M., and H. Green (1967), "Human-mouse hybrid cell lines containing partial complements of human chromosomes and functioning human genes," *Proc. Nat. Acad. Sci.* **58**:1104–1111.

Westergaard, M. (1948), "The relation between chromosome constitution and sex in the offspring of triploid *Melandriun*," *Hereditas* **34**:257–279.

———— (1958), "The mechanism of sex determination in dioecious flowering plants," *Adv. in Genet.* **9**:217–281.

Westin, M., H. Perlmann, and P. Perlmann (1967), "Immunological studies of protein synthesis during sea urchin development," *J. Exptl. Zool.* **166**:331–345.

White, M. J. D. (1946), "The cytology of the Cecidomyidae (Diptera). II. The chromosome cycle and anomalous spermatogenesis in *Miastor*," *J. Morph.* **79**:323–369.

Whitehouse, H. K. L. (1969), *Toward an Understanding of the Mechanism of Heredity* (2nd edition). London: Edward Arnold.

Whiting, P. W. (1943), "Multiple alleles in complementary sex determination in *Habrobracon*," *Genetics* **28**:365–382.

Whitt, G. S. (1970), "Developmental genetics of the lactate dehydrogenase isozymes of fish," *J. Exptl. Zool.* **175**:1–36.

Wilkie, D. (1964), *The Cytoplasm in Heredity*. London: Methuen.

———— (1970), "Reproduction of mitochondria and chloroplasts," *Organization and Control in Prokaryotic and Eukaryotic Cells.* Cambridge: Cambridge University Press, pp. 381–399.

Wilt, F. (1965), "Regulation of the initiation of chick hemoglobin synthesis," *J. Mol. Biol.* **12**:331–341.

Wimber, D. E., and D. M. Steffensen (1970), "Localization of 5S RNA genes on *Drosophila* chromosomes by RNA-DNA hybridization," *Science* **170**:639–641.

Winge, Ö. (1937), "Goldschmidt's theory of sex determination in *Lymantria*," *J. Genetics* **34**:81–89.

Witkin, E. M. (1969), "Ultraviolet-induced mutation and DNA repair," *Ann. Rev. Genetics* **3**:525–552.

Witschi, E. (1936), "Studies on sex differentiation and sex determination in amphibians. VII. Experiments on inductive inhibition of sex differentiation in parabiotic twins of salamanders," *Anat. Rec.* **66**:483–503.

———— (1960), "Genetic and postgenetic sex determination," *Experientia* **16**:274–278.

Wollman, E. L., and F. Jacob (1955), "Sur le méchanisme du transfer de matériel génétique au course de la recombinaison chez *E. coli* K-12," *Comptes Rend. Acad. Sci.* (Paris) **240**:2449.

———— (1958), "Sur les processus de conjugaison et de recombinaison chez *E. Coli*. V. Le méchanisme du transfer de matérial génétique," *Ann. Inst. Pasteur* **95**:641.

Wolstenholme, D. R., C. A. Vermeulen, and G. Venema (1966), "Evidence for the involvement of membranous bodies in the process leading to genetic transformation in *Bacillus subtilis*," *J. Bacteriol.* **92**:1111–1121.

Woodward, V. W., J. R. DeZeeuw, and A. M. Srb (1945), "The separation and isolation of particular biochemical mutants of *Neurospora* by differential germination of conidia, followed by filtration and selective plating," *Proc. Nat. Acad. Sci.* **40**:192–200.

Wright, S. (1934), "The results of crosses between inbred strains of guinea pigs, differing in number of digits," *Genetics* **19**:537–551.

Yamada, T. (1961), "A chemical approach to the problem of the organizer," *Adv. Morph.* **1**:1–53.

Yanofsky, C., B. C. Carlton, J. R. Guest, D. R. Helinski, and U. Henning (1964), "On the colinearity of gene structure and protein structure," *Proc. Nat. Acad. Sci.* **51**:266–272.

Ycas, M. (1969), *The Biological Code*. New York: Wiley (Interscience).

Zinder, N. D., and J. Lederberg (1952), "Genetic exchange in *Salmonella*," *J. Bacteriol.* **64**:679–699.

Zuckerkandl, E., and L. Pauling (1965), "Evolutionary divergence and convergence in proteins," *Evolving Genes and Proteins*, V. Bryson and H. F. Vogel (eds.). New York: Academic Press, pp. 97–166.

Index

*All page numbers in boldface type refer to illustration appearing on that page.